Seeking the Truth

How Science Has Prevailed over the Supernatural Worldview

RICHARD H. SCHLAGEL

an imprint of Prometheus Books
59 John Glenn Drive, Amherst, New York 14228–2119

Published 2010 by Humanity Books, an imprint of Prometheus Books

Inquiries should be addressed to
Humanity Books
59 John Glenn Drive
Amherst, New York 14228–2119
VOICE: 716–691–0133
FAX: 716–691–0137
WWW.PROMETHEUSBOOKS.COM

14 13 12 11 10   5 4 3 2 1

Library of Congress Cataloging-in-Publication Data

Schlagel, Richard H., 1925–
Seeking the truth : how science has prevailed over the supernatural worldview / by Richard H. Schlagel.
p. cm.
Includes bibliographical references and index.
ISBN 978–1–59102–774–4 (pbk. : alk. paper)
1. Religion and science—History. 2. Religion—Controversial literature. I. Title.
BL245.S35 2010
201'.65—dc22   2009039176

Printed in the United States of America

Dedicated with love and gratitude to my wife, Jody (Josephine) who, although a theist, graciously read the entire manuscript correcting innumerable typos, but, more importantly, recommended grammatical and stylistic revisions enhancing the literary quality of the treatise.

ACKNOWLEDGMENTS

Given the prevailing uncertain economic conditions that have diminished the market for lengthy scholarly publications, I am very grateful to Steven L. Mitchell, Editor in Chief of Prometheus Books, along with his Acquisitions Committee, for agreeing to publish the present work. In addition, I want to extend my utmost gratitude to the following members of The George Washington University for supporting the grant that also made the publication possible: Professor Carol K. Sigelman, then Associate Vice President for Graduate Studies and Academic Affairs; Marguerite P. Barratt, Dean of the Columbian College of Arts and Sciences; Professor Geralyn M. Schulz, Associate Dean for Research; and Professor David D. DeGrazia, Chair of the Department of Philosophy, also of the Columbian College of Arts and Sciences. I want to thank Yonah Levy, too, for his assistance when I encountered computer glitches when formatting the manuscript. Finally, I want to express my indebtedness to all the scholars whose previous research and scholarship made this book feasible.

CONTENTS

INTRODUCTION

In an effort to be tolerant of deeply held religious beliefs there is a mistaken tendency to conflate religious tolerance with the condoning of religious convictions. This is shown in the deference given to religious beliefs which are asked to be taken on faith when confronted with disconfirming evidence, whereas non-religious beliefs are expected to be discarded in such cases. For instance, the craze a few years ago over alleged sightings of flying saucers died when other explanations for "crop circles" and reports of strange creatures descending and entering illuminated space ships were found.

Yet countless Christians hold religious beliefs just as implausible, such as Mary's immaculate conception of Jesus by mediation of the Holy Spirit; that at communion the consecrated bread and wine are actually transmuted into the flesh and blood of Christ; that Jesus miraculously multiplied loaves of bread on command, walked on water, raised Lazarus from the dead; and that following his crucifixion and entombment several days in a cave he rose from the dead and bodily ascended into heaven. Because none of these beliefs are credible today they are asked to be taken on faith, evoking St. Paul's doctrine of "salvation by faith" as the justification. But there remains the question of whether this faith itself is justified in light of what we have learned since Paul's day.

As a result of religious persecutions throughout history governments have erected legal defenses protecting individuals from being attacked for their religious beliefs, however implausible they may be in terms of what is generally considered reliable evidence and rational knowledge. At least this is true as long as such beliefs are not personally or socially threatening or destructive. But this commendable treatment does not obviate the responsibility of critically evaluating and revising *all beliefs* in light of newly discovered facts and well tested explanations, otherwise no intellectual advances would have been made throughout history. Thus as fair-minded as religious tolerance is, it has the adverse effect of shielding incredible beliefs from criticism.

Aside from the essential personal consolations and moral and spiritual guidance that Christianity provides, the main reason its followers are not troubled by the implausibility of their beliefs is that they have little understanding of their historical origin and development. This is reinforced by being told that they should be taken on faith which deflects any critical appraisal. As any study of early beliefs reveal, they were crude, animistic, anthropomorphic, and mythical. Devoid of knowledge about what caused natural events, but drawing an analogy with intentional human acts, primitive man conceived natural occurrences to be purposely willed by evil or benign spirits, demons, angels, and gods. Rather than trying to understand *how* natural occurrences were produced, they asked *why* they occurred, the form of the question itself evoking an intentional answer. This is the reason that in ancient times supernatural causes and mythical narratives were the typical forms of 'explanation' and why various sacrifices, ascetic practices, and supplications or prayers were the predominant means of attempting to attain some control over threatening or destructive events.

However, endeavoring to elude the inadequacies of these early beliefs a rational mode of thinking was gradually developed to attain a more realistic understanding of the universe, usually at considerable psychological strain and opposition from the guardians of traditional beliefs. The psychological strain arises from the necessity of replacing more familiar and emotionally satisfying beliefs with those that are initially counterintuitive and demystifying, though more explanatory, while the opposition stems from those who have a vested interest in preserving enshrined beliefs and institutions. But though the innovators are compensated for the emotional and cognitive dissonance by the greater explanatory insights the new conceptualizations bring, this is not true of the general public who usually finds them incomprehensible and disturbing.

In the history of Western thought there were three outstanding examples of this. The first was the endeavor of the Greek philosophers to dispel *mythos* with *logos* to attain a more rational conception of the universe. It was they who began critically questioning and replacing the narrative accounts of natural occurrences characteristic of early myths and epics by a progressive tradition of rational inquiry. Rejecting previous mythical accounts despite their dramatic and emotional appeal, Greek *thinkers* strove to formulate more credible explanations based on empirical evidence and logical reasoning, introducing such prescient theories as Anaximander's account of land animals evolving from aquatic creatures, Democritus' atomic theory, Aristarchus' heliocentricism, and the purely naturalistic cosmology of Anaxagoras, along with the innovative formalisms of Parmenides' logical cosmological criteria, Aristotle's logic, Euclid's geometry, and Archimedes' method of exhaustion, a precursor of calculus. As Nietzsche observed: "it is all over with priests and gods when man becomes scientific."

The second example of a scientific development that greatly improved our understanding of the universe but was opposed by the dominant religions of the time, was the modern heliocentric theory. Its primary creators, Aristarchus, Copernicus, Kepler, Galileo, and Newton, were persuaded of its truth because it offered a simpler, more harmonious conception of the solar system. Convinced that the centrality of the sun was the key to explaining more exactly the orbital motions of the planets, Kepler was able to formulate the first correct planetary laws while Galileo's telescopic observations of the sun's surface and Jupiter's satellites undermined the distinction between the celestial and terrestrial worlds and supported heliocentricism. But because it conflicted with ordinary observations, refuted Genesis, and displaced humankind from the center of the universe, the theory was ridiculed by Martin Luther and opposed by the Catholic Church which forced Galileo to abjure his belief in heliocentricism.

The third example of a scientific theory that aroused the opposition of Christians especially because it refuted the myth of special creation in Genesis was Darwin's theory of evolution. Instead of being accepted as a much more plausible account of the origin of species, including homo sapiens, than the verbal creation by God ("let there be"), Darwin's theory was vilified. Though based on the surprising indigenous diversity of species he had observed on isolated atolls during his voyage on the Beagle to South America and supported by the recent discovery of transitional fossils embedded in different geological periods, his evolutionary theory was resoundly rejected. Christians found the biblical claim that God "brought forth . . . living creatures according to their kinds" and created man in "God's image" (1: 14–26) more credible than Darwin's naturalistic explanation by the theory of evolution. Yet his conception that living creatures had evolved due to their advantageous genetic modifications being competitively selected for survival and therefore propagated more abundantly, opened up whole new areas of further explanation.

While Christians in the nineteenth century may be forgiven for accepting the biblical account at that time, it is dismaying that with all the subsequent supporting evidence from geology, paleontology, zoology, embryology, microbiology, and the recent deciphering of the human genome evangelical and fundamentalist Christians still reject it for theories of special creation and intelligent design. Although John Paul II apologized for the Catholic Church's disgraceful treatment of Galileo and declared its present acceptance of evolution, he did so with the unwarranted proviso that God was the ultimate cause of the process which adds nothing to the understanding of it.

As traumatic as those developments seemed at the time, they were just a prologue to later advances in science and technology that would drastically change our living conditions and conception of the universe. Except perhaps for Leonardo da Vinci or Jules Verne, who would have foreseen the introduction of electricity, radios, telephones, automobiles, airplanes, television, nuclear reactors, computers, organ transplants, and cloning, or the launching of earth satellites and the lunar landing? How trivial the heliocentric revolution seems as compared to the telescopic discovery by Edwin Hubble in the 1920s that our Milky Way galaxy consists of about 100 billion stars and that it is just one galaxy among 100 billion others in the universe.

Or consider our present estimate of the origin and age of the universe, in contrast to the biblical account, based on the spectroscopic discovery of the red shift of the light spectrum from outer space indicative of an inflationary universe expanding from an original singularity or explosion called the Big Bang. This initial discovery was strengthened with the detection by Arno Penzias and Robert Wilson of the background microwave radiation left over from the Big Bang which recently was confirmed by the astrophysicists John C. Mather and George F. Moot in their COBE satellite experiments that measured faint temperature variations in the microwave background. Rather than the 4004 BCE date of the creation of the universe the Irish prelate and scholar James Ussher computed by tracing biblical genealogies in the middle of the seventeenth century and still generally accepted in the early nineteenth century, we now have evidence that its age is 13.7 billion years. Just considering our brief history on this insignificant planet compared to the age of the universe, the enormity of its size, and that even the known universe may be just a bulge of a larger universe should be humbling enough to extinguish any pretensions of a special creation or intentional design.

Moreover, evolution is no longer just a "theory" given the extensive conciliatory evidence mentioned previously and now the sequencing of genomes which show the similarity of our genetic coding to organisms as primitive as worms and mice and only a 1.2 percent difference between the human genome and that of the pigmy chimpanzee and bonobo gorilla. Yet either because religionists are ignorant of these developments or just ignore them they continue to maintain convictions that are no longer sustainable.

But wanting to have the benefits of both cultural developments religionists attempt to reconcile them, conceding that while science has been tremendously successful in investigating the physical world and improving the material conditions of life, religion has its own divine reality revealed in scripture and corroborated by religious experiences such as mysticism, and therefore must be uncritically accepted on faith. Although the proponents of this view claim that the difference in these belief systems precludes any conflict between them, there is the obvious difficulty of reconciling religious belief with what we have since learned about the universe and human existence. This continuing conflict is illustrated in the recent Catholic Document, *Dignitas Personae* or "Dignity of the Person," that condemns embryonic stem cell research, in vitro fertilization, and the "morning-after pill," along with many other practical technologies which it opposes on dogmatic religious grounds, despite their offering hope of alleviating some of our most acute problems.

The worldviews of science and religion are too disparate to permit a facile reconciliation and the burden of proof has passed from science to religion, for the achievements of science are too obvious and substantial to be ignored. Moreover, as will be discussed later, religious experiences such as mysticism do not have the sacred status they once had because now there is considerable evidence that they are experienced by schizophrenics and temporal lobe epileptics, as well as correlated with other abnormal electrical stimulations within the temporal lobe, called temporal lobe transients. Detaching it from its historical setting and empirical evidence, defenders of religion shield it from contemporary knowledge and criticism as if it had an eternal validity denied past cultural institutions. That children are taught religious beliefs as a child before having developed a critical attitude or historical perspective and later made to feel guilty if the beliefs are questioned, insisting they should be held on faith, is a major factor in their still being held despite their implausibility. How prevalent would religious convictions be if they were not inculcated at an early age but confronted later in life?

Yet yearning for the solace of religion, such as trust in a beneficial and caring God, anticipating the reunion with loved ones in an afterlife, and hoping that human destiny has a transcendental meaning, but also desiring the practical benefits of science and technology, such as curative medical procedures, protection from natural disasters, a higher standard of living, and greater educational and occupational opportunities, religionists accept the latter because of its significant practical value while retaining the supernatural framework underlying the former. The difficulty with this divisional argument, as was true of the ancient distinction between the celestial and terrestrial worlds, is that there is no *evidential* basis for it.

The reason scientific inquiry has become more influential is that we now realize that biblical claims that a spirit, angel, or divine being can intervene in natural processes or human affairs has no factual basis nor explanatory value, in contrast to scientific investigations and theories that explain how such occurrences are caused and are experimentally confirmed. There has never been any extension of factual knowledge as a result of alleged biblical revelations and though the belief in God has been consoling and predominantly inspirational and morally edifying, promoting ethical behavior, charitable work, and social solidarity, this is not a confirmation of its truth. Four of the major five religions, Buddhism, Judaism, Christianity, and Islam (omitting Hinduism), were founded on the charismatic lives and inspired teachings of Buddha, Abraham and Moses, Jesus and Paul, and Muhammad, yet it is no accident that they originated when oracular dreams, angelic visitations, divine revelations, extraordinary miracles, and deifications were intrinsic to the culture, which is less true today in more advanced societies.

The most successful attempt to establish Christianity on a secure cosmological foundation was Thomas Aquinas' thirteenth-century synthesis of Christianity with Aristotelianism, but that philosophical foundation was completely demolished by the scientific contributions of Copernicus, Kepler, Galileo, and Newton. One forgets how fanciful the medieval setting was with its satyrs, cherubs, unicorns, dragons, demons, and angels. Any viewing of medieval and early Renaissance art with its countless paintings of the archangel Gabriel's annunciation to the Virgin Mary, angels standing guard over the tomb of Jesus, along with Mary, Jesus, and martyrs being lifted into heaven on pink clouds presents a glaring discrepancy between that era and ours.

Despite their aesthetic value, these portrayals are representative of a primitive culture and vanished world. Nor is the emotional satisfaction derived from still believing them a vindication of their truth, since history is strewn with the relics of discarded rituals, icons, and beliefs which also had their emotional attractions at the time. Consider the spiritual gratification derived in the ancient world from the cult worship of such gods as Persia's Marduk and Ahura Mazda, Egypt's Osiris and Isis, Greece's Zeus and Athena, or Rome's Jupiter and Venus which today have largely been disavowed. A derivative of Judaism, Christianity is a legacy of this earlier deification tradition which is shown in medieval literature and Renaissance paintings where the classical and Christian figures and motifs are intermingled. While Christians like to think of Christianity as the only true religion, this is pure chauvinism.

If instead of Christianity being vindicated by divine revelation and faith, it were viewed more realistically as a refined continuation of more primitive religions with their sustaining miracles, angels, and gods, we could still appreciate the wonderful pageantry, literature, art, music, and architecture that it has inspired while discarding its supernatural scaffolding. One can also acknowledge the tremendous appeal of Jesus' support of the afflicted and downtrodden as depicted in the Gospels or of the qualities of gentleness, kindness, and friendliness in countries like Thailand and Cambodia where the people have embraced the teachings of Buddha, yet this does not justify worshiping Jesus or Buddha. Confucius had a similar impact in China, yet he was not normally deified.

Though science cannot provide the consoling beliefs of religions because the evidence is lacking, it has done much to alleviate the poverty, disease, suffering, and ignorance that tormented human existence in the past, as well as today in much of the world. The technological applications of science in agriculture, engineering, industry, medicine, pharmacology, electronics, nuclear energy, transportation, and computers are mainly responsible for whatever improvements have been achieved in the living conditions of human beings, not reliance on God's benevolence or wisdom. One only has to contrast the standards of living in parts of Africa, Southeast Asia, China, and Central and South America with Europe and North America to realize the significance of the benefits of modern science and technology. Despite the consolations of their religions, it is the Muslims, Africans, Asians, and Latinos who want to emigrate to the United States or Europe to seek a better way of life, not the reverse.

It is this difference in the historical origins and contrasting justifications of the worldviews of religion and science, along with the reasons for the ascendancy of the latter, that will be described in this book, drawing on my previous publications *From Myth to Modern Mind: A Study of the Origins and Growth of Scientific Thought*, Vol. I, *Theogony through Ptolemy* (1995) and Vol. II, *Copernicus through Quantum Mechanics* (1996), along with *The Vanquished Gods* (2001). The reason for the present publication is that it makes more accessible in a single volume the perspective and arguments of the other three volumes which is especially relevant given the resurgence of the regressive religious right in the United States and the Taliban movement among the Muslims that has created the worldwide terrorist threat.

Our European friends are usually puzzled (and dismayed) that a country as advanced in science and technology as the United States can have such a large proportion of evangelicals and fundamentalists among its population. Although our graduate programs in mathematics and the physical sciences are the envy of the world, our primary and secondary educational systems are generally deficient in teaching these disciplines which are essential for instilling respect for objective evidence, rigorous thinking, and the testing of beliefs, along with addressing such pressing problems as global warming, energy sources, demographic increase, and the present economic crisis. My hope is that by reconstructing the tremendous achievements of science and showing how it has brought about fundamental conceptual revisions in the past in contrast to the archaic, reactionary nature of religions, those confronting their different claims will be better able to evaluate the two worldviews and understand why scientific inquiry is prevailing.

Chapter 1

OUR GREEK HERITAGE

INTRODUCTION

Western civilization was forged by two radically contrasting cultures: the Hellenic and Hellenistic civilization of the ancient Greeks and the Hebraic religion of the Israelites. The former left an astonishing legacy: a pantheon of gods (Zeus, Hera, Athena, Apollo, etc.), myths such as Hesiod's *Theogony* and Prometheus bringing fire to man, the two greatest and most influential philosophers in history, Plato and Aristotle, along with engendering other schools of philosophy, such as Epicureanism, Stoicism, Skepticism, and Neoplatonism. While there is some evidence that a few Greek philosophers such as Thales and Democritus voyaged to distant lands where they came into contact with Near Eastern scholars, the Greek city-states scattered in areas such the Peloponnesus, Attica, and Boeotia, as well as the Ionian coastline of the Near East, were too isolated for their incredible mathematical and scientific contributions not to have been mainly indigenous.

Among their prominent mathematicians were Pythagoras, Hippocrates of Chios, Archytas of Tarentum, Theaetetus, Eudoxus (the latter three connected with Plato's Academy), Appolonius, and Eratosthenes. But the most famous were Euclid whose *Elements* was the foundation of geometry and Archimedes (a Greek although he spent most of his life in Syracuse in Sicily) who, with Eudoxus, created the mathematical "method of exhaustion," a forerunner of modern calculus. The most prominent name in medicine in the ancient world is Hippocrates of Cos who is referred to as the "father of medicine," while significant biological investigations were carried on in the Museum in Alexandria by Herophilus and Erasistratus. Herodotus' history of the Greek triumph at the Persian Wars and Thucydides' history of the defeat of Athens by the Spartans at the end of the tragic twenty-seven year Peloponnesian War were the first attempts to write factual histories which are our only reliable source of these momentous events.

Also among their significant contributions are the epics of Homer and the dramatic works of the tragedians, the beautiful columned temples so imitated in Washington, DC, the realistic sculptures copied by the Romans, and perspective drawings in murals, frescoes, and mosaics from Pompeii and the Villa of Livia. Add to these superlative achievements the instillation in Athens of the first constitutional, democratic government in the world by the founding father Cleisthenes shortly after 507 BCE and reinforced by Solon's laws and one realizes what a tremendous debt we owe to the Greeks. It is because we have assimilated so much of this legacy in our own culture that we take it for granted.

In contrast, the Hebrews gave us the Talmud; the Old Testament containing the Genesis myth of creation, the story of Adam and Eve with their disobedience of Yahweh bringing sin on all humankind; the story of Noah's flood; and sublime scripture such as the Book of Psalms, along with quasi historical figures such as Abraham, Moses, and David. Moreover, through the *dramatis personae* of the New Testament it should not be forgotten that John the Baptist, Mary, and Jesus, along with disciples such as Peter and James, were Hebrews. So too was Saint Paul, who did more to spread the word of Christianity throughout the Levant than anyone, followed by the unknown authors of the Gospels of the New Testament. That Christianity succeeded to grow was due therefore to the influence of Jesus' life and teachings as spread by his disciples, the mission of Saint Paul, the four Gospels, and its acceptance as the state religion by Constantine.

Given the obvious contrast between the two cultural contributions, the secular achievements of the Greeks and the Hebraic-Christian religion, the astonishing thing is that the revelatory tradition of the Hebrews and the life and words of Jesus should have spread so rapidly and had such a continuing dominant influence on Western civilization, especially following Constantine's adoption of Christianity and transfer of the capital of the Roman Empire to Constantinople. Foremost among the reasons for this success was the devout and exemplary lives of the Christians when being persecuted and during the plagues, inspired by the example and teachings of Jesus who taught the brotherhood and love of mankind, along with the promise of an everlasting life in a beatific heaven.

THE MILESIANS

While all civilizations have a prehistory of epics, legends, and myths, the Greeks were alone in being able to emerge from this mythical past to achieve a more realistic understanding of the universe, the actual origins of their city-states, and an explanation of the capricious forces determining the fate of societies and individuals.[1] Thus the emergence of rational thought began in the sixth century BCE with the Milesian philosophers Thales, Anaximander, and Anaximines, so named because they were from the city of Miletus in Ionia, now on the south-Western coast of Turkey on the Aegean Sea. Although Aristotle said he had many predecessors, Thales was the first we know of to ask the question, what is the first principle of things, the reality behind phenomena? Designated as one of the seven wise men of antiquity, he was acclaimed as a statesman, astronomer, meteorologist, physicist, engineer, mathematician, businessman, philosopher, and world traveler.

Nothing remains of his writings, but his main doctrines preserved in the doxography or historical records of ancient Greek manuscripts were that water is the first principle (archê) of things; that the earth floats on water like a log; that whatever attracts or repels, such as magnets and perhaps electrified objects, do so because they possess souls, and that "all things are full of Gods." It is believed that he arrived at his first tenet, that water is the prime element from which everything originated, because it can exist in the three basic states, liquid, solid (as ice), and gas (when evaporated), that it supports life, and that it can generate fire due to fermentation. The notion that water is the first principle and that the earth floats on water like a log is a reasonable inference from Greece being surrounded by water while the last two tenets show the influence of animism and mythology, in that moving things must be animate and that the universe is pervaded by gods. Nevertheless, as Simplicius, a Neoplatonist philosopher of the sixth-century AD states, Thales' significance consists in his being "traditionally the first to have revealed the investigation of nature to the Greeks."

The second Milesian philosopher to speculate about the original state of the universe was Anaximander, referred to as a pupil, associate, or successor of Thales, along with being "the most outstanding of his generation." Although he begins his speculation with Thales' question regarding the first principle of things, his answer seems to resemble more that of Hesiod's *Theogony* written a century or so earlier. Serving the Greeks much as the Old Testament has for the Hebrews, the *Theogony* contains a narrative account of the origin and structure of the universe, along with the genealogy of the gods. It was Hesiod's conception of the original form of the universe as "Chaos," described as a "yawning gap," that appears to have influenced Anaximander, along with "Strife" as the generating force.

Fortunately, there exists a fragment of Anaximander's philosophy recorded by Theophrastus who is one of the best sources of the writings of early Greek philosophers and who succeeded Aristotle as head of the Lyceum which the later founded. The fragment indicates that it was neither water nor any of the other elements that first existed, but an indestructible and endless entity variously named the "Indefinite," "Infinite," and "Boundless," an original state from which everything comes about through a process of "separating out" (a concept also used in Genesis).This unending transformation apparently caused by the conflict of opposites, such as Hot and Cold and Wet and Dry, is guided by the moral principle of equity, for things necessarily come to be and pass away to "pay penalty and retribution to each other according to the assessment of time." That moral precepts guided creation was characteristic of early mythical cosmologies and religions.

There were many attempts to interpret exactly what Anaximander meant by these fragments, as well as expand on how the present nature of the universe could have arisen from such a chaotic antecedent state, but whether they are based on information regarding Anaximander's system that is now lost, or just conjecture, is unknown. In any case, most commentators agree that the Hot and the Cold separated from the Boundless, the Hot forming a "sphere of fire" surrounding the air formed by the condensing Cold like "bark around a tree." The Cold further condenses the moist core of the air producing the earth in the center of the sphere. The outer fiery circumference surrounding the air then separates into successively distant rings of fire each of which is covered by a concealing mist. The misty covered rings have holes that reveal the inner fire creating the appearance of the sun, moon, and stars. There is some evidence that he speculated about the relative sizes of the sun and moon compared to the earth, along with their distances from the earth. For some reason he claimed the ring of the stars was closer to the earth than those of the sun and moon despite their distant appearance. The earth has the shape of a truncated Greek column with "a height one-third of its width," retained in the center of this spherical universe by a kind of equilibrium, "there being no reason for it to move in any direction." This may be the first use of "the principle of sufficient reason" that a particular explanation suffices if there is no justification for supplementing it.

His speculations about the origins of life in general and of human life in particular are further evidence of his advance over previous mythical thought. According to Aëtius, another of our doxographical sources, "[t]he first animals were generated in moisture and enclosed in thorny bark [similar to sea urchins]; as they grew older, they came out onto the drier land and, once their bark was split and shed, they lived in a different way for a short time." In contrast, "man was generated from living things of another kind, since the others are quickly able to look for their own food, while only man requires long nursing." Despite some uncertainty of the authenticity of these ascriptions, what is obvious is that Anaximander, using familiar examples and empirical analogies, attempted to devise a cosmology based on plausible empirical descriptions and explanations, in contrast to earlier mythical narratives such as the account in Genesis where the verbal commands of God, "Let there be. . . . ," portray creation. The fault in explanations by magical commands is that they foreclose all further inquiry without really offering any explanation. To say, as one even does today, that evolution was caused by God or by intelligent design is to terminate inquiry without having actually contributed anything to the understanding. The confirmation of this is that there is no evidence the biblical accounts ever resulted in any additional investigations or furtherance of knowledge, in contrast to Anaximander's description of the origin of the solar system, life, and man that stimulated further inquiry and allowed for emendations and improvements.

The last in the trilogy of Milesians is Anaximines, younger than Anaximander and probably his pupil and associate. His contribution illustrates what was just said about building on and revising previous theories. Rejecting Anaximander's Boundless as the archê, like Thales he selected a more specific element, Air, although not our common atmospheric air, but an "infinite Air-substrate" from which the existing elements originated. Also dismissing "separating out" as an explanation of how the present state of the universe came into being, he used the more explicit processes of "rarefaction and condensation," rarified air becoming fire or conversely condensing or "felting" into wind, clouds, water, earth, and stones. The concept of "felting" also explains how "exhalations" arising from the earth produce the sun and moon as well as the stars by added heat. These fiery "heavenly bodies" are also described as flat, floating in the air like leaves: "the sun is flat like a leaf" while "the stars are fiery leaves like paintings." They do not rotate under the earth but circle eternally around it as "a felt cap turns around the head."

As with Thales, there is a trace of animism and divinity in the original Air which begets "gods and things divine; the rest aris[ing] from the progeny of the Air." Moreover, he identifies the Air with breath and therefore with life and soul: "As our soul being air holds us together and controls us, so wind [or breath] encloses the whole world." Yet despite these traces of more primitive forms of thought in his system, using the concept of condensation to explain the generation of the universe allowed him to make significant contributions also to meteorology. Aëtius reports that "Anaximines said that the clouds occur when the air is further thickened; when it is compressed further rain is squeezed out, and hail occurs when the descending water coalesces, snow when some windy portion is included together with the moisture." He also offered credible explanations of lightening as similar to the flash of the sea when cleaved by oars, rainbows as caused by light reflected through clouds, and earthquakes produced when the earth "dries up and cracks after excessive rain." Obvious to us, these explanations were unusual for the time.

Although all traces of mythical thought have not been extruded from his philosophy, it is clear that we are further along the way to a rational explanation of the universe. Storm clouds no longer are considered the extended wings of the gigantic bird Imdugud, as in Babylonian myths, nor lightening explained as a thunderbolt thrown by Zeus, nor creation as due to divine commands. As Francis Cornford states:

> It was an extraordinary feat to dissipate the haze of myth from the origins of the world and of life. The Milesian system pushed back

> to the very beginning of things the operation of processes as familiar and ordinary as a shower of rain. It made the formation of the world no longer a supernatural, but natural event. Thanks to the Ionians, and to no one else, this has become the universal premise of all modern science.[2]

HERACLITUS

After this remarkable breakthrough and progression of thought by the Milesians, the following contributions by the Presocratic philosophers, though not constituting a similar linear progression, do exhibit distinctive conceptual innovations that often display prescient anticipations of the crucial aspects of science, if still in a rudimentary form. The first to be discussed, Heraclitus, about forty years younger than Anaximines, spent his life in Ephesus, a city near Miletus in Ionia, but boasted of not being a disciple of anyone. In fact his approach to cosmology did not display the scientific outlook of the Milesians, but was more mystical and metaphysical. Known as an unusually "haughty," "obscure," and "prophetic" individual, his philosophy exhibits an enigmatic, cryptic, and syncretic style.

The point in discussing him is to contrast his manner of thinking with the Milesians to show how speculative thought, when too disjoined from empirical constraints becomes fanciful, but also because he chose fire as the basic stuff of things. As previously indicated, "opposition and strife" were commonly evoked in the systems of the Milesians to explain change and becoming which was known as the problems of "the One and the Many" and "Change and Identity." But rather than focusing on the conflicting elements themselves, it was the feature of change itself that Heraclitus abstracted as the crucial principle in his system. As Plato recounts in the *Cratylus* (402A), "Heraclitus is supposed to have said that all things are in process and nothing at rest; he compares this to the stream of a river, into which you could not step twice into the same water." Just as time is fleeting prohibiting occuping more than one moment of time, so one cannot "step twice into the same waters of a flowing river."

Yet, because the water in a river can be still or stagnant, even this simile is inadequate. So one can imagine Heraclitus asking: what constantly changes while still remaining the same? concluding that the only possible answer is fire. Who has not been entranced watching flames dance and flicker as they consume a log in a hearth? So it must have occurred to Heraclitus who said "[t]his world order did none of the gods or men make, but it always was and is and shall be an ever living fire." Yet not satisfied with fire as the complete explanation, he adds Hesiod's principle of "Strife" as another generating principle: "It is necessary to know that war is common and right is strife and that all things happen by strife and necessity." Yet despite this inherent conflict and diversity producing change, there also is an underlying unity or "measured transformation" whereby fire is "extinguished" to become the sea and the sea the earth, which eventually are "exchanged" back into fire. It is this additional unifying process that he describes as "the path up and down as being one." Finally, along with fire and Strife constituting change, he adds the "Logos," a divine influence which "steers the process," although acknowledging the obscurity of the conception: "Of the Logos men always prove to be uncomprehending. . . . For although all things happen according to this Logos men are like people of no experience [and hence fail to comprehend it]."

Unlike the Milesians, Heraclitus does not offer rational explanations based on empirical analogues, but a diffuse system that disdains logical distinctions, conceptual clarity, and empirical evidence. This is especially obvious when one contrasts his meteorological explanations with those of Anaximines, claiming the sun is the size it appears to be, "the breadth of a man's foot" and "is new each day." He is said to have described the heavens as containing "bowls turned with their hollow side towards us, in which the bright exhalations are collected and form flames," of which the "brightest and hottest is the flame of the sun." Eclipses and phases of the moon were due to different turnings of the bowls, the "sun and moon . . . eclipsed when the bowls face upwards; and the monthly phases of the moon occur as its bowl is gradually turned."

However, unlike his convoluted cosmology and implausible meteorological explanations, his ethical aphorisms have a Shakespearean ring showing astute insight into the human condition: "A lifetime is a child playing with draughts"; "Man like a night light is kindled and put out"; "It is hard to fight desire, what it wants it buys with the soul"; "Nature loves to hide"; "If one does not expect the unexpected one will not find it out"; "Men who love wisdom must inquire into many things." But the one I consider the truest and the most profound is the following: "Man's character is his destiny."

THE PYTHAGOREANS

Little is known about the founder of the next school of Presocratics, Pythagoras, other than his spending his early years on the island of Samos off the coast of Ionia across from Miletus, later leaving for Croton in southern Italy to escape from the tyrant Polycrates. There, in the last quarter of the sixth century, he founded a religious order whose members were sworn to secrecy regarding their rituals, mathematical discoveries, and cosmological speculations. As was true of other religious leaders (e.g., Christ), he was alleged to have performed various miracles and to have been divinely conceived from a mortal woman by Apollo. It was not until a later follower Philolaus, in the following century, violated the vows of secrecy by publishing accounts of the Pythagorean system that their contributions became known.

What makes the speculations of the Pythagoreans so perplexing and intriguing is that the distinction between mathematics (especially geometry) as a purely axiomatic formal system and physical space had not yet been made. It was not until the nineteenth century when Carl Friedrich Gauss, Nicolai Lobachevsky, Janos Bolyai, and Bernard Riemann constructed non-Euclidean geometries that the question of the difference between the two became apparent. Considering alternatives to Euclid's crucial maxim that parallel lines never meet, they discovered that other valid geometries could be derived from different maxims where the lines do intersect.

Originally these non-Euclidean geometries (so named by Gauss) were dismissed as irrelevant because it was thought that only Euclidean geometry was a true description of objective space. However, when Einstein later incorporated Riemannian geometry to represent curved space in his General Theory of Relativity that it was recognized that whether geometry was considered to be a true account of space had to be decided by empirical investigation, not as an a priori given. In a cosmological system such as Einstein's where space is spherical, a different geometry is required than where it is rectilinearly infinite as in Newton's cosmology.

It was this distinction that was not apparent at this time. Yet paradoxically, it was their inadvertently conflating the two that enabled the Pythagoreans to construct their unusual cosmology. Today we understand that numbers are necessary for counting, adding and subtracting, and dividing quantities, that geometry and trigonometry are the formalisms for constructing and computing the properties of geometrical figures, and that equations represent the functional correlations of physical properties discovered in nature (e.g., $v = d/t$ or $E = mc^2$). Assigning magnitudes to the known variables in the equations and then solving them enables scientists to calculate the values of the unknown quantities, making possible their prediction and testing. It is the task of mathematicians to provide valid mathematical functions underlying the computations, while it is the role of scientists to utilize the correct formalism in solving a problem.

As the theorem, $a^2 + b^2 = c^2$, which was both discovered and proved by Pythagoras demonstrates, given the dimensions of the two equal sides of a right triangle, the magnitude of the unknown third side can be calculated. As a method of triangulation, it is very useful in astronomy. A simple example, it nonetheless exemplifies the function of the most complicated equations and computations in mathematics and science. In quantum mechanics, for example, the differential equations of Newton were superceded by Schrödinger's wave and Heisenberg's matrix mechanics along with Dirac's transformational equations (which surprisingly proved mathematically equivalent in their solutions to quantum problems). Unlike language whose symbols have an established conventional meaning, intentional (conceptual) and extensional (denotative), mathematics is an uninterpeted formalism until its symbols are assigned quantities.

The purpose of elucidating this distinction between the mathematical formalism and its empirical applications is to show how limitations of knowledge can distort the interpretation of important discoveries. It may have been Pythagoras himself using a monochord, an ancient musical instrument, who discovered that the intervals of the musical scale correspond to numerical ratios, the basis of his number cosmology. By plucking a single string stretched over a sound box having a movable bridge with a graduated scale, it was found that the different tones produced by the various lengths of the string are in numerical ratios expressed by the first four integers. That is, dividing the string by halves (1/2) produces the next higher octaves, division by two-thirds (2/3) produces the major fifth, and by three-fourths (3/4) the major fourth. Then, in what must be one of the boldest generalizations in history, the Pythagoreans concluded that the numerical ratios underlying the musical scale must exemplify the origin and structure of the universe itself!

Although seemingly very far sighted and actually extremely influential, what prevented this discovery from being as innovative as it appears to be was the inability to distinguish between mathematical ratios as formal notations and the world itself. Perhaps it was the common practice of representing numbers diagrammatically as physical marks on papyrus scrolls or in the sand (Archimedes was allegedly killed by an angry Roman soldier because he was so engrossed in a diagram drawn in the sand that he ignored the latter's command) that led to their conflating numerical properties with spatial and material configurations. Just as geometrical figures can be drawn from points and lines, three-dimensional pyramids constructed from four identical triangles, and rectangular solids from six rectangles, so the Pythagoreans believed that the physical universe could be constructed from arithmetic-geometrical units, analogous to crystallography.

This is succinctly described by Speusippus, another important doxographer: 'For 1 is the point, 2 the line, 3 the triangle and 4 the pyramid . . . and the same holds in generation too; for the first principle in magnitude is the point, the second the line, the third the surface and the fourth the solid.' But it is Aristotle who provided an exact summary of the Pythagorean position.

> And the Pythagoreans, also, believe in one kind of number—the mathematical; only they say it is not separate but sensible substances are formed out of it. For they construct the whole universe out of numbers—only not numbers consisting of abstract units: they suppose the units to have spatial magnitude. But how the first I was constructed so as to have magnitude, they seem unable to say. (*Metaphysics,* 1080b 16–22; Ross translation)

Not only is this an excellent summary of the Pythagorean position, Aristotle adroitly identified the crucial weakness in their system, that of explaining how the original unit acquired spatial dimensions.

From their original numerical discovery the Pythagoreans abstracted the principles of their cosmological system. Just as the movable bridge imposed "limitations" on the "unlimited" continuum of sound with the tonal variations corresponding to even or odd ratios, they claimed that the universe arose from the imposition of the "Limiting" on the "Unlimited," which in turn generated diverse opposites under the categories of the "Odd" and the "Even." And since oddness and evenness clearly are features of numbers, while the limited and the unlimited apply to spatial distinctions and thus are geometric, the Pythagoreans were able to construct an original arithmogeometric cosmology.

Moreover, they enlarged the categories of the Limited and the Odd and the Unlimited and the Even to include under the former "one, right, male, resting, straight, light, good, and square," while under the latter "plurality, left, female, moving, curved, darkness, bad, and oblong." Much ingenuity has been expended in trying to make sense of these groupings, with little success. With his usual acuteness, Aristotle asked how "weightiness and lightness," or sensory qualities such as "white, sweet, and hot" can be considered an "assemblage of units?" They considered ten to be the perfect number because the addition of the four integers composing the ratios in their initial discovery added to ten, $1+2+3+4=10$, while the total number of each of the two categories also equaled ten. Even in their astronomy they added a counter-earth so that the total number of the celestial bodies, Mercury, moon, Venus, sun, Mars, Jupiter, and Saturn, along with the fixed stars, the earth, and the counter-earth, would add up to ten. According to Aëtius, the number ten was so important they used to invoke the "tetrad" as their most binding oath: "Nay, by him that gave to our generation the tetractys, which contains the fount of eternal nature."

Returning to the question raised by Aristotle, how the unit 1 acquired the spatial properties essential for generating the three-dimensional universe, an answer was proposed by Philolaus. According to Kirk and Raven, in their excellent source book on the Presocratics, his answer drew on an analogy with biological creation.

> Just as the sperm . . . is deposited in the womb, so also in cosmogony, the first unit, which represents the principle of the Limit, is somehow implanted in the midst of the surrounding Unlimited; and just as the child, immediately after birth, inhales the breath outside, so also the first unit, immediately after it is generated, proceeds to draw in the void from the surrounding Unlimited.[3]

Hardly convincing, this germinal explanation shows how intractable the problem was and how primitive were their attempted explications.

This is evident also in their astronomical speculations in which they arrived at a true conclusion which had significant consequences, but for fanciful reasons. Having introduced the counter-earth to make the number of the astronomical bodies add up to ten, in a similar a priori fashion they displaced the earth from the center of the universe by Fire, calling it the "Hearth of the World." As Aristotle again relates: "At the center, they say, is fire, and the earth is one of the stars, creating night and day by its circular motion about the center. . . . Their view is that the most precious place befits the precious thing but fire, they say, is more precious than earth. . . ." (*De Caelo*, 293a21–a32, Stocks translation) Based on his own empiricist philosophy, Aristotle correctly criticizes the reasons given for their conclusions: "In all this they are not seeking for theories and causes to account for observed facts, but rather forcing their observations . . . to accommodate them to certain theories and opinions of their own" (293a25).

Still, it was an audacious and prescient move to replace the central earth by fire vaguely foreshadowing the heliocentric conception of the universe. Further exhibiting their remarkable though fanciful originality, as with their discovery that numerical harmonies underlie the musical scale they attributed a musical harmony to the motions of the celestial bodies. Assuming that because of their size and speed they should emit sounds, they concluded that "their speeds, measured by their distances, are in the same ratios as musical concordances," thus conceiving the extraordinary idea of the music of the spheres, though a refrain too distant to be audible to human ears.

This conjecture that the speeds and distances of the heavenly bodies are in ratio (similar to musical harmonies) would be vindicated two millennia later by Kepler. In his *Harmonice Mundi* (*World Harmony*) he introduced his Third Law that the ratio of the orbital periods of the planets squared is proportional to the cube of their mean distances from the sun. Thus the imaginative speculations of the Pythagoreans received precise mathematical expression by Kepler, who considered himself a neo-Pythagorean.

Copernicus, too, in the prefatory letter to the *De Revolutionibus*

(*Revolutions of the Heavenly Spheres*), credited Philolaus with attributing to the earth an obliquely circular motion around the central fire and Heraclides of Pontus and Ecphantus with adding a diurnal rotation on its axis from West to East. Since these were the two motions attributed to the earth by Copernicus, he cited these Pythagoreans as justifying his own advocacy of the heliocentric universe. Galileo also praises the Pythagoreans when, in response to Sagredo who, in *Dialogues Concerning the Two Chief World Systems* expressed surprise that there were so few adherents of the Pythagoreans over the centuries, has Salviati reply,

> my surprise is very different from yours. You wonder that there are so few followers of the Pythagorean opinion, whereas I am astonished that there have been any up to this day who have embraced and followed it. Nor can I ever sufficiently admire the outstanding acumen of those who have taken hold of this opinion [the fire or sun centered universe] and accepted it as true; they have through sheer force of intellect alone done such violence to their own senses as to prefer what reason told them over that which sensible experience plainly showed then to the contrary.[4]

As Galileo pointed out, apart from the conflict with scripture, the underlying issue in the controversy between the geocentric and heliocentric universe was whether to accede to the evidence of our senses as the Aristotelians advocated or to aspire to a simpler mathematical system of the universe even though opposed to ordinary observations.

Yet the greatest tribute to the speculative foresight of the Pythagoreans is to be found in contemporary string theory. As described by Michio Kaku,

> if you had a supermicroscope and could peer into the heart of an electron, you would see not a point particle but a vibrating string. . . .
>
> If we were to pluck this string, the vibration would change; the electron might turn into a neutrino. Pluck it again and it might turn into a quark. In fact, if you plucked it hard enough, it

> could turn into any of the known subatomic particles. In this way, string theory can effortlessly explain why there are so many subatomic particles. They are nothing but different "notes" that one can play on a superstring. . . . In fact, all the subparticles of the universe can be viewed as nothing but different vibrations of the string. The "harmonies" of the string are the laws of physics.[5]

The resemblance to the Pythagoreans in these passages is astonishing, as noted by Kaku.

> Historically, the link between music and science was forged . . . when the Greek Pythagoreans discovered the laws of harmony and reduced them to mathematics. . . . Not surprisingly, the Pythagoreans' motto was "All things are numbers." Originally, they were so pleased with this result that they dared to apply these laws of harmony to the entire universe. Their effort failed because of the enormous complexity of matter. However, in some sense, with string theory, physicists are going back to the Pythagorean dream. (p. 198)

What greater recognition could these early thinkers have wished for! Yet it must be added that just as the mathematical cosmogony of the Pythagoreans lacked empirical support, after several decades of research there still is no empirical evidence for string theory.

PARMENIDES

The next Presocratic to be discussed was a younger contemporary of Pythagoras who was born in Elea, not far from Metapontium where Pythagoras died, and though it is said that he was a Pythagorean in his youth there is no evidence that they ever met.

Although his system does resemble the Pythagoreans in that it is based on formal principles, it is unlike the Pythagoreans in not being derived from a mathematical discovery (that the musical scale consists of mathematical ratios), but on logical maxims that are entirely devoid of empirical content. Indeed, like Plato, Parmenides denied any credence to sensory evidence or physical properties in the preferred "Way" of his two proposed systems. Because his ontology is so farfetched having no scientific relevance, its consideration will be as brief as possible. The purpose in discussing it is to show how unrealistic it is to impose on the world a conceptual framework independent of observable evidence or empirical tests, the essential criteria of scientific inquiry that took several millennia to establish. One can forgive Parmenides for not recognizing this at such an early period, but not those who currently want to impose "intelligent design" or "special creation" on the empirical data of evolution which is unjustified for the same reasons.

Parmenides left only one work, an oracular poem which begins with a Prologue followed by two parts, the first called "The Way of Truth" and the second "The Way of Opinion." Claiming that together they exhaust the possible conceptions of reality, the first is extremely cryptic and obscure, yet it had considerable influence on the theorizing of the following Presocratics. The second (that I will not discuss) presents the more familiar representation of the world for those incapable of following the logical prescriptions of the first Way. The following is a typical statement in the "Way of Truth" setting forth the logical criteria for establishing his ontology.

> Come, I will tell you—and you must accept my word when you have heard it—the ways of inquiry which alone are to be thought; the one that IT IS, and it is not possible for IT NOT TO BE, is the way of credibility, for it follows Truth; the other, that IT IS NOT, and that IT is bound NOT TO BE: this I tell you is a path that cannot be explored; for you could neither recognize that which IS NOT, nor express it. . . . For it is the same thing to think and to be. . . . One should both say and think that Being Is; for To Be is possible and Nothingness is not possible.[6]

What this enigmatic passage maintains is the equivalence of Knowing and Being because, according to the argument, the logical principles of knowing establish that only Being exists, along with the nonexistence of Not-Being and the impossibility of knowing it or of deriving Being from it. What he is attacking is the conceptual framework of his predecessors who assumed that the universe arose from an antecedent element or state and that its present condition was brought about by a becoming or transformation. Yet it is not their actual explanations that he is critiquing, but the logical presupposition of the framework itself. Expressed in today's logical symbolism, what he maintains is that BEING (P) IS and NOT BEING (-P) IS NOT -(-P), nor CAN IT COME TO BE, hence either P or -P (the exclusive P v -P), and since -(-P) is excluded, therefore P. As he says: "Being has no coming-into-being . . . for what creation of it will we look for? How, whence [could it have] sprung? Nor shall I allow you to speak or think of it as springing from Not-Being; for it is neither expressible nor thinkable that What-Is-Not Is" (p. 43).

This restrictive logical schema that only Being Is and that Not-Being is Not, nor can there be any Coming-to-Be, excludes all change, becoming, motion, time, degrees of being, spatial separation, and qualitative alterations, the fundamental explanatory concepts of the previous philosophers. According to Parmenides, they all presupposed a coming to be from what did not preexist, that is from nothing. Hence the choice is between accepting a worldview precluded by his a priori logical criteria of what can exist or pursuing an explanation of the world based on inferences from its observed structures and processes.

There is little doubt as to what we would select today, which is obvious when we confront the conception of the world implied by these logical constraints.

> There is only one . . . description of the way remaining, that [What Is] Is. To this way there are very many sign-posts: that Being has no coming-into-being and no destruction, for it is . . . without motion, and without end. . . . Nor is Being divisible, since it is all alike. Nor is there anything . . . which could prevent

> it from holding together, nor any lesser thing, but all is full of Being. Therefore it is altogether continuous . . . in the limits of mighty bonds, without beginning, without cease . . . therefore all things that mortals have established, believing in their truth, are just a name: Becoming and Perishing, Being and Not-Being, Change of position, and alteration of bright colour. (pp. 43–44)

Thus the typical categories used in interpreting the external world are only nominal, having no objective validity. And since he has not admitted corporeality into his system, characterizing Being as "continuous, complete, unchanging, and timeless," his ontology seems to reduce to the single property of spatial extension. This is reinforced by his assertion that "remaining the same in the same place, it [Being] rests by itself and thus remains there fixed" (p. 44). It would appear at first that Being, the quintessence of existence, could not possibly be equated with spatial extension which is usually thought of as emptiness. Yet this seems to be the only consistent conclusion allowed by his logical principles and description of Being.

Furthermore, claiming that if Being were unlimited it would lack fullness or completeness, he deduces that it must have a Limit, but one that would make Being self-contained. So in addition to its spatial extension, it has a spherical shape: "Being shall not be without boundary. For . . . if it were [spatially infinite] it would be lacking everything. . . . [And] since there is a Limit, it is complete on every side, like . . . a well-rounded sphere. . . ." (p. 44). Curiously, his conclusion is not unlike Einstein's who also concluded that the universe is spherical and therefore both finite and infinite: finite in the sense that the dimensions are bounded, but infinite in the sense that it is continuous such that one could never go beyond it, its global shape forcing one to continue in a spherical trajectory. The difference is that Einstein's universe is four-dimensional, three dimensions of space and one of time, the latter excluded by Parmenides.

Parmenides' disciple, Melissus of Samos, wrote a treatise *On Being* defending Parmenides' system except for his description of Being as spherical. Arguing that whatever has a limit, even if equal in every direction as defined by Parmenides, must be limited by Non-Being, Melissus concluded that Parmenides' Being must be without limits or infinite. As he says: "For it is impossible for anything to Be, unless it Is completely [self-contained and everlasting]. . . . But as it Is always, so also its size must be infinite." In addition, "If it were not One, it will form a boundary in relation to something else [Non-Being]. If it were infinite, it would be One" (p. 48).

As these arguments indicate, they depend solely on the meanings of the terms (as in theology and metaphysics) because there are no empirical facts that one could refer to in attempting to settle the matter. Just as in Christianity where the doctrine of the Trinity as three persons in one, the Father, Son, and Holy Spirit, exist as one Godhead or consubstantial is purely definitional, there being no analogue in nature, so the doctrines of Parmenides depend solely on definitions resulting from his prescriptive logical ontology.

In conclusion, we have seen how Parmenides' system depended on his discovery of the logical maxim, "BEING IS AND NOT BEING IS NOT," which formally precludes any change and becoming. Just as the Pythagoreans conflated mathematics with spatial existence in their cosmology, so Parmenides conflated logical principles with ontological possibilities. As our previous discussion of pure mathematics distinguished geometry and numerical computations from their empirical interpretations or applications, so the use of logical formalism to test the logical validity of arguments must be distinguished from whether the arguments are factually true or false. Today we take for granted such distinctions, but as everything else we know, they had to be established at one time.

Yet despite the overall implausibility of Parmenides' ontological conclusions, his rationalistic principles had a decisive impact on later philosophers, such as Empedocles, Anaxagoras, and Democritus. Though they rejected his dismissal of all sensory and empirical evidence, they nonetheless felt compelled to abide by the following logical canons: (1) that not-being cannot be the source of being or becoming; (2) that because whatever exits cannot come from not-being it must be ultimate; (3) that since plurality cannot arise from a unity it too must be ultimate; (4) that as change or transformation cannot be a coming to be from nothing or perishing, it must be explained by what exists; and (5) that if there is motion it too must be an original property of being. So while rejecting Parmenides' specific logical precepts as a priori criteria of what can exist, those that followed nonetheless found them essential for guiding further theorizing.

EMPEDOCLES

A few years ago (May 2005) when visiting Sicily my wife and I discovered the Port of Empedocles just outside of Agrigento, the present name for the ancient city of Akragas where Empedocles was born of a noble family in the middle of the fifth century. It was gratifying to find that recognition, usually reserved for generals or statesmen was given to an ancient scholar. A complex personality, Empedocles combined wide ranging scientific interests with the temperament of a poet, mystic, and prophet. As founder of the Sicilian school of medicine his reputation as a physician was so outstanding that he was said to have performed miraculous cures, such as bringing a woman back to life who had been dead (probably in a coma) for thirty days. Although his physical, cosmogonic, and astronomical writings were strongly influenced by the works of the Milesians, especially the metaphysical outlook of Heraclitus and the logical canons of Parmenides, he is primarily distinguished for his original research in biology, embryology, and physiology.

While he wrote other works, like Parmenides he is especially known for two poems, one titled *On Nature* containing his scientific theories and the other, *Purifications*, showing the mystical influence of the Orphic religion. He begins *On Nature* by emphasizing the importance of sensory evidence, declaring that though the "eyes are more accurate witnesses than ears," one must utilize "whatever way of perception makes each thing clear." Relying on empirical evidence rather than logical canons in constructing his worldview, he rejected Parmenides' ontological monism selecting the combined elements of his predecessors, fire, air, earth, and water, as the eternal components of the universe. Yet illustrating the continued influence of myth on his thought, he introduced the four elements as "the four roots of things: bright Zeus, life-bearing Hera, Adonis, and Nestis which causes a mortal spring of moisture to flow with her tears." It is generally agreed that Zeus represents fire, Hera (goddess of the heavens) air, Adonis (a god of the underworld) earth, and Nestis water.

Regarding the four elements as ultimate, he explains change and becoming as a mixing of these four entities, with the defining characteristics of compounds determined by their predominant components. To explain change he returns to the notion of an eternal cyclical process which, in his account, consists of the combining and separating of the elements caused by the action of Hesiod's Love and Strife. This occurs in four phases, two when either Love or Strife is dominant and two transitional states brought about by their contrasting waxing and waning. When Love reigns and Strife is abated there is a fixed uniform blending of the elements while the mixing occurs when Strife is dominant and Love inactive, the whole composition having Parmenides' spherical shape. Combining this empirical explanation with his mysticism, Empedocles identified the spherical universe with God, declaring that God "is equal in all directions to himself and altogether eternal, a rounded Sphere enjoying a circular solitude."

Permeating the Sphere consisting of transitional mixtures of the elements, in concomitant contrasting directions Love and Strife "stream" back and forth from "the lowest depth of the whirl" to "the outer circumference," the conflicting results producing the motion and segregation of the elements. As described by Simplicius:

> When Strife had reached to the lowest depth of the whirl, and Love was in the middle of the eddy, under her do all these things come

> together so as to be one, not all at once, but congregating each from different directions. . . . And as they came together Strife began to move outwards to the circumference. . . . And in proportion as it was ever running forth outwards, so a gentle immortal stream of blameless Love was ever coming in. [Thus] what had been unmixed before was now mixed, each exchanging its path. . . . [7]

It is difficult to make sense of this, especially as to how the opposing streams of Love and Strife could concurrently flow in reverse directions to produce at one extreme the uniform blending when love is in the center and at the other a differentiated mixture when hate reaches the periphery. Yet despite its obscurity, his intent is to conform to Parmenides' logical maxims: "From what in no wise exists, it is impossible for anything to come into being; and for Being to perish completely is incapable of fulfillment and unthinkable. . . . Nor is there any part of the Whole that is empty or overfull" (p. 329). While complying with each of Parmenides' logical canons, this cosmological cycle also produces the astronomical structure of the universe, along with the generation of plant and animal life.

His description of the origin of astronomical phenomena draws on the explanations of Anaximander and Anaximines. Borrowing Anaximander's term "separating off" to explain how the celestial bodies were formed, he asserts that when the outflow of Strife began separating the elements, the first to appear was air, then fire, with the earth and water later "coming together" in the center of the Spherical universe. Forming a band around the circumference of the Sphere, the outer edge of the air solidifies creating a "crystalline" border, perhaps the origin of the later concept of "crystalline spheres." Then adopting Anaximines explanation, he asserts that fire subsequently separates off rising to displace and surround the air beneath the solidified border. A portion of the fire mixes with the air and sinks to the lower half of the sphere while the remaining pure fire fills the upper half, forming two distinct hemispheres foreshadowing the later distinction between the terrestrial and celestial worlds.

The upper hemisphere of pure fire and the remaining air is the source of light, while the loss of fire and mixture of air in the lower hemisphere causes night, the imbalance of the two producing a circular motion creating day and night. Like Anaximander, he says the earth is kept in the center of the universe by the circular motion of the hemispheres whose centripetal force also produces water: "From the earth, as it was exceedingly constricted by the force of the rotation, sprang water." The fixed stars are made of fire and attached to the crystalline periphery similar to Anaximines' description that they are "like nail heads on the vault of the heavens." The sun and moon are not fiery as such, the sun deriving its light by reflection from the fire of the upper hemisphere, while the moon "gets its light from the sun," causing eclipses by rotating between the sun and the earth. As imprecise as this cosmological explanation may sound to modern ears, at least we have the initial formation of the ancient geocentric universe with the earth in the center of a sphere bounded by a crystalline orb and the sun and moon rotating within.

Disregarding Anaximander's remarkably prescient conception of the evolutionary generation of living organisms, including human beings, Empedocles' explanation involves the previously described "double process" of the coming together and separating under the influence of Love and Strife. Confining our discussion to the origin of humans, according to Aëtius this dual process is divided into four stages: the first consists of the grotesque creation of disjointed limbs, the second the combining of these limbs into monstrous creatures, the third the emergence of "whole-natured forms," and the fourth the normal sexual generation of organisms.

That he rejected Anaximander's more sensible explanation of the origin of advanced forms of life from aquatic creatures for the following surrealistic description is disappointing: "Here sprang up many faces without necks, arms wandered without shoulders, unattached, and eyes strayed alone, in need of foreheads" (p. 336, fn 442). This is succeeded by a next phase just as grotesque: "Many creatures were born with faces and breasts on both sides, man-faced ox-progeny, while others sprang forth as ox-headed offspring of man, creatures compounded partly of male, partly of the nature of female, and fitted with shadowy parts" (Ibid.). Yet when we turn to his more empirical investigations as a physician we find a genuine scientific curiosity and pursuit of sensible explanations. In the kinds of questions he raised, as well as his explanations in terms of empirical analogies, we can see the early signs of a truly scientific mind.

To cite a few examples, asking what determines the sex of a child he answered that it depends upon whether the father's seed enters the warmer or colder part of the mother's womb, the colder resulting in females and the warmer in males. Regarding the source of inherited features, he claims that they too are determined by the temperature of the parents' seed: if they are equally hot the child resembles the father and if equally cold the mother, while if the father's seed is hotter than the mother's a boy with his mother's features is born and if the mother's is hotter a girl with her father's features. He also attempts to explain sterility, the birth of twins, triplets, and deformed fetuses. After describing the embryological formation of different organs, he noted that embryos are encased in a "sheepskin" (amnion) and that the blood and bones are composed of different proportions of the elements. While understandably crude, what is significant is his attempt to provide naturalistic explanations of these familiar but puzzling phenomena.

His description of respiration and perception is more astute, attributing respiration not only to the nostrils but also to the inward and outward flow of air in tubes stretched beneath the skin but extending to "pores" on its surface. Inhalation occurs when the blood draws inward sucking the air with it and exhalation when the blood flows outward forcing the air out. As an illustration he cites the example of the klepsydra, a vessel with a perforated bottom and narrow neck commonly used by woman in Greece to draw water from the wells. To fill the vessel the opening in the neck is left uncovered to allow the air to escape when the klepsydra is pressed into the water forcing it to flow inwards through the perforations, then after the vessel is filled the opening is closed to prevent the air from entering and the water from escaping when the klepsydra is withdrawn from the well. An excellent example of an analogical explanation, the forces in the vessel illustrate how the air pressure in our bodily tubes causes the respiration.

In another analogy he uses the example of a lantern to illustrate how perception occurs. Despite the advances made in understanding the structure and function of the brain, it still is largely a mystery as to how the stimulations of the sense organs transmitted to neurological processes in the brain produce our perceptions of an external world. We thus can appreciate the difficulty these early thinkers faced in trying to explain the same experience. Just as the reflection from the enclosed fire in a lantern illuminates the surrounding world, Empedocles claims that perception occurs when the fire behind the pupils of the eye streaming outward through tiny passages encounters the 'effluences' emanating from external objects. But aside from not explaining how this interaction with effluences results in the perception of a visual image, this does not explain how we perceive elements other than fire, since in his system like perceives like.

Furthermore, there seems to be confusion as to whether it is the fiery projections from the eyes interacting with the effluences that produce vision or whether it is the emanations from the objects entering the appropriate senses that cause the perceptions. According to Aristotle, "Empedocles speaks sometimes as if sight were due to the fire in the eye shining out, and sometimes as if emanations from objects entered the pores of the eye." Smell, at least, is explained as due to effluences taken in as we breathe and thus is adversely affected by colds, while the ear functions as a bell clapper when struck by the pressure of the outer air. So perhaps the apparent ambivalence is due to the perspectival differences among the senses, sight and to a lesser degree hearing, portraying distant objects, while smell, taste, and touch are more obviously caused by immediate contact with their respective stimulations.

He also is concerned to explain the nature of thought. Consistent with his earlier identification of the Spherical universe with God, he asserts that "all things . . . have intelligence and a portion of Thought," yet he locates thought in the blood, "for the blood round the heart is Thought in mankind." This attribution of thought to the blood is based apparently on the fact that since blood is composed of a nearly equal mixture of the four elements, plus Love and Hate, this conformity enables us to think about the different effluences coming from the various phenomena in the external world. Combining his physiological explanation of perception and thought in *On Nature* with his religious view of the incarnation and immortality of the soul maintained in the *Purifications* suggests that the two works are complementary. Thus his assertion in the former that "all things have a portion of Thought" is consistent with his claim in the latter that "God is Mind, holy and ineffable . . . which darts through the whole universe with its swift thoughts." This is reinforced by the story that to substantiate his claim in the *Purifications* that he was an immortal god having completed his cycle of Purifications on the earth, he leapt into Mt. Etna to destroy any trace of his mortal ending.

In conclusion, one finds in Empedocles' physical theory of the four elements, his astronomical speculations containing the rudiments of a geocentric universe, and his astute biological and physiological investigations the definite traits of a scientific mind. That one also finds a residue of primitive thought in his use of Hesiod's mythical Love and Strife to explain the motion of the elements, the identification of God with the universe, and his bizarre description of the origin of human bodies is to be expected at this early stage of cognitive development. But in contrast to mythical or theocratic cosmogonies in which gods are believed to be the controlling forces in the universe, precluding scientific inquiry and explanation, Empedocles displays an avid curiosity about nature along with the belief that natural explanations are attainable. This is especially evident in the kinds of factual questions he raised about human organisms, such as the sexual origin of human traits and his use of empirical analogies to explain respiration and perception. But it was the doctrine of the four elements that was his most significant and lasting contribution, facetiously referred to as the Greek "periodic chart." Adopted by Aristotle, this conception of the basic elements remained the dominant theory of the composition of physical reality for over two millennia, until the chemical discovery of the atomic-molecular structure of matter in the eighteenth century.

Thus Empedocles' investigations illustrate several prerequisites of scientific inquiry: that one possesses a curiosity about natural phenomena, that one ask the right questions, and that by conducting the appropriate empirical inquiries one eventually will arrive at the correct answers. As none of these characteristics are true of religious worldviews, they are what distinguish science from religion and account for the tremendous advances in knowledge of the former, in contrast to the latter which teaches that the ultimate answers are to be found in a higher power inscrutable to human reason, thus discouraging or terminating inquiry.

ANAXAGORAS

As were most of his predecessors, Anaxagoras was from Ionia born in the city of Clazomenae not far from Miletus, Ephesus, and Samos, with the date of his birth given as 500 BCE. At the age of twenty he left for Athens where he remained for the greater part of his life pursuing philosophy and becoming the teacher of such prominent Athenians as Pericles, Euripides, and Archelaus who in turn taught Socrates, who instructed Plato, who was a teacher of Aristotle, who taught Alexander the Great and Theophrastus who succeeded him as head of the Lyceum in what surely must be the most distinguished philosophical dynasty in history. Although Anaxagoras wrote only one book, it was extremely important because of its complete disavowal of the earlier mythopoetic tradition and acceptance of scientific rationalism. Sadly, his emancipated thought was not appreciated by some Athenians who brought charges of impiety against him for maintaining that the sun and moon were material bodies rather than deities. This led to his being prosecuted, condemned, and exiled after which he left for Lampsacus where he died a few years later in 428. This sorry episode prefigured the trial and death of Socrates also on charges of impiety, although in his case fabrications because he did not speculate about the nature of astronomical bodies.

Like his younger contemporary Empedocles, according to Simplicius (who is the source of nearly all Anaxagoras' fragments), he rejected Parmenides' ontological monism while acknowledging his logical maxims in the assertion that "nothing comes into being nor perishes, but is rather compounded or dissolved from things that are. So they would be right to call coming into being composition and perishing dissolution." But while agreeing with Empedocles that becoming and change should be interpreted as a decomposition or compounding of what already exists, he denies that the great diversity of things can be derived from just four elements, fire, air, earth, and water. As he asks: "How could hair come from what is not hair or flesh from what is not flesh?" So believing that the four elements were insufficient to explain the great variety of things and that perhaps they too were composite, he claimed that the original mixture contains an infinitude of particles representative of all that exists, each of which in turn is infinitely divisible, so infinitesimal as not to be discernable, except for air and aither (fire).

Such was the composition of the original cosmos before the present universe was created by the "separating off" process.

> All things were together, infinite in respect of both number and smallness; for the small too was infinite. And while all things were together, none of them were plain because of their smallness; for air and aither covered all things, both of them being infinite; for these are the greatest ingredients in the mixture of all things both in number and size. (p. 268, n. 365)

In this conception that the original mixture contained an infinite number of elements each of which was infinitely small, Anaxagoras rejected not only Parmenides' ontological monism but his disciple Zeno's argument that if there were a plurality of entities each would have to be both infinitely small and infinitely large, a logical impossibility precluding the pluralism proposed by Anaxagoras.

In our previous discussion of Parmenides we had mentioned his disciple Melissus, but not his more well-known disciple Zeno who is famous for his reductio ad absurdum arguments proving the impossibility of motion and pluralism in defense of Parmenides' static monism. The most well-known of these arguments allegedly refuting motion involves Achilles and the tortoise: in a race in which the tortoise is given a head start, according to Zeno Achilles could never catch up to the tortoise no matter how fast he ran because during the time it took to reach the tortoise the latter would have advanced some distance farther. So while Achilles could continue to reduce the distance between himself and the tortoise by endless increments, he could never breach the gap to overtake the tortoise demonstrating that any motion is logically excluded.

Similarly, to refute the notion that there can be a plurality of elements he argued that if discrete spatial entities existed they would have to be both infinitely small and infinitely large, a logical impossibility. If the universe consisted of particles of various dimensions, then theoretically they would be infinitely divisible and therefore ultimately composed of infinitesimal units. But to be composed of an infinite number of units each particle would have to be infinitely large. Since the concept of discrete particles leads to a contradiction, that concept, too, must be rejected.

The fault in these arguments, like those of Parmenides, is that the logical principles preclude the existence of separate elements only because they (the principles) are not suitable for analyzing their status. By making certain logical assumptions one can deny the existence of anything. As every scientist realizes, any deductions from a formal system, whether logical or mathematical, are only empirically valid if the formalism is suitable to or fits the data. Apparently ignoring Zeno's arguments, Anaxagoras adopted both horns of his dilemma declaring that the original elements being infinitely divisible are both infinitely small and infinitely large, in size as well as number.

Even more perplexing, he claimed that while the original mixture did not manifest any discernable differences, it nonetheless contained such opposite qualities as "the moist and the dry, the hot and the cold, the bright and the dark, and of much earth . . . and . . . seeds countless in number and in no respect like one another" (p. 368, n. 496). His motive clearly was to ensure that whatever exists does not come to be from nothing, hence he endowed the original cosmos with a bewildering mixture of countless elements, states, qualities, and seeds. This conception of the original nature of the universe before the present world was created drew heavily on Anaximander. It is as if Anaxagoras decided to fill Anaximander's Unbounded, which also was imperishable, unlimited, and composed of a homogenous blending of the elements so that nothing was distinguishable, with an infinite number of elements too small to be discernable. Moreover, he replaced Anaximander's "steering principle," along with Heraclitus' "Logos" and Empedocles' "Love and Strife" with the conception of "Nous" or "Mind," defined as "self-ruled" and possessing "knowledge about everything" and the "greatest power controlling all things." Being unmixed and "the same throughout," it is "the finest of all things and the purest."

Then, analogous to the "whirl" of Empedocles' Sphere, this Mind initiates a "rotary motion" in a small area of the original mixture which then spreads throughout without blending with it. But in contrast to Empedocles' centripetal force, this increasing rotary motion creates a centrifugal force inaugurating the separating off process in the next phase of creation. Though having identified the second of the two physical forces generated by circular motion, Anaxagoras apparently felt the need to introduce 'Mind,' a nonphysical force, as an additional explanation of motion. He will claim later that all living things have a portion of Mind and that the soul "causes motion in the organism just as Mind did in the Whole." Once the rotary motion began the pairs of opposites in the original mixture, "the hot and the cold, the bright and the dark, the dry and the moist, and the rare and the dense," began to separate off. Emerging from each other, this opposition is intended to avoid the charge that they came to be from nothing.

He then includes a further statement that greatly adds to the confusion: "But there are many portions of many things, and nothing is altogether separated off nor divided one from the other except Mind" (pp. 372–73, n. 503). So in contrast to the original mixture in which the separate elements (in what I believe is the most plausible interpretation following Aristotle) were homogeneous or "homoeomerus," he now indicates that after the separating off process "everything contains a portion of everything else" except for Mind, distinguished by whatever portion is the largest. As he states:

> And since the portions of the great and of the small are equal in number, so too all things would be in everything. . . . Since it is not possible that there should be a smallest part, nothing can be put apart nor come-to-be all by itself, but as things were originally, so they must be now too, all together. In all things there are many ingredients, equal in number in the greater and in the smaller of the things that are being separated off. (pp. 375–76, n. 508)

Here we confront new paradoxes for if all things contain "ingredients equal in number in the greater and in the smaller," how can anything be distinguished from anything else? That in the original mixture all elements were "all together," as he stated, does not imply that each element was composed of a portion of everything else rather than being identical or homoeomerous, as he now seems to claim of the composition of elements that have been separated off. This, if true, would obviate things being distinguishable because if everything, from the largest to the smallest contains a portion of everything else, and each of these portions also are composed of everything else, then the position is self-refuting for there are no distinctive things that, in larger number in some portions than others, could give them their distinguishing qualities or properties.

Furthermore, there is the problem of interpreting what Anaxagoras meant by the assertion quoted previously that "air and aither covered all things, both of them being infinite . . . are the greatest ingredients in the mixture, both in number and size." But again, how can there be "a greatest number of ingredients" in a mixture in which each portion contains an infinite diversity of ingredients? Does it make any sense to refer to greater or lesser infinite quantities? Furthermore, what gives the air and aither their distinguishing qualities? Are they simply two different aggregates of identical particles of air and aither or do they also contain an infinite diversity of things?

Throughout history there have been numerous ingenious attempts to answer these questions without, however, dispelling the ambiguity. As Kirk and Raven assert:

> No Presocratic philosopher has given rise to more dispute, or been more variously interpreted, than has Anaxagoras. . . . It is actually very doubtful whether any critic, ancient or modern, has ever fully understood him, and there are some points on which certainty is now unattainable. (p. 367)

While some commentators have maintained that Anaxagoras could not have been that confused, I believe that with such a complex and intricate system the kinds of conceptual distinctions and clear explanations that we find easy to make after 2500 years of theorizing were not yet explicit or evident. We shall have to await his successors, especially Aristotle, for a clearer discrimination and elucidation of these problems. Yet the fact that Anaxagoras raised these questions with the need for clarification is much to his credit. Even the necessity of positing an underlying imperceptible domain of particles to explain the experienced world, the solution that will be adopted by the Atomists and become the foundational framework of modern physics, chemistry, and molecular biology, was just being recognized and introduced.

Concluding with his astronomical conjectures since his contributions to biology were less significant than those of Empedocles, here his break with the ancient mythopoetic tradition and commitment to a worldview based on empirical evidence and scientific rationalism is especially evident. As the separating out process continued, the increasing vortical motion forced the denser elements to the center and the lighter to the periphery, as asserted by his predecessors. "The dense and the moist and the cold and the dark came together here, where the earth now is, while the rare and the hot and the dry went outwards to the farther part of the aither." Thus air is a collective name for two pairs of the contrasting qualities (the dense and moist, the cold and dark) and aither for the other two pairs (the rare and dry, the hot and bright). The continuing pressure of the water causes the moist to condense into clouds, the clouds into water, and the vortical motion into the earth in the center.

Considering the earth to be flat, he says that it is supported by the dense air with the rivers and the seas formed from the water remaining in its "hollows." Solidified by the cold, stones from the earth are cast outward by the increasing rotary motion becoming the sun, moon, and the stars, the basis of the charge of impiety against him. The sun and the stars are heated by coming in contact with the aither, but the moon being closer to the earth is unheated. There are additional stones in the sky, all of which are kept from falling by the centrifugal force generated by the speed of the rotation. The sun is described as a "red-hot stone" larger than the Peloponnesus and the moon as having "plains and ravines on it," the latter description probably inferred from his having seen a large meteorite which fell in Aegospotami in 476. Given the commonly held belief that the sun, moon, and stars were celestial, heavenly, or divine bodies, his descriptions were extremely far sighted. It was not until Galileo's telescopic observations begun in 1609 that these inferences about the surface of the moon were confirmed, while it is only since the lunar landing provided samples of the moon's crust that we now know that it is composed of a different material.

In addition, Anaxagoras held that the moon is nearer to the earth than the sun and that its phases are produced by the light from the sun. "Eclipses of the moon are due to its being screened by the earth . . . those of the sun to screening by the moon when it is new." The stars are carried east to west by the aither that circles under the earth causing darkness or night. He suggests that the Milky Way, which is obscured by the light from the sun, consists of starlight shining through the shadow projected on the sky by the earth as the sun passes below it. "Comets are a concatenation of planets" while "shooting stars are broken off from the aither like sparks and immediately quenched." Lightning is a streak of fire bursting through the clouds while thunder is due to the fire being suddenly extinguished. The sun's rays reflected from the clouds produce rainbows.

Again, all traces of mythical fantasy or religious deification have been eliminated from these explanations. Even though three centuries or so separated the writing of Genesis from Anaxagoras, the striking difference in the treatment of cosmological and astronomical questions has to be explained by more than a lapse of time. With Anaxagoras there is no appeal to God creating anything or that light and darkness preceded the existence of the sun as in Genesis. Instead, Anaxagoras relied entirely on empirical data and terrestrial analogies to explain astronomical phenomena. What is even more significant is that while Genesis never gave rise to any further inquiry, as is true of all explanations in terms of God's commands or will, Anaxagoras was one link in the chain of Greek empirical inquiries and speculations that eventually led to modern classical science. This is why it is so astonishing that people should still believe in the ancient religious cosmogony of the Old Testament which never included or produced any significant knowledge, although it does contain some very beautiful literature and uplifting moral exhortations.

LEUCIPPUS AND DEMOCRITUS

We have now reached the coda of Presocratic thought with Leucippus and Democritus having originated the most powerful scientific theory of all time, the atomic theory, while Democritus' prescient conception of a universe consisting of an infinite number of solar systems in various stages of existence would not be confirmed until the telescopic discovery of vast numbers of extragalactic universes in the 1920s by Edwin Hubble. That Leucippus and Democritus were able to conceive such an extraordinarily effective physical theory, along with anticipating the results of one of the most significant telescopic discoveries ever made, is a tribute not only to the genius of the Greeks, but also to the capability of some human beings of overcoming tradition by initiating cognitive change. It is sad that while nearly everyone has heard of Galileo, Newton, Bohr, and Hubble, the originators of two of the most significant developments in the physical sciences are unknown. This is especially disgraceful considering that most people in the West know of Genesis which has no validity, while ignorant of the contributions of Leucippus and Democritus.

Although Aristotle is justifiably acclaimed because of the originality and comprehensiveness of his writings, in the fourth century BCE he was rivaled by Democritus who was one of the most prolific authors among the ancient Greeks. Reputed to have written fifty-two works on such diverse subjects as art, music, ethics, psychology, politics, mathematics, medicine, physics, and astronomy, he nevertheless remains unknown outside the cloistered halls of academe. Moreover, unlike Aristotle's physics or astronomy which has no significance today (in contrast to his biological investigations and treatises on ethics, poetics, aesthetics, and politics), practically all the theoretical speculations of Leucippus and Democritus have been vindicated by later developments in physics and astronomy.

There are two main reasons for this disparity in reputation: first, although atomism was prominent in Greece from 430 to 280 BCE owing largely to its adoption by Epicurus and was well known in

Rome due to Lucretius' popularization of the theory in his poem *On Nature*, the conception of the universe it presents, unlike Aristotle's, was so advanced for the time and divergent from ordinary experience that it descended into oblivion after the fall of Rome, not surfacing again until the seventeenth century when Newton adopted the corpuscular framework. The second reason is due to the tragic loss of Democritus' original works, even though much of it has been retained in the doxography, while only one fragment from Leucippus' book has been preserved.

Although it is generally agreed that Leucippus originated Atomism in a book titled the *Great World-system*, little else is known about his writings while his life remains a mystery. What can be said with confidence is that he developed atomism in response to the logical strictures of Parmenides and the paradoxical implications of Anaxagoras' system. Much more is known about Democritus who was from Abdera in Thrace which was slightly northeast of Stagera, the birthplace of Aristotle. In his treatise modestly titled the *Little World-system* after Leucippus' book, he states that he was forty years younger than Anaxagoras which places his birth at 460 and indicates he was ten years younger than Socrates and Leucippus.

Born in a family of wealth, when his father died he asked for his patrimony in money so that he could travel, whereupon he is said to have journeyed to Egypt to "visit the priests and learn geometry," as well as to Babylon and the Red Sea, possibly going as far as India and Ethiopia. Through these travels he acquired a broad background evident in his prolific writings earning him the sobriquet of "wisdom" among his contemporaries. He later visited Athens where he met Socrates, but apparently his reputation had not preceded him because he said "I came to Athens, but no one knew me." Generally cheerful in disposition, he acquired the title of "the laughing philosopher" due to his good-natured amusement at "the vain efforts of men."

Confronted with the same problems as his predecessors of explaining the changes and motions in the world, along with the compounding and destruction of things without violating Parmenides' canons and yet avoiding Anaxagoras' paradoxes of infinite divisibility, Leucippus conceived of physical reality as an indefinite number of imperceptible, indestructible elements. Defined as so dense as to be devoid of internal structure they were given the name "atmos" or "atoms," meaning compactness, a conception not refuted until the discoveries in the late nineteenth and early twentieth centuries of radioactive substances by Röntgen, Becquerel, and the Curies and subatomic particles such as the electron by J. J. Thomson, the proton by Rutherford, and the neutron by Chadwick. Though internally unalterable, the seemingly endless possibilities in their sizes, shapes, configurations, and positions, as well as their diverse entanglements with other atoms, were thought sufficient to explain the formation of macroscopic objects. As described by Simplicius: "They (Leucippus, Democritus, Epicurus) said that the first principles were infinite in number, and thought they were indivisible atoms and impassible owing to their compactness, and without void in them; divisibility comes about because of the void in compound bodies . . ." (pp. 407–408, n. 556).

In addition to the atom's primary qualities of solidity, size, shape, and position, the Atomists made the astute decision to dispense with any psychic or animistic sources of motion, such as the Love and Strife, Logos, or Nous of their predecessors. They simply endowed the atoms, as Newton will, with an inherent kinetic motion. For this they were criticized by Aristotle who claimed they did not specify what motion was 'natural' to the atoms nor what their 'original source' of motion was, reflecting his own erroneous presuppositions. According to Aristotle, the celestial and terrestrial realms were ontologically distinguished not only by the kinds of substances they were composed of, the former aither (ether) and the latter the four Empedoclean elements, fire, earth, air, and water, but also by their 'natural' motions, the celestial being circular and uniform while the terrestrial were vertically rectilinear and accelerating. Furthermore, Aristotle posited the Prime Mover as the ultimate source of motion which, though itself unmoved, "moves" the universe owing to the empathetic attraction exerted on the celestial bodies. Dispensing with these unnecessary properties and explanations by endowing atoms with an inherent kinetic motion, the Atomists again foreshadowed modern classical science. Thus in addition to connoting "compactness" or "fullness," the atom was also called "rhymos" or "onrush" because of its inherent mobility.

With this added property the Atomists were able to explain change, becoming, and alteration as due to the impact, interaction, compounding, and separation of the atoms. Using the model of motes randomly interacting in a sunbeam, they described the atoms as "colliding," "rebounding," "scattering," or becoming "entangled" because of their various shapes. The image of motes was particularly apt because the atoms were too tiny to be observed by the naked eye, so the mote image provided indirect evidence to reinforce the inferential arguments, analogous to Einstein's famous article on Brownian motion. As in modern chemistry, when compounded to form substances the individual atoms do not lose their identity, though the combined substances acquire new properties. Aristotle describes how the Atomists illustrated the possible generation of shapes with letters of the alphabet: "Leucippus and his associate Democritus . . . say the differences in the elements are the causes of all other qualities. For these differences . . . are three — shape and order and position. For . . . A differs from N in shape, AN from NA in order and I from H in position" (*Metaphysics*, 985b4; Ross translation).

This theory that the ceaseless changes and motions in the world could be accounted for by the unlimited variations in the physical properties and motions of the atoms required rejecting another one of Parmenides' maxims, that there could be no separation, gaps, or void in Being. But like Newton, the atomists realized that the discreteness, motion, and rearrangement of the atoms could not occur in a universe that was densely filled with matter, but required a medium which allowed for the unimpeded motions and interactions. Thus they were not only the first to introduce indestructible elements, they also added a kind of nonbeing or 'void' for the atoms to move in. With his usual astute ability to describe accurately other philosophical positions though not agreeing with them, Aristotle presents their rationale as follows:

> The void, they argue, "is not:" but unless there is a void with a separate being of its own, "what is" cannot be moved — nor again can it be "many," since there [would be] nothing to keep things apart. . . . The "many" move in the void . . . and by coming together they produce "coming-to-be," while by separating they produce "passing-away." (*De Generatione et Corruptione*, 325a5–30; brackets added, Joachim translation)

While the Pythagoreans and Parmenides had posited the existence of a spherical space, they did not consider it as void or empty, but as having some kind of quasi physical reality. So in positing an empty space, the Atomists had to distinguish between the existence of the atoms and a form of nonexistence of the void separating the atoms. Leucippus addressed the problem by drawing a distinction between the real and existing, declaring that both the atoms and the void exist, but only the atoms are real: "though empty space is not in a sense a real thing, it nonetheless exists." However, there still remained the question of whether space, as in Newtonian science, existed in itself as an independent container of the atoms, or whether it was merely the empty intervals between the atoms. Cyril Bailey, an outstanding historian of the Atomists, suggests that Democritus may have held the notion of space as "the whole extent of the universe some parts of which were occupied by matter," while Leucippus tended to think of it merely as "the empty parts, the intervals between bod[ies]."[8] There is no evidence that they attempted to explain the origin of the atoms and the void, apparently assuming their eternal existence.

Another problem pertains to whether the atoms possessed weight. After Galileo's limited and Descartes' generalized definition of inertial motion and Newton's conception of gravity as a force which is directly proportional to the product of the masses of objects and inversely proportional to the square of the distance between them, within Newtonian mechanics weight is considered a derived property created by a gravitational force that is relative to the masses of objects and the distances separating them. Beyond the gravitational perimeter of a massive body objects still possess mass but no weight, as the weightlessness of astronauts and objects in space ships has demonstrated. In outer space objects vary in their resistance to being pushed because of their densities or masses, but owing to the lack of gravity they do not fall when unsupported.

Because the Atomists had no notion of gravitational attraction, they had to explain weight by other means. Being homogeneous or made of the same material atoms did not vary in mass or density because of what they were made of, but they did differ in shape and size, along with bulk when conjoined in compounds, properties that explain their different weights. Since there is no record of Leucippus having discussed weight, one has to turn to Democritus for an answer. Various interpretations of his position have been offered, but the one that has gained the most consensus is that of Aëtius who maintained that the atoms in the infinite void do not possess weight since they do not evince an inherent downward motion. Endowed with kinetic motion, they move in empty space in a straight line until deflected by striking other atoms. However, their different sizes and shapes, along with their varying bulks or densities when compounded into larger objects, do contribute to differences in mass. Therefore, it is only *after* the formation of individual worlds, when the whirl of the vortices force bulkier or more massive objects towards the center and smaller and lighter objects towards the periphery, that motion downward or upward is considered evidence of an object's weight.

This bring us to their conception of the formation of innumerable worlds which presents the most striking evidence of the extraordinary originality and contemporaneity of their views, with all trace of myth eliminated from their system. In reading their account of the formation of worlds one could imagine perusing the earlier modern "nebular hypothesis" explanation of the formation of individual solar systems and galaxies, along with the "steady-state theory" of the universe that preceded the Big Bang theory. Originally, the cosmos consisted of an endless number of kinetic atoms scattered randomly throughout an infinite void. Given the enormous number of atoms in an eternity of time, occasional chance collisions would result in some coalescing because of their converging "hooked shapes," their increasing bulk attracting more atoms creating the vortical whirling motion of unique worlds located in a particular area of the void.

This whirling motion forced the larger of the compounded atoms to fall to the center of the vortex eventually coalescing as the earth, while the smaller "round and smooth" fiery atoms are "squeezed" to the periphery. Evidently aware of the inclination of the earth's axis, the Atomists explained this as a tilt to the south due to an imbalance of the earth's bulk. Like their predecessors, they erred in describing the earth's shape as "oblong," its length given as one and a half times its width. Apparently influenced by Anaxagoras, they asserted that eventually some huge "earth stones" are thrust into the periphery forming the sun, moon, and the stars. Originally cold, the sun and the stars become "molten" bodies acquiring their heat and brightness either by their rapid movement through the heavens or by coming in contact with the outer layer of fiery atoms. In contrast, the moon's surface resembles the earth's consisting of rocky expanses along with mountains and valleys, acquiring its light from the sun. They claimed that although the velocities of the planets and the stellar bodies increase as their distance from the earth increases, they appear to move more slowly owing to their greater distance, thus the moon seems to revolve more rapidly than the sun, and the sun more than the fixed stars, an illusion due to their being closer to the earth and having smaller orbits. In this manner, the void came to be filled with an endless number of vortical worlds occupying their individual regions of space.

This theory that a chance aggregation of the atomic particles resulted in their condensing into the material bodies of the earth, moon, sun, and fixed stars forming a solar system is roughly analogous to the "nebular hypothesis" used today to explain the origin and formation of similar astronomical systems. Furthermore, the Atomists' conception of the universe consisting of different solar systems in various stages of creation and disintegration, along with their divergent environmental conditions either supporting or prohibiting multiple forms of life, again resembles the modern conception of the universe. As described by Hippolytus,

> there are innumerable worlds, which differ in size. In some worlds there is no sun and moon, in others they are larger than in our world, and in others more numerous. . . . The intervals between the worlds are unequal; in some parts there are more worlds, in others fewer; some are increasing, some at their height, some decreasing; in some parts they are arising, in others failing. They are destroyed by collision one with another. There are some worlds devoid of living creatures or plants or any moisture. (Kirk and Raven, p. 411, n. 564)

Except for their conception that the worlds are not distributed uniformly, as they were in the steady-state theory, there is nearly perfect agreement between the two cosmological systems. When one considers that the Atomists arrived at theirs about twenty-four centuries ago the achievement is astonishing. For those seeking a rational conception of the universe, it certainly is a tribute to the Greek genius that in just two centuries their philosophers were able to replace the prevailing mythological accounts with a cosmological system comparable to one accepted in the early twentieth century.

I find it difficult to decide whether their conception of accidental causes signify some kind of inherent indeterminacy as exists in modern quantum mechanics, or perhaps more likely that some causes are too complex to be completely understood and therefore are described as accidental or due to chance. As Bailey describes their view:

> The Atomic conception of "chance" then is . . . the purely subjective conception which is proper to a scientific view of nature. "Chance" is no external force which comes in to upset the workings of "necessity" by producing a causeless result; it is but a perfectly normal manifestation of that "necessity," but the limits of the human understanding make it impossible for us to determine what the cause is. (Bailey, p. 143)

The consistency of their explanations even extends to their conception of the soul which is defined as spherical "soul-atoms" similar to the mobile fiery atoms mentioned earlier, thereby avoiding the notorious mind-body problem that has preoccupied so much of modern philosophy. According to Aristotle, "Democritus says that the spherical is the most mobile of shapes; and such is mind and fire." Breathing draws in these spherical atoms from the air when one inhales, thus unconsciousness and death are due to a depletion of the soul atoms. The mind consists of pure soul atoms and exists in the breast, while thought is a motion of the soul atoms when activated by sensory stimulation or images, another example of their empirical thought. It is only in the twentieth century that developments in neurophysiology and brain research have discovered evidence tending to explain consciousness and thought as produced by purely chemical-electrical discharges of complex neural structures in the brain, precluding a dualistic explanation.

This theory that all physical phenomena as well as human experience is explainable entirely by the atomic framework raised the same epistemological questions that modern philosophers have confronted with the rise of modern science; namely, how can one justify accepting the reality of such an abstract theoretical framework so different from the phenomenal world of experience. While the primary properties, such as hardness, size, shape, and motion, can be abstracted from objects large enough to be perceived and therefore assumed also to apply to the insensible atoms, how does one explain the origin of the additional sensory qualities displayed by ordinary objects which do not characterize the atoms? Since the objects we experience are not the naked atoms with their primary properties, but a world camouflaged by colors, sounds, smells, tastes, and tactual sensations, how do these latter properties arise?

The explanation depended on the realization that ordinary macroscopic objects with their sensory qualities appear as they do because of the particular sense organs of the perceiver, and thus would not appear as they do either for beings with different sense organs or none at all. This is not as obvious for sight and hearing, which initially seem just to disclose an independently existing world as it is, in contrast to sensations such as pain and tickling or those of touch, taste, and smell which evince more dependence upon the interaction between the stimulus and the appropriate sense organ. Yet even the reliance of colors and sounds on our eyes and ears is obvious in cases of color blindness and deafness. This is what led to the distinction between the nonsensory physical causes of our experiences and their effects on the sense organs that produce our sensations and sensory perceptions of colors and sounds.

While these early thinkers did not know anything about the neurophysiological processes of perception, they still were aware that an intervening bodily process justified distinguishing the physical causes producing the seeing and hearing from what was seen and heard. In the eighteenth century Immanuel Kant introduced his famous distinction between the world as it is experienced and as it exists in itself, "the world of appearances," and the unknowable world of "things in themselves." The atomists, in contrast, like modern scientists drew the opposite conclusion claiming that knowledge of the sensory world was "obscure," while knowledge of the atoms and the void was "genuine": "To the obscure belong all the following: sight, hearing, smell, taste, touch. The other [that of the atoms and the void] is genuine, and is quite distinct from this." As stated in a famous fragment preserved by Sextus: "By convention are sweet and bitter, hot and cold, by convention is colour; in truth are atoms and the void. . . . In reality we apprehend nothing exactly, but only as it changes according to the condition of our body and of the things that impinge on or offer resistance to it" (p. 422, n. 589).

Yet like modern scientists the Atomists tried to infer what the unobservable properties of the atoms must be to produce subjective sensations in contrast to the more objectively appearing visual images and sounds: "Bitter taste is caused by small, smooth, rounded atoms, whose circumference is actually sinuous; therefore it is both sticky and viscous. Salty taste is caused by large, not rounded atoms, but in some cases jagged ones . . ." (p. 423, n. 591). It is apparent that they were trying to correlate sensations with what they believed were the most likely primary qualities of the atoms producing them. This is true also of their explanation of the perception of elements such as fire, air, and water, defining fire as composed of round, mobile atoms while the others "they distinguish by magnitude and smallness" whose effects produce our perception of them.

Democritus' explanation of vision is more detailed and complex. Following Empedocles' explication of sight based on the analogy with the lantern, Democritus claimed that objects give off 'effluences' having the primary qualities of the respective objects (so in that sense resembles them) which are met in the intervening air by a projection from the eyes that "stamps" an "eidola" or "image" of the object on the air which is then conveyed to the eye. As described by Theophrastus:

> Democritus explains sight by the visual image, which he describes in a peculiar way; the visual image does not arise directly in the pupil, but the air between the eye and the object of sight is contracted and stamped by the object and the seer; for from everything there is always a sort of effluence proceeding. So this air [image], which is solid and variously coloured, appears in the eye. . . . (p. 421, n. 587; brackets added)

If Theophrastus' account is correct, color would not be a subjective quality added to the objective image when it enters the eye, but a feature of the image owing to the interaction of the effluences with what is projected by the eye and therefore colors the image before it enters the eye. Hearing differs from sight in that it is not mediated by an image, but produced when "sound-particles" leaving the sound source strike the air breaking it up into particles of smaller size but similar shape, propelling them to a sentient body which they penetrate throughout but particularly through the receptacle of the ears: "Democritus says that air is broken up into bodies of like shape and is rolled along together with the fragments of the voice." It would seem that as sound is conveyed by the air and spreads, its likeness is conveyed by sound-particles while its dispersal is due to their breaking up into smaller particles which are facsimiles of the original. Why the impact should be distributed throughout the body, rather than limited to the ears, is not mentioned. But what is significant about the Atomists' attempt to explain perception is not the understandably crude explanation itself, but the fact that they tried to provide an explication at all. It is this conviction that ordinary physical phenomena and conscious experiences could be explained by inferred imperceptible particles that was so admirable and far-sighted.

As indicated previously, all sensation, perception, and thought are eventually produced by the impact of the effluences or images on the appropriate sense organ which then sets in motion the soul atoms that constitute consciousness and thought. Although thought is generated by the impact of effluences or images, it is a more "genuine" form of knowledge than perception, yet it should not forget its origin: "Wretched mind, do you, who get your evidence from us [the senses], yet try to overthrow us? Our overthrow will be your downfall." This empiricism was even extended by Democritus' to his conception of the gods and the origin of the belief in them. He conceived of the gods as corporeal beings composed of fiery atoms similar to the soul atoms in ourselves and to those at the periphery of the spherical worlds. The source of the belief in the gods he attributed to visions, some beneficial and some harmful, that occur in dreams and are reinforced by fear of the threatening forces of nature. However, since all events occur of necessity there is no divine creator who intervenes in natural processes, therefore prayer is useless in influencing events. Thus his conception of the gods had nothing to do with an explanation of natural events, but was an acknowledgement of the prevalence of the human experience of them.[9]

CONCLUSION

To conclude, we have seen how the Presocratics, using empirical models in their reasoning, progressively devised or anticipated the methods of modern science to achieve a more rational understanding of the universe despite the initial primitive presuppositions of their thinking. In summary, how the Milesians began rational speculations about the origin of the universe, dispersing the fog of myth in Hesiod's *Theogony,* to arrive at a more realistic cosmology; how Heraclitus introduced fire as the *archê* but then reverted to a more primitive form of thought except for his wise ethical maxims; how the Pythagoreans' discovery of a numerical harmony underlying the musical scale but inability to abstract mathematics from its spatial representations enabled them to conceive of a mathematically created universe; how Parmenides' logical canons produced a completely implausible conception of the universe yet imposed important logical strictures on later cosmological speculations; how Empedocles introduced fire, air, earth, and water as the basic elements that was not refuted until the development of modern chemistry, along with important contributions to physiology; how Anaxagoras conceived the notion of an infinitely divisible substructure of elements as the basic physical reality, while his claim that the sun and the stars were made of earthen stones and that the moon had plains and ravines resembling the earth were extraordinary for the time; and how this culminated in the remarkable foresight of Leucippus and Democritus who, in introducing the atomic theory and setting the example that empirical inquiry guided by reason could explain all natural phenomena, were the sole thinkers in antiquity to completely replace the prevalent mythical and religious worldviews with a naturalistic unified theory that was a forerunner of modern classical science. As Bailey states:

> In reading or writing the history of Greek Atomism it is impossible not to have an eye on modern science. Yet the temptation to see closer parallels than actually exist and to read modern ideas into ancient speculation is very real and must be resisted. . . . But even a layman may assert without contradiction that it was the Greeks who put the questions which modern science still endeavors to answer: had their problems been different, the whole course of European scientific thought might never have existed. (pp. 4–5)

While Bailey's cautionary advice is correct, it still is true that we owe to the ancient Atomists the unique example in the ancient world of trying to understand and explain the universe and human existence in purely empirical-rationalistic terms. More generally, had not those audacious Greeks, including such later Hellenistic thinkers as Euclid, Archimedes, Aristarchus, Galen, and Ptolemy, following the Golden Age of philosophy of Plato and Aristotle, not existed, we still might be living in the cultural and intellectual morass of the Middle or Dark Ages when Christianity was dominant. I wonder how many Christians would want to return to this period of religious domination when illiteracy, superstition, rigid hierarchical social structures, repression of the vast number of people without any individual rights, and the wealth and imperialism of the church were the prevailing culture?

NOTES

1. The supporting references for this chapter can be found in my book, *From Myth to Modern Mind: A Study of the Origins and Growth of Scientific Thought*, Vol. I, *Theogony through Ptolemy* (New York: Peter Lang, 1995) and

other doxographical sources which I have listed. I have omitted individual citations to make the book more readable and to save space.

2. F. M. Cornford, *Principium Sapientiae,* ed. by W. K. C. Guthrie (New York: Harper & Row, 1965), p. 145.

3. G. E. Kirk and J. E. Raven, *The Presocratic Philosophers* (Cambridge: At The University Press, 1962), p. 313.

4. Galileo Galilei, *Dialogue Concerning the Two Chief World Systems,* trans. by Stillman Drake (Berkeley: University of California Press, 1967), pp. 327–28.

5. Michio Kaku, *Parallel Worlds: A Journey through Creation, Higher Dimensions, and the Future of the Cosmos* (New York: Doubleday, 2005), pp. 196–97.

6. Kathleen Freeman, *Ancilla to the Pre-Socratic Philosophers* (Oxford: Basil Blackwell, 1962) pp. 42–43. The immediately following quotations are from this work.

7. Simplicius, *De Caelo*, 529, I and *Physics*, 32, 13. Quoted from G. S. Kirk and J. E. Raven, op. cit., pp. 346–47, f.n. 464. Unless otherwise indicated, all the following quotations are from this work.

8. Cyril Bailey, *The Greek Atomists and Epicurus* (Oxford: At The Clarendon Press, 1928), p. 77.

9. Cf. Kathleen Freeman, *The Pre-Socratic Philosophers*, 2nd ed. (Oxford: Basil Blackwell, 1959), pp. 311–12.

Chapter 2

THE WORLDVIEW OF PLATO

It was reading Plato's dialogues that enticed me into studying philosophy after having completed an undergraduate premedical major. I have always retained a great admiration for him as an extraordinarily gifted writer, original thinker, and visionary philosopher, despite his worldview being despoiled by later developments in science. His conception of a meritocracy ruled by Guardians who had attained knowledge of the Good, his notion of ethics and mathematics as based on a transcendent realm of Forms, along with his own contributions and that of the Academy to mathematical investigations and astronomical research had a tremendous impact on Western civilization. Such was the originality of his thought that he had a strong influence on thinkers as diverse as Plotinus, Kepler, Heisenberg, and Karl Popper. It is his remarkable imagination, as well as his vision of a transcendent realm of eternal values and truths culminating in the Good, that have inspired and enthralled readers throughout the ages. Yet there is little doubt that his idealized conception of knowledge and mythical cosmology would have impeded the development of modern science if taken as the true model of inquiry.

Despite some uncertainty, Plato's birth and death are given as 428 to 348–7 BCE. He came from a distinguished aristocratic family with some ancestors on both sides having been leaders of Athens. But he grew up after its Golden Age when the city had suffered a series of calamitous plagues and eventual military defeats during the last third of the fifth century. These include the tragic twenty-seven year Peloponnesian war brought on by Sparta's invasion of Attica in 431 followed by the dreadful plagues that ravaged Athens, particularly the one from 430 to 429 when the city also was under siege by Sparta.[1] Probably carried to the harbor of Piraeus by incoming sailors, it spread throughout the terribly overcrowded city whose population had swelled to about 200,000 as a result of Pericles' policy of bringing into the fortifications the agrarian families evacuated from the farmlands of Attica before the invading army of 60,000 Spartans. Due to the unsanitary conditions tens of thousands of people died in the first plague alone. Pericles, who had been the outstanding leader of Athens for the previous thirty years, died from the plague near the outset in 429.

Then Athens suffered disastrous military defeats in the misguided attempt to conquer Syracuse whose coalition annihilated the Athenian land forces while the Spartan fleet destroyed most of Athens' triremes in the Great Harbor of Syracuse in 414. Yet despite those losses the Athenians in the following decade were able to rebuild their fleet and achieve a number of naval victories during the Ionian war culminating in the stunning destruction of the Spartan triremes at Arginusea in 406. But then in a war that had seen so many unexpected triumphs and defeats reflecting the fickleness of history, the entire Athenian fleet was caught off guard the succeeding year at Aegospotami, near the Hellespont, by the Spartan admiral Lysander and completely annihilated in one of the greatest naval disasters in history. Followed by a six month siege by the Spartans, the Athenians finally surrendered in 404 bringing an end to Athens' former prosperity, great empire, dominance of the sea, and Golden Age. From its position as the grandest and most powerful polis in Greece before the war, it was reduced to an impoverished provincial city-state, though it regained some of its prosperity, prestige, and especially its cultural influence during the fourth century owing to the schools founded by Plato and Aristotle, the most influential centers of learning at the time.

Socrates served in the war and it is likely that Plato also served in his youth. His experience of the war and its aftermath must have influenced his idealized conception of the state ruled by the Guardians presented in the *Republic*. Because of his illustrious family he was well acquainted with the outstanding individuals of his time, traveled widely (he twice visited Syracuse in Sicily attempting unsuccessfully to convert either the despot Dionysius I or his brother-in-law Dion to a philosopher-king) and founded the Academy at the age of forty, which he directed until his death. Lasting for nine hundred years until its closing by the Byzantine Emperor Justinian, it had a longer history than any other academic institution in the West. Many leading scholars visited or taught at Plato's Academy or Aristotle's Lyceum in the succeeding several centuries.

Though a contemporary of Democritus, under Plato's direction the Academy disdained the Presocratic tradition of seeking empirical explanations of natural phenomena, instead becoming famous for its research in mathematics and astronomy. Influenced in early life by Socrates' dialectical search for the essential meaning of such concepts as truth, beauty, goodness, justice, and piety, Plato's conception of knowledge and of reality took an entirely different turn from that of the Presocratics. Rather than attempting to discover the physical reality underlying and causing the appearances of nature, which he deemed impossible because of the inchoate status of the material world, his dialogues continued Socrates' unresolved dialectical quest for the ultimate meanings of these axiological terms. But while Socrates avowed (or feigned) ignorance of their essential definitions, content to unmask the pretentious answers of others, Plato posited a supersensory realm of absolute meanings as their referents, though modestly claiming that his solution was nothing more than a "likely story." As he has Socrates (who speaks for him) assert in the *Republic*, the "ascent to see the things in the upper world you may take as standing for the upward journey of the soul into the region of the intelligible. . . . Heaven knows whether it is true, but this, at any rate, is how it appears to me."[2]

So rather than seeking a naturalistic worldview, his "likely story" consisted of positing a world of absolute, eternal "Forms," a reification of meanings, mathematical concepts, and values. Subsisting beyond the shadowy world of appearances which he named the "Receptacle," these ideal Forms were the ultimate referents and answers to Socrates' search for essential meanings, as well the perfect exemplars of genus and species, differentia, mathematical concepts and relations (which is why so many mathematicians throughout history have been Platonists), the ideal archetypes of the imperfect entities "mirrored" in the Receptacle, and absolute truths. At the pinnacle of the Forms is the Form of the Good which irradiates and illuminates all knowledge.

While the dialogue method leads to a "recollection of the Forms," the actual knowledge of them was acquired by the soul in its existence before incarnation, but lost due to the trauma of birth. Thus in place of empirical inquiry, Plato offers the strange doctrine of "anamnesis" or "recollection" as the origin of knowledge. As he states in the *Phaedo*: "Our present argument applies . . . to absolute beauty, goodness, uprightness, holiness, and . . . characteristics which we designate . . . by the term 'absolute.' So that we must have obtained knowledge of all these characteristics before our birth."[3]

The third component of his cosmological system, in addition to the Receptacle and the intelligible realm of Forms, is the "Demiurge." In the *Sophist* Plato dismisses the view of Democritus that everything occurs by causal necessity arguing, as religionists do today, that the order in the universe, as imperfect as it obviously is, exhibits an intelligent design and therefore must have been imposed by a supernal mind according to a perfect model. Rejecting the empirical explanations of the Atomists, he reverts to the primitive notion of a "maker" of the universe, not in the sense of a first cause, since each of the three components of the universe is eternal, but as a "Divine Craftsman" who strives to impose as much order on the intractable elements in the Receptacle as possible.

Thus Plato's Demiurge is neither a creator *ex nihilo* nor a religious figure to whom one can pray, but a Supreme Craftsman who attempts to "overcome Necessity with Reason" by "persuading" the errant, imperfect contents of the preexistent Receptacle to "participate in" the world of perfect Forms. However, since this conception is presented as a "likely story" it is hard to know how literally or figuratively to take it. As Timaeus says in the dialogue by that name, "it is a hard task to find the maker and father of this universe, and having found him it would be impossible to declare him to all mankind."

After his theory of the Forms and conception of the state, Plato is best known for his contribution to astronomy, either indirectly owing to the astronomical research pursued by others in the Academy or directly by his own theories and conception of the proper method of astronomical investigation. While Plato's own contribution is more contentious, there is no disputing the significance of two of Plato's students and/or associates: that of Eudoxus whose theory of concentric planetary orbits was adopted by all the leading astronomers until Kepler introduced elliptical orbits and that of Heraclides of Pontus who held that the apparent diurnal rotation of the heavens is actually due to the daily rotation of the earth in the opposite direction. What makes Plato's own contribution contentious is his conception of the role of observation in astronomy which, as a discipline, became more significant as he grew older.

In the earlier dialogues he stresses the instrumental function of astronomy in leading the intellect away from the irregular appearances of the celestial bodies to their more perfect mathematical correlations in the World of Forms. This view is dramatically presented in the *Republic* where he has Socrates rebuke Glaucon for saying that astronomy is especially suited to direct "the mind to look upwards, away from the world of ours to higher things," with Socrates replying that it depends on whether what is meant by "higher things" is the visible appearances of the heavenly bodies or the "unseen reality" of Forms subsisting beyond. As he further states in the *Republic*:

> These intricate traceries in the sky are, no doubt, the loveliest and most perfect of material things, but still part of the visible world,

> and therefore they fall far short of the true realities—the real relative velocities, in the world of pure number and all perfect geometrical figures, of the movements which carry round the bodies involved in them. These, you will agree, can be conceived by reason and thought, not seen by the eye. (p. 248)

He apparently held that because astronomers cannot attain an exact mathematical representation of the observed irregularities in the motions of the "wanderers" (the Greek meaning of planets) because of their imperfections, they should seek a "revelation" of the more perfect mathematical model on which their creation was based. He has no conception of mathematics as an applied science. Unlike Kepler who formulated the first astronomical laws by replacing the concentric orbits of Eudoxus with elliptical orbits to fit the precise astronomical observations of Tycho Brahe, Plato believed that the search for exact astronomical laws based on observations of the celestial bodies could take one only so far, claiming that it is "absurd to study them in all earnest with the expectation of finding in their proportions the exact ratio of any one number to another" (p. 248).

So it seems that it was not that Plato believed that the astronomical observations at the time were too imprecise to yield an exact mathematical description of the planetary motions, but that it was the *inherent imperfection* of the planetary bodies themselves that precluded their irregular orbital motions and changes conforming to the ideal mathematical model used in their creation. As he again states in the *Republic*:

> The genuine astronomer . . . will admit that the sky with all that it contains has been framed by its artificer [the Demiurge] with the highest perfection of which such works are capable. But when it comes to the proportions of day to night, of day and night to month, of month to year, and of the periods of other stars to Sun and Moon and to one another, he will think it absurd to believe that these visible material things go on forever without change or the slightest deviation, and to spend all his pains on trying to find exact truth in them. (p. 248; brackets added)

He then adds the statement which confirms the above interpretation: "If we mean, then, to turn the soul's native intelligence to its proper use by a genuine study of astronomy, we shall proceed, as we do in geometry, by means of problems, and leave the starry heavens alone" (pp. 248–49). This suggests that in the final stage of astronomical inquiry one should look beyond the astronomical observations to their geometrical representations because "every soul possesses an organ better worth saving than a thousand eyes, because it is our only means of seeing the truth . . ." (p. 245).

It is this view of pure mathematics as the discovery of real abstract relations representative of a transcendent reality that caused so many leading mathematicians in the past to become Platonists (despite the necessity of the empirical confirmation of scientific theories). As expressed by Plato in the *Epinomis* which, although some scholars have questioned its authenticity as a work of Plato, is certainly stylistically and doctrinally consistent with his other writings:

> To the man who pursues his studies in the proper way, all geometric constructions, all systems of numbers, all duly constituted melodic progressions, the single ordered scheme of all celestial revolutions, should disclose themselves, and disclose themselves they will, if, as I say, a man pursues his studies aright with his mind's eye fixed on their single end. As such a man reflects, he will receive the revelation of a single bond of natural interconnection between all these problems.[4]

It is this vision of a "unified mathematical theory" that has been uppermost in the thinking of the greatest scientists, from Newton to Einstein to Steven Weinberg.

So far a general description of Plato's tripartite worldview has been presented based on the earlier dialogues. But it is the *Timaeus*, perhaps his last dialogue, that contains the most comprehensive description of his cosmological system, including the creation and structure of the solar system and the geometric composition and interchange of the four elements in the Receptacle, that displays his remarkable creative imagination. In his highly original cosmology he combines many of the salient conceptions of his predecessors in an exceedingly novel way. On the other hand, as a narrative system or "likely story" it resembles earlier mythological accounts more than the rational cosmologies of the Presocratics or of his student Aristotle. Perhaps no other thinker possessed such a "multifaceted intellect" combining the attributes of a mythologist, mathematician, astronomer, and mystic.

Having previously characterized the Demiurge as a Divine Craftsman about which nothing more can be said, nothing more *will* be said. Plato does, however, elaborate his descriptions of the intelligible world of Forms and the Receptacle, the former now referred to as the "intelligible Living Creature" and the latter simply as the "Living Creature." The first he continues to describe as the eternal referent of both human and divine thought whose archetypal, generic, and axiological Forms constitute the transcendent model of reality attainable by the "science of dialectic." He now adds that it embraces the "main families of living creatures," the heavenly gods (which include the earth, planets, and stars), birds, fish, animals, and humans. As a group, they represent all the species of living creatures.

Turning to the Receptacle, his expanded description in the *Timaeus* was foreshadowed in his famous "Allegory of the Cave" presented in the *Republic*. In that allegory human beings are described as existing in a cave in which they are accustomed to seeing only reflected images on the wall, cast by a hidden source of light, which they mistake for the real world. When they do occasionally venture outside the cave and are exposed to the actual world they are blinded by the illumination and withdraw to the cave. In the allegory the cave stands for the Receptacle, the images reflected on the cave walls to the world of appearances, while the emergence from the cave represents the ascent to the ideal world of Forms which initially seems to be blindingly unreal, but in truth is the actual reality as revealed to our intelligence.

In the *Timaeus* the allegory is developed more precisely. The Receptacle, as one of the eternally existing components of reality along with the Demiurge and Forms, is described as a preexisting spatial matrix that initially is empty, but which the Divine Craftsman fills with various unruly components. So while the Receptacle itself is "invisible and characterless," reminiscent of Parmenides' 'One,' when perceived it is filled with ephemeral transient qualities, as in the system of Heraclitus. Furthermore, it includes the four Empedoclean elements composing the ordinary objects in the world due to the Divine Craftsman "persuading" the Receptacle to "mirror" their "Archetypes" in the realm of Forms. The Receptacle, therefore, is "Space, which is everlasting, not admitting destruction; providing a situation for all things that come into being, but itself apprehended . . . by a sort of bastard reasoning. . . ."[5] Although the Receptacle or "Mother of Becoming" appears filled with external qualities, these appearances are caused by "powers" within the Receptacle striking the sense organs (just as the perceived external world appears to us to be independently real), so that if we did not exist they would not exist. As Cornford quotes, the Receptacle has

> "every sort of diverse appearance to the sight," in the sense that, if there were a spectator with eyes to see it, it would cause in him sensations of various colours. But in the absence of any spectator, there are, strictly speaking, no colours—only changes capable of causing such sensations. Space is . . . filled with "powers" whose motions are in unordered and unbalanced agitation. (p. 205)

Just as he tried to explain how the empty spatial Receptacle appears to have all the qualities it does, so he also attempted to explain how, without recourse to Democritus' atoms, the four elements, fire, air, earth, and water, appear in the Receptacle as substances and objects. First, the four elements acquire their sensory appearances from the four qualities, hot or cold and dry or wet. Secondly, in an ingenious explanation drawn from the Pythagoreans he attributes the solidity of each element to its geometrical structure: the pyramid or tetrahedron (four sided) constituting fire, the octahedron air, the icosahedron (twenty sided) water, the cube the earth, with the dodecahedron (twelve sided) constituting the form of the universe as a whole. As he states:

> Now, taking all these figures, the one with the fewest faces (pyramid) must be the most mobile, since it has the sharpest cutting edges and the sharpest points in every direction, and moreover the lightest, as being composed of the smallest number of similar parts; the second (octahedron) must stand second in these respects, the third (icosahedron), third. Hence . . . we may take the pyramid as the element or seed of fire; the second in order of generation (octahedron) as that of air, the third (icosahedron) as that of water. (pp. 222–23)

The cube was assigned to the earth because the "earth is the most immobile and the most plastic of bodies." The fifth solid, the dodecahedron, was added because it "approaches the sphere in volume and therefore was used by the 'god . . . for the whole.'"

These geometrical solids had a considerable fascination for later astronomers or physicists, such as Euclid, Kepler, and Heisenberg. Plato then shows how the four geometrical figures can be constructed from two types of triangles: the equilateral triangle forming the faces of the first three geometrical figures and the square forming the cube. From these he describes how the first three figures can be transformed into one another by their equilateral triangles being broken down into half-equilateral or right angled triangles and recombined in different ratios to form different arrangements of elements (cf. pp. 224–25).

For example, the eight surfaced octahedron as air can be decomposed with its triangular surfaces regrouping to form two four surfaced fire pyramids; water as the twenty surfaced icosahedron can be broken down and recombined as two eight-surfaced air octahedrons and one four surfaced pyramid, producing warm air; five four-surfaced fire pyramids can recombine to form one twenty-surfaced water icosahedron, and so forth "in an endless diversity." Thus Plato replaced the Atomists' physical descriptions of the four elements with geometric structures. The difference, of course, is that Plato's theory was entirely a priori while the Atomists' conjectures and the sterochemical structures of modern chemistry are based on experimental evidence (cf. p. 225).

Before describing the Receptacle further along with the formation of the solar system, Plato presents a brief sketch of "the World's Body" and "the World's Soul." He adopts the Parmenidean conception that the world or living creature is "a sphere, without organs or limbs, rotating on its axis." Then, in contrast to the Atomists who simply endowed the atoms with a kinetic motion, Plato argues in the *Phaedrus* and in the *Laws* that because the only conceivable source of motion is the "self-moving soul," the spherical universe must be animated by a soul (an argument that Aristotle will also adopt). Thus this World-Soul, which pervades the universe from the center to the periphery, is the source of the various motions in the universe, such as the rotation of the World Sphere itself and the specific revolutions of the celestial bodies or "heavenly gods." In what is too fanciful a conception to be fully described, in the *Sophist* Plato describes the World-Soul itself as compounded of three kinds of Forms, "Existence," "Sameness," and "Difference," to explain positive and negative judgments (cf. pp. 59–60).

Given these semantic components composing the nature of the World-Soul, the Demiurge formed this into a long band with two numerical series, 1, 2, 4, 8 and 1, 3, 9, 27 (cf. pp. 67–69). Further reflecting the approach of the Pythagoreans, each series is further subdivided into arithmetical and harmonic segments corresponding to the intervals of the diatonic musical scale, so that every segment has an arithmetical or harmonic relation to every other. The two numerical series are then divided into two separate strips joined at the ends to form two circular bands, one band making up the outer circle positioned horizontally to correspond to the sidereal equator, the second placed inside the other with an inclination representing the Zodiac (cf. p. 76).

The strip forming the sidereal equator is given a circular motion from east to west around its axis representing the diurnal rotation of the spherical world, designated the "movement of the Same." This causes the common movement of the fixed stars with each star also revolving on its axis to face the earth. The band of the Zodiac revolves in the opposite direction from west to east producing the orbits of the planets. Called the "movement of the Different," this band revolves in the opposite direction to the motion of the Same, yet participates in the diurnal rotation of the whole universe (imagine an armillary sphere). These motions are those of the World-Soul before the heavenly bodies were enshrined in it, explaining how the Divine Craftsman might have organized the world from preexistent materials to give it a rational structure with a purpose.

The two major astronomical movements accounted for, Plato is ready to explain how the individual planets were given their specific orbital motions. The Zodiacal strip is now divided "into seven unequal circles" whose diameters are supposed to correspond to the two series of numbers making up the original long band, 1, 2, 3, 4, 8, 9, 27 while comprising the successive orbits of the seven planets. Their centers coinciding with the center of the spherical universe, the planet's orbital strips are nested within each other according to their increasing size and inclined at different angles:

> . . . the Moon in the circle nearest the Earth; the Sun in the second above the Earth; the Morning Star (Venus) and the one called sacred to Hermes (Mercury) in circles revolving so as, in point of speed, to run their race with the Sun, but possessing the power contrary to his: whereby the Sun and the star of Hermes and the Morning Star alike overtake and are overtaken by one another. (p. 105)

The three outer planets have a "proper" motion in addition to the dual motion of the Same and Different, but he declares that "where he [the divine Craftsman] enshrined them and for what purpose is too difficult to be explained" (brackets added). Although this system certainly seems arbitrary and obscure, Plato did attempt to describe accurately the relative positions and motions of what are now called the "inferior planets," along with noting the apparent retrograde motion of five of the planets (except for the sun and moon, the other planets appear to stop, briefly reverse their motion, and then move forward again) showing that he tried to take into account their observed motions. In addition, like the stars each planet revolves on its axis to present the same face towards the earth. Just as the vortical motion of the Spherical World is derived from the World-Soul, so the proper motions of the planets or heavenly bodies are explained by their being "living creatures with a soul" and thus "self-moving."

Plato now adds to the orbital motions of the planets their composition and "well-rounded shape," stating they are predominately made of fire with a small portion of the other elements, earth, air, and water. Although they are "living creatures everlasting and divine," they are not unchanging and eternal like the intelligible Forms, but have existed since the creation of the World-Soul and the World-Body. Their heavenly revolutions created time, as marked by days, months, and years: "For there were no days and nights, months and years, before the Heaven came into being . . . but these have come into being as forms of time, which images eternity and revolves according to number" (p. 98).

This creation by the Demiurge was intended to have the heavenly bodies "mirror" as much as possible the unchanging perfect motions of the Intelligible Living Creature to inspire in man the mathematical study of nature and the pursuit of philosophy. As he wrote, the

> sight of day and night, of months and the revolving years, of equinox and solstice, has caused the invention of number and bestowed on us the notion of time and the study of the nature of the world; whence we have derived all philosophy, than which no

> greater boon has ever come or shall come to mortal man as a gift from heaven. (pp. 157–58)

Although exceedingly fanciful, this cosmological theory had considerable influence on later astronomers because Plato's Divine Craftsman could be identified with the Christian God. For example, in the Preface and dedication to Pope Paul III preceding his *On the Revolutions of the Heavenly Spheres*, Copernicus says that it was the "lack of certitude in the traditional mathematics concerning the composition of movements of the spheres" that led him to seek a more exact model to explain the "machinery of the world, which has been built for us by the Best and Most Orderly Workman of all," while Kepler initially was (mis)guided by the platonic solids in his search for the fundamental laws describing planetary motion.

Although there is evidence that Plato was somewhat influenced by astronomical observations in creating his "likely story" and certainly realized the crucial role of mathematics in astronomical theorizing, like Genesis it is a most "unlikely" account given what we now know. But even Plato seemed to be aware of this as he modestly acknowledges in the *Timaeus.*

> If then, Socrates, in many respects concerning many things—the gods and the generation of the universe—we prove unable to render an account at all points entirely consistent with itself and exact, you must not be surprised. If we can furnish accounts no less likely than any other, we must be content, remembering that I who speak and you my judges are only human, and consequently it is fitting that we should, in these matters, accept the likely story and look for nothing further. (p. 23)

He knew what he referred to as the 'shortsighted' astronomical systems of Anaxagoras and Democritus, but dismissed them as 'no less likely' than his own despite their obvious superiority to us today. As he says in the *Sophist*:

> For on a shortsighted view, the whole moving contents of the heavens seemed to them [Anaxagoras and Democritus] a parcel of stones, earth, and other soulless bodies. . . . It was this that involved the thinkers in those days in so many charges of infidelity [one is reminded also of Bruno and Galileo] . . . but today the position has been reversed. . . . No son of man will ever come to a settled fear of God until he has grasped the two truths we are now affirming, the soul's dateless anteriority to all things generable, her immortality and sovereignty over the world of bodies, and moreover that presence among the heavenly bodies of a mind. . . . [6]

Those who advocate "intelligent design" as the cause of the universe are just as misguided as Plato. What was unknown in the past usually was attributed to the soul or to God, the final refuge of ignorance. That the Presocratics refused to indulge in this deception is what makes them so remarkable.

NOTES

1. For an excellent account of the tragic Peloponnesian War and the horrifying consequences of the plagues, see Victor Davis Hanson, *A War Like No Other* (New York: Random House, 2005).

2. Francis M. Cornford, *The Republic of Plato* (New York: Oxford University Press, 1945), p. 231. Unless otherwise indicated, the immediately following quotations are to this work.

3. Plato, *Phaedo*, trans. by Hugh Tredennick, *The Collected Dialogues of Plato*, ed. by Edith Hamilton and Huntington Cairns, Pantheon Books (New York: Bollinger Foundation, third printing, 1964), p. 58.

4. *Epinomis*, trans. by A. E. Taylor, *The Collective Dialogues of Plato,* op. cit., p. 1532.

5. F. M. Cornford, *Plato's Cosmology* (New York: The Liberal Arts Press 1957), p. 192. This discussion along with the immediately following quotations and page reference are from this work.

6. Plato, *Laws* XII, trans. by A. E. Taylor, *Collected Dialogues of Plato*, op. cit., p. 1512 (brackets added).

Chapter 3

THE WORLDVIEW OF ARISTOTLE

My respect for Aristotle even surpasses that for Plato because of the originality and comprehensiveness of his writings that addressed all the areas of philosophy or knowledge (which came to the same thing) at the time.[1] His extensive published treatises and edited lecture notes covered such diverse subjects as physics, astronomy, metaphysics, ethics, rhetoric, poetry, and politics, along with more unusual topics dealing with the soul, generation and corruption, the history, parts, and generation of animals, memory and reminiscence, constitutions, as well as dreams and prophesying. In the *Prior Analytics* of his *Organon* he even created the first systematic treatment of logic, known as sentential or Aristotelian logic, preceding the later development of mathematical logic in the nineteenth and twentieth centuries. Having said of Plato there are few who could match his "multifaceted intellect," there are even fewer who can approach the tremendous range, analytical powers, and originality of Aristotle's extraordinary intellect—never equaled by another thinker in my opinion—which earned him the accolades "the master of them that know" and "the philosopher."

Having said that, it must be admitted that in contrast to the worldviews of the Milesians and the Atomists, who laid the foundations of later scientific inquiry, there is little of value today in Aristotle's specific theories of physics and astronomy. While one can still marvel at his insights when reading his works on ethics, aesthetics, politics, and even embryological development, his explanation of physical changes and the nature and source of the motion of the planets have no validity today. The reasons for this is that his explanatory framework, unlike the ancient Atomists and modern scientists, relies on the superficial appearances of things and that he followed Plato in believing that only the soul or mind could be an ultimate source of motion. Yet his uncompromising commitment to sensory observation and its essential role in scientific inquiry, in contrast to the fanciful explanations of Plato and the dialectical arguments of many of the Presocratics, was necessary to advance empirical inquiry.

Born in 384 BCE in the city of Stagira located in Thrace on the northern coast of the Aegean Sea, he is often referred to as "the Stagirite." His father was a member of the Asclepiad clan or guild named for the renowned physician Asclepios. Though he died while Aristotle was still young, the fact that Aristotle did extensive research in biology and that it was that branch of science that provided his model of scientific explanation can be ascribed to the influence of his father. At the age of seventeen or eighteen he entered Plato's Academy where he remained for twenty years until Plato's death in 347–48. He must have been an outstanding student because Plato referred to him as "the reader" and "the mind of the school." His early *Dialogues* especially show the strong influence of Plato, though his later emphasis on sensory observation and interest in all natural phenomena, along with his using logic rather than mathematics in his method of scientific explanation, contrasted sharply with Plato.

Shortly after Plato's death he left Athens settling among a small circle of Platonists for three years in Assos on the coast of Asia Minor just north of Lesbos. Unlike Plato who had the sexual preference of many male Greeks at the time, in Assos Aristotle married and fathered a daughter named after his wife Pythias. Following his brief stay in Assos he moved to Mitylene, in Lesbos, where he pursued his interest in biology. As W. D. Ross states in his book *Aristotle*, "[t]o his stay in Assos, and even more to his stay at Mitylene, belong many of his inquiries in the field of biology. . . ."[2] Then in 343–42 at the invitation of King Philip of Macedon he left for the court at Pella to become tutor to Prince Alexander, later to become "Alexander the Great."

Although little is known about the tutelage, it is tantalizing to imagine what mutual effect this auspicious relationship might have had on the great philosopher and on one of the greatest conquerors and political visionaries in the history of the world. It seems probable, however, that Aristotle wrote *Monarchy* and *Colonies* with Alexander in mind, while Alexander reciprocated by helping to fund and supply with items of scientific interest a museum of natural history in the Lyceum later established by Aristotle in Athens. Then in 335–34 following Alexander's accession to the throne after Philip's death, Aristotle returned to Athens and because Athenian law prevented a foreigner from owning land he rented some buildings in a grove sacred to Apollo Lyceius with an adjoining garden, lecture halls, rooms for his large collection of manuscripts and maps, and the museum of natural history. He lectured in the morning on more difficult subjects such as logic, physics, and metaphysics in the Peripatos, a loggia, with students strolling behind giving rise to the name "peripatetic" or "follower of Aristotle." He lectured to larger audiences in the afternoon on rhetoric, ethics, and politics.

In addition to these lectures, during the twelve years he headed the Lyceum he supervised the dissection of animals and insects, compiled the history and classification of plants and animals, pursued historical research in philosophy, collected 158 constitutions, and resolved the question of the cause of the flooding of the Nile (due to melted snow from the mountains) in the *Meteorology*. While biological research was conducted also at schools of medicine, such as the Hippocratic School at Cos, Aristotle was the first we know of to organize such research on so broad a scale, which set an example for the later investigations pursued in the Museums and Libraries at Pergamum and Alexandria. As Werner Jaeger states in his book, *Aristotle*:

> We do not hear of many disciples of Aristotle by name, but what Greek is there who wrote during the next hundred years on natural science, on rhetoric, or literature, or on the history of civilization, and was not called a Peripatetic? Lavish as the grammarians are with this title, it is easy to see that the intellectual influence of the school soon extended over the whole Greek speaking world. . . . [T]he Lyceum . . . became a university in the modern sense, an organization of sciences and of courses of study.[3] (Brackets added)

It was during this period in Athens following his first wife's death that he formed an "illegalized" relationship with Herphyllis who bore him a son named Nicomachus, after whom he titled his well-known work on ethics. Then in 323 the anti-Macedonian feeling that erupted with the news of the death of Alexander forced him to leave Athens because of an 'absurd charge of impiety' lest, as he said, 'the Athenians sin twice against philosophy' (a reference to the previous false charge of impiety brought against Socrates that forced him to take the hemlock). Leaving Theophrastus in charge of the Lyceum, he withdrew to the paternal property of his deceased mother in Chalcis on the Euboean peninsula where he died a year later at the age of sixty-three, reportedly from an intestinal disorder. In his will he provided for his daughter Pythias and his son Nicomachus who were still minors; left property to Herphyllis with permission that she should be "given to one not unworthy" if she chose to remarry; and arranged for the freedom of some of his servants while leaving bequests to others. Respecting the wishes of his first wife he was buried with her. These Provisions attest to his outstanding character, the traits of a just, kind, generous, and considerate person whose actions were in conformity with his ethical principle that 'the end of man is virtue.'

Because I sacrificed some expository space to present a fuller biographical account of Aristotle's life, believing that it would be of special interest to the reader, in describing his philosophy I shall limit the exposition to his two major contributions: (1) his deductive method of scientific explanation and (2) his fundamental principles of explanation exemplified in his physics, astronomy, and cosmology. Starting with the prior conception, it comprises two essential components: (1) that scientific knowledge consists of necessary facts proved or demonstrated by syllogistic deduction from general premises that are "true, primary, immediate, better known than and prior to the conclusion," the proven fact; and (2) that the premises, though originally dependent on inductions from sensory observations, have to be "rationally intuited by the soul" to be known with certainty.

These two components of his system, that scientific knowledge consists of logical demonstrations from premises that are dependent initially on sensory observations but actually known by "rational intuition," are presented in his *Organon*: the first component, the logic of syllogistic demonstrations in his *Prior Analytics* and the second, his explanation of how premises are attained, in his *Posterior Analytics*. It is the first that presents his method of syllogistic reasoning or sentential logic consisting of three statements demonstrating that an essential quality necessarily inheres or does not inhere in an object because the object belongs to a class or genus which has or does not have that property as part of its essence or definition. As stated in Book I of the *Posterior Analytics*: "there are three elements in demonstration: (1) what is proved, the conclusion—an attribute inhering essentially in a genus; (2) the axioms . . . which are premises of demonstration; (3) the subject-genus whose attributes, i.e., essential properties, are revealed by the demonstration" (p. 121).

Thus if asked "Why are men mortal?" the answer consists in the demonstration that since "All animals are mortal," and "Man is an animal" (belonging to the genus animal), "Man too is mortal." Or to take a negative example, "Why do the planets (in contrast to the stars) not twinkle?" the answer is that "No proximate celestial bodies twinkle," and since "The planets are proximate celestial bodies," it follows that "The planets do not twinkle." Or to take a causal example, "What causes a lunar eclipse?" the answer is that "All shutting out of the sun's light by the imposition of the earth causes an eclipse," and "All lunar eclipses have the imposition of the earth between the sun and the moon," therefore "It is the imposition of the earth that causes the lunar eclipse." The proof is dependent on the two premises being connected by a common or "middle" term, in the above examples "animal," "proximate celestial body," and "the earth's imposition between the sun and the moon shutting out the light" that he considers not only as the connecting link in the syllogism, but as the actual 'cause' of the conclusion. As he states in Book II of the *Posterior Analytics*, "demonstrative knowledge must be knowledge of a necessary nexus, and therefore must clearly be obtained through a necessary middle term. Otherwise its possessor will know neither the cause nor the fact that his conclusion is a necessary connection" (p. 120).

The underlying assumption is that the world is organized in terms of species and genus, as in biology, whose essential qualities not only define but *cause* things to be what they are. Knowing the genus of an object or event proves why it necessarily is the kind of thing it is or why it acts as it does. This is quite different from the modern conception of scientific explanation as discovering the concomitant underlying causes of phenomena based on observation and experimentation, devising theories to explain their modus operandi, and inferring and testing predictable consequences, and thus explains why Aristotle's method proved unsatisfactory. In fact, it was Aristotle's method based on the meanings of terms as used by the Scholastics in the Middle Ages that led to Aristotelianism falsely acquiring the reputation as a disputatious, sterile method of inquiry.

While this deductive conception of scientific knowledge certainly did not prove correct in terms of what we now know about scientific explanations, his claim in the *Posterior Analytics* that sensory observations rather than dialectical arguments are the primary bases of all scientific knowledge proved a turning point in the development of a preliminary conception of scientific inquiry. In Book I of *On Generation and Corruption* he harshly criticizes the Pythagoreans,

Parmenides, and Plato for accepting "dialectical" arguments over the evidence of their senses.

> Reasoning in this way . . . they were led to transcend sense perception, and to disregard it on the ground that "one ought to follow the argument:" and so they assert that the universe is "one" and immovable . . . [A]lthough these opinions appear to follow logically in a dialectical discussion, yet to believe them seems next door to madness when one considers the facts. For indeed no lunatic seems to be so far out of his senses as to suppose that fire and ice are "one". . . . (p. 498; brackets added)

This severe criticism of dialectical reasoning and defense of sensory evidence reverberates throughout his writings, an apparent reaction to the kind of philosophy pursued in Plato's Academy during the twenty years that he was a student there and the reason empirical research was pursued in the Lyceum which he later established in Athens. As he states in Book I of the *Metaphysics*: "All men by nature desire to know. An indication of this is the delight we take in our senses; for even apart from their usefulness they are loved for themselves; and above all the sense of sight . . ." (p. 689). It also explains Galileo's admiration of him "considering him as I do a man of brilliant intellect"[4] and Darwin's assertion that "from quotations which I had seen, I had a high notion of Aristotle's merits, but I had not the most remote notion what a wonderful man he was. Linnaeus and Cuvier have been my two gods . . . but they were mere school-boys, compared to old Aristotle."[5]

Aristotle even traces and describes the perceptual and cognitive reasons that make reliance on sensory evidence true. First, the sensory capability of differentiating among sensory qualities and configurations is "a congenital discriminative capacity" possessed by all animals. But once these discriminations are made the ability to organizing them into classes of species and genus depends upon identifying and abstracting their common, distinguishing, universal features that he attributes to a capacity of the human soul. It is these "manifest qualities," particular to each science such as biology, physics, or astronomy, that are apprehended by the soul's rational intuition that is the foundation of scientific knowledge. Thus the ultimate goal of modern science to arrive at a "unified theory" was denied by Aristotle because each science has its unique defining qualities and first premises.

Having discussed his conception of scientific knowledge, we turn now to his principles of explanation as applied in physics and astronomy or cosmology. Because his cosmological system drew a qualitative distinction between the terrestrial and the celestial worlds which in turn involved different explanatory principles, we will begin with the better known terrestrial world. As he states in *Parts of Animals*:

> Of things constituted by nature some are ungenerated, imperishable, and eternal, while others are subject to generation and decay. The former are excellent beyond compare and divine, but less accessible to knowledge. . . . The scanty conceptions to which we can attain of celestial things give us, from their excellence, more pleasure than all our knowledge of the world in which we live. . . . On the other hand, in certitude and in completeness our knowledge of terrestrial things has the advantage. (p. 656)

The science that deals with the terrestrial world is physics which he depicts as discovering the underlying "principles, conditions, or elements" of explanation. However, if we expect this to resemble the modern scientist's inquiry into the atomic-molecular structure of matter, as well as the forces that bring about change and motion, and that function according to universal causal laws, we are surprised to discover that instead of atoms and forces his principles of explanation involve concepts such as 'subject,' 'form,' 'contrary,' 'privation,' 'potentiality,' 'becoming,' and 'actuality.' Equally bewildering are such statements as "if 'man' is a substance, 'animal' and 'biped' must also be substances," or that "imagination and thought are motions of the soul," or that motion is "the fulfillment of what exists potentially, insofar as it exists potentially."

The possible principles of explanation that he inherited in the fourth century were as divergent as the number cosmology of the Pythagoreans, the logical maxims and the 'One' of Parmenides, the four elements of Empedocles, the atomic system of Leucippus and Democritus, and the Forms of Plato. Also, like his predecessor he was determined to explain change, generation or destruction, and motion without violating Parmenides' logical cannons. But he makes it clear that he is going to be guided by experience, not by abstract logical arguments, declaring in Book I of the *Physics* that "to investigate whether Being is one and motionless is not a contribution to the science of Nature," nor is there any point refuting arguments "which deny the existence of common phenomena, such as change and motion: accept one ridiculous proposition and the rest follows." Who can seriously deny that leaves change color, that the sun and the moon revolve in their orbits, that children grow, that wood burns, or that the city of Miletus was destroyed by King Artaphernes during the Persian wars? Thus instead of denying such phenomena, Aristotle devises the principles by which they can be understood without violating Parmenides' maxims that something cannot come to be from nothing nor entirely cease to be.

He concludes that in all such occurrences three principles of explanation are involved: (1) the underlying "substance" or "matter" which persists throughout all change and therefore is the subject of the changing forms, the leaves, the sun and the moon, the children, the wood, and the city of Miletus in the previous examples; (2) the initial quality, form, or state that is replaced, such as the original color of the leaf, the previous positions of the sun and the moon, the unburned wood, the intact state of Miletus, and (3) the new form, quality or state that replaces the previous one. To exclude a coming to be from nothing, he adds that whatever properties an object originally lacked but changes into do not arise from Not-Being because they existed previously in the object's substance as a 'privation' or 'potentiality.' Moreover, he adds that because the original forms change into their contraries, change is not random or arbitrary: "change proceeds from a contrary to a contrary or to something intermediate: it is never the change of any chance subject in any chance direction. . . ."

As he summarizes: "The causes and the principles, then are three, two being the pair of contraries of which one is definition and form and the other is privation, and the third matter." Thus one color changes from one color to another, not to a sound; humans beget humans, not another animal; burnt wood changes into ashes, not metal; a lute produces music, not smoke, and so forth. These inherent privations give objects the potential for certain kinds of changes and not for others, so that when the appropriate cause is present the potential opposite will be realized.

Thus his principles of explanation are not as puzzling when applied to the changes around us. Even in "substantial change" involving "coming-to-be and passing-away" such that the original substance is no longer identifiable, as when wood burns, dough is baked into bread, or cities destroyed, there still persists throughout the change the "prime matter," the basic matter of the universe which has always existed and will continue to exist. Similarly, because the opposite or contrary form that replaces the original form does not arise from nothing, but preexists in the substance as a privation, it does not come to be from nothing. Furthermore, although the original form may cease to exist in a particular object, yet preexists in the realm of forms, the forms themselves do not cease to be. Thus Aristotle has achieved his objective of accounting for change without violating Parmenides' principles nor denying that change does occur. As he says, "neither the matter nor the form comes to be—and I mean the last matter and form" along with ensuring that "no state of not-being ensues." Novelty is a result of the unique actualization of contrary forms in an object, not because of their transformations.

No description of Aristotle's principles of explanation would be complete without including his celebrated doctrine of the four causes, each of which must be accounted for in any complete understanding of change. The following description in Book I of *On Generation and Corruption* is probably his most concise statement of the four causes.

> There are four causes underlying everything: first, the final cause, that for the sake of which a thing exits; secondly, the formal cause, the definition of its essence (and these two we may regard pretty much as one and the same); thirdly, the material; and fourthly, the moving principle or the efficient cause. (p. 665)

Here he presents the final cause first because he is considering the generation of animals in which purpose plays such a crucial role, though normally he presents the material and formal causes first because they are theoretically prior or more basic.

Because these causes follow from his previous discussion of explanation and are so well known they will only be briefly discussed. The material cause is the substance or matter of which anything is made and that underlies all change. In addition, the material cause includes the privation or potentiality for whatever can come to be. What is curious are the examples Aristotle uses to illustrate the material cause and the kinds of entities he includes under this category of cause: bronze has the potentiality for becoming a statue and silver of becoming a bowl, "but because bronze and silver are species of the genus metal, the genus is also a material cause. . . ." As he says in the *Physics*: "In one sense, then, that out of which a thing comes to be and persists, is called 'cause,' e.g. the bronze of the statue, the silver of the bowl, and the genera of which the bronze and the silver are the species" (p. 240). That material causes are relative is not perplexing; what is puzzling is that they include such abstract classifications as species and genus.

Yet this is less surprising when we recall that his previous discussion of the basic 'principles' to be used in understanding change involved syllogistic demonstrations showing that an object possesses a certain property because it belongs to a higher or more general class of objects, resembling biological classifications. This taxonomic mode of explanation, rather than explaining things in terms of their atomic-molecular structures, is the source of the puzzlement. The formal cause is the 'essence,' 'definition,' 'form' or even 'formula' that determines what an object is. Initially existing in the substance as a potential cause, when actualized the form constitutes the kind of object it is. In the previous examples, what the statue represents is the actualization of the bronze while the shape is the actualization of the silver bowl. As he says in the *Physics*, in this second type of cause, "the form or the archetype, i.e. the . . . essence and its genera are called 'causes' (e.g., of the octave the relation 2:1, and generally number), and the parts in the definition" (p. 240). His famous Categories stating the ultimate types of classifications is another example of the significance of formal distinctions in his system.

The efficient cause is the most familiar to us because it closely resembles the modern scientific sense of cause, that which brings about the change or actualization of things, along with their motions. As the third type of cause it is defined in Book II of the *Physics* as "the primary source of the change or coming to rest; e.g., the man who gave advice is a cause, the father is cause of the child, and generally what makes of what is made and what causes change of what is changed" (p. 241). This definition is incomplete, however, in that it does not include an explanation of motion. As we shall see, it is the 'Prime Mover' which is the cause of the heavenly spheres producing the circular motion of the stars and the planets, including the sun or "generator," that is the efficient cause of the changing seasons which in turn is the efficient cause of the meteorological, botanical, organic, and biological alterations in the terrestrial world.

> Thus, from the being of the "upper revolution" [i.e., 'the eternity of the revolution of the heavens'] it follows that the sun revolves in this determinate manner; and since the sun revolves *thus*, the

> seasons in consequence come-to-be in a cycle . . . and since they come-to-be cyclically, so in their own turn do the things whose coming-to-be the seasons initiate. (p. 531; brackets added)

Lastly is the final cause, the 'actualization,' 'fulfillment,' or 'end result' of any change. Once again the influence of his biological model is apparent in that all organisms and plants grow for a purpose, along with the persistent form of primitive explanationsas due to intentional causes. As he states in Book II of the *Physics* in one of the most famous passages in all of his extensive writings:

> If then it is both by nature and for an end that the swallow makes its nest and the spider its web, and plants grow leaves for the sake of the fruit and send their roots down (not up) for the sake of nourishment, it is plain that this kind of cause is operative in things which come to be and are by nature. And since "nature" means two things, the matter and the form, of which the latter is the end, and since all the rest is for the sake of the end, the form must be the cause in the sense of "that for the sake of which." (p. 250)

These four causes are exemplified in his conception of the Prime Mover which as a thinking Being is the *material* cause, as the eternal source of motion the *efficient* cause, as pure actuality the *formal* cause, and as the culminating mover in the universe the *final* cause. This is not meant in the sense of an ultimate creator since all the components of the universe, such as substance, form, and motion, are eternal, but as a raison d'être for all that exists.

To give concreteness to the preceding principles of explanation, among all the possible conceptions of the composition of the actual world proposed by his predecessors, Aristotle chose the four elements of Empedocles, fire, earth, air, and water. It was this choice that led to his theory of the four elements being predominant until its replacement by Newton's corpuscular theory in the seventeenth century. But he amends Empedocles' theory with a deeper analysis illustrating how, consistent with his explanatory principles, their interaction brings about physical changes. Thus the elements are analyzed into prime matter and pairs of the primary "opposites" or "contrarieties," the hot and the cold and the wet and the dry, that endow them with their particular qualities or properties. As he states in Book II of *On Generation and Corruption*: "For Fire is hot and dry, whereas Air is hot and moist (Air being a sort of a aqueous vapour), and Water is cold and moist, while Earth is cold and dry" (p. 511).

He then provides an astute explanation of how the interactions of the four elements plus the exchanges among the "couplings" of the contrary qualities can account for the various transformations we see in nature. Those elements will convert most easily that have one quality in common, such as water and air which share the quality moist and can exchange their coupled qualities, cold and hot, to convert from one to the other when water is heated or air cooled. Those elements, such as fire and water and earth and air which have no qualities in common are the most difficult to transform, yet such changes can occur under extreme conditions. Other qualities possessed by the elements, such as "heavy-light," "hard-soft," "viscous-brittle," "rough-smooth," and "coarse-fine," are formed by adding additional qualities to the four original ones; for example, heavy is an additional property of earth and water due their 'tendency' to fall to the center of the universe, while lightness is a further property of fire and air owing to their 'tendency' to rise. From these examples one can understand why the atomic-molecular conception of substances being composed of inner particles and forces, accounting for their external and interactive properties, subsequently replaced the ancient theory of the four elements.

We turn now to his astronomical and cosmological views which are further proof of his reliance on sensory evidence, his extensive knowledge of astronomy, and his tremendous ingenuity in fitting this information into his unique system of explanation. As for the importance of sensory evidence, he reaffirms that "in the knowledge of nature is the unimpeachable evidence of the senses as to each fact." As for his cosmological system, it required (1) delineating and explaining the difference between the two most prominent forms of motion, terrestrial and celestial; (2) describing the different substances involved in the two kinds of motions; and (3) ascertaining the necessary causes or reasons for these motions. It was these entrenched theoretical assumptions underlying his astronomical explanations that were the initial basis of astronomical research from the fifteenth to the seventeenth centuries that provoked such controversy and eventually had to be entirely replaced. His explanations occur primarily in Book VIII of the *Physics*, the *De Caelo* (*On the Heavens*), and book Lambda of the *Metaphysics*.

Beginning with the terrestrial world because it is the most familiar and accessible, as previously described it exhibits two contrary motions: the downward direction of the earth and water, and the upward thrust of fire and air. His explanation of these contrasting movements goes no further than the verbal assertion that they have a 'natural tendency' respectively to move in those directions correlated with the 'potency' of their natural places to receive them. Thus the earth rests immobile in the center of the universe surrounded by water, air, and fire which successively fill the gap between the earth and the innermost side of the moon's orbital sphere, the division between the terrestrial and the celestial worlds.

His defense of the spherical shape of the earth and arguments for why it remains stationary are considerably advanced over his predecessors. For example, he attributes the previous misconceptions of the earth's shape as drum-like and flat to the perceived flatness of the earth's surface and that the horizon line of the earth *across the sun* at sunset appears straight. In defending its sphericity, he points out that the ordinary horizon line of the earth appears curved, not straight, that during lunar eclipses the edge of the earth reflected on the moon also is curved, and that the hulls of sailing ships passing beyond the horizon disappear before their masts, which would occur if the earth were spherical but not if it were flat.

He adds that the earth's size is insignificant because as one travels north and south different stars become visible "which goes to show not only that the earth is circular in shape, but also that it is a sphere of no great size, for otherwise the effect of so slight a change of place would not be so quickly apparent." He gave the accepted calculation of the earth's circumference as 400,000 stades or 9,987 geographical miles, a figure much too small since the actual dimension is 24,990 miles. Owing to the earth's small size he claimed "there is continuity between the . . . pillars of Hercules and . . . India, and that in this way the ocean is one," a major reason Columbus believed that by sailing westward he could reach India. All of this shows the immense learning and astuteness of Aristotle's reasoning in contrast to his usual reputation.

He refutes the Pythagorean arguments that the earth has two circular motions (as Galileo will rebut *his* arguments two millennia later), one an orbital revolution around the central fire and the other a westward diurnal rotation on its axis. If the earth revolved in an orbit, he claims, then "there would have to be passings and turnings of the fixed stars [i.e., their displacements relative to the earth's motion]. Yet no such thing is observed" (brackets added). Against the claim that the earth rotates on its axis, he introduces the contentious argument in Book II of *De Caelo* that the earth must be immovable since "heavy bodies forcibly thrown straight upward return to the point from which they started" (p. 434) that would be impossible if the earth rotated because during the time the object was in the air it would have revolved eastward. Each of these arguments will be used by the Scholastic Aristotelians during the Copernican revolution to oppose the heliocentric system that interchanged the positions of the sun and the earth.

Thus the supporters of heliocentricism had to refute Aristotle's conclusions that were based on direct perceptions and common sense arguments by counterarguments and new evidence. To deny his argument disproving the earth's rotation, Galileo demonstrated that an object dropped from the masthead of a sailing ship fell parallel to the mast whether the ship was moving or at rest (if it move uniformly), therefore no conclusion could be drawn regarding the motion of the ship from the trajectory of the falling object. But Aristotle's crucial qualitative distinction between the terrestrial and celestial worlds posed equally serious problems. Because "no change appears to have taken place . . . in the whole scheme of the outermost heaven" (p. 403), according to the records, they must be "ungenerated and indestructible," hence "immutable," "eternal," and "divine." How was it possible, then, for a terrestrial object with a natural rectilinear motion to exist in the heavenly realm composed of weightless ethereal bodies having a natural circular motion or for the ethereal weightless sun with a natural circular motion to be at rest at the center of the territorial world?

One can see what a radical cognitive change was necessary to shift from the geocentric to the heliocentric universe, one that required two millennia to achieve. Yet the transformation was aided by new arguments and discoveries in the sixteenth and seventeenth centuries, such as Tycho Brahe's startling discovery in 1572 of a previously unseen luminous body (a Nova) located in the fixed stars that was brighter than Venus. Then in 1577 he observed an elongated brilliant newcomer (a comet) which he determined was at least six times as distant in space as the moon.[6]

These discoveries by the "greatest naked eye astronomer in history," along with Kepler's laws describing elliptical rather than circular orbits and Galileo's revealing telescopic observations brought new challenges to the ancient conception that the heavenly realm was too perfect to undergo any changes, while also providing further evidence for revising the venerable geocentric system. Yet given such caveats as 'at least with human certainty' and 'so far as our inherited records reach,' as well as Aristotle's steadfast adherence to what 'we can see,' Galileo was correct in asserting "I declare that we do have in our age new events and observations such that if Aristotle were now alive, I have no doubt he would change his opinion" (op. cit., p. 50).

Returning to our exposition, the natural state of things would be for each of the four elements to attain their natural places and remain there, but "violent" or "unnatural motions" occur due to the interaction of the elements or their displacement owing to natural causes, as when fire moves downward in a flash of lightening or water and physical objects are swept upward by strong winds or hurricanes. Another common example of an "unnatural motion" is the propelled movements of objects contrary to their natural tendency to fall. Explicitly rejecting an inertial tendency in the projected object to remain in motion when the propellant ceases because all moving objects tend to come to rest, Aristotle explains projectile motion as due to the impact of a contiguous mover or, when that stops, to the compressed pressure of the air behind the object until it is exhausted. Recognizing the weakness of the explanation of natural motions in terms of a "natural tendency" and of projectile motions as due to contiguous "motive causes," the founders of modern classical science replaced these explanations by various forces, such as gravity and inertia, within a completely revised explanatory framework.

Just as terrestrial motion had two contrary rectilinear directions, upward and downward, this natural rectilinear motion also had a contrary, the natural circular motion of the celestial world. In the *Physics* orbital motion was described as continuous, eternal, and prior, hence the perfect motion of the heavens. Furthermore, being eternal and perfect it had to have a substance other than the four terrestrial elements, hence the introduction of a fifth element, the "aither," a weightless substance finer than fire. Confronted with describing the various heavenly motions without utilizing mathematics, Aristotle adopted the "concentric" or "homocentric spheres" developed by the famous Pythagorean mathematician and astronomer, Eudoxus, while also refining Eudoxus' system with the help of another astronomer, Callippus.

To account for the various motions of the celestial bodies Eudoxus had posited twenty-seven homocentric spheres: a single sphere for the fixed stars, three conjoined spheres each for the moon and the sun, and four each for the remaining planets in their order of ascendance: Mercury, Venus, Mars, Jupiter, and Saturn. These spheres had their various axes positioned so that the contiguous rotations would produce each of the orbital periods. The two outermost spheres of the seven planets convey similar motions to each of them, the first sphere with its axes oriented towards the north and south poles replicates the motion of the fixed stars thereby producing the planets' east to west diurnal rotation. The axes of the second sphere is tilted 23 1/2° from the north-south poles so that its equator bisected the Zodiac forming an elliptic with the sphere of the stars. The various velocities of these second spheres determine the particular orbital speeds of each of the planets in their common revolutions from west to east opposite the diurnal rotation, from roughly eighty-eight days for Mercury to the twenty-nine years for Saturn.

Thus the celestial realm consisted of successive nesting spheres rotating and revolving in various inclinations and speeds around a common center occupied by the earth (hence the term "geocentric"). A third sphere was added to the moon and the sun to adjust for the variations in their orbital periods. As for the five remaining planets, along with their orbital irregularities they presented an additional problem because during their periodic revolutions they can be seen to have a retrograde motion in which they reverse their eastward direction moving some distance westward and then looping back to their regular forward motion. So while having the motions of the first two spheres, Eudoxus added two additional spheres to the orbits of the remaining five planets with their axes so positioned to explain their observable irregularities and retrograde motions. In an effort to further "save the phenomenon," Plato's famous admonition to astronomers, Aristotle states that Callippus added additional spheres to bring their total to thirty-four, including the sphere of the fixed stars.

While the systems of Eudoxus and Callippus were limited to describing the orbital periods of each planet separately, Aristotle realized that this did not take into account the effects of the motions of the planets above it. Thus he introduced "counteracting spheres" to negate the influence of the preceding motions of the planets. Yet this still left two major anomalies in this geocentric system: that at various times or positions in their orbits the sun and moon can be seen to be at different distances from the earth and to vary in brightness, neither of which could be explained by a system of concentric spheres which place the planets always at an equal distance from their common center. Later Apollonius and Hipparchus will attempt to remove these discrepancies by replacing the several spheres with a single concentric sphere, a "deferent," and an epicycle with its center on the circumference of the deferent conveying the planet. The alternating forward and backward circular motion of the epicycle would explain the variations in the distances and brightness of the moon and the sun because at different points in their epicyclical revolutions they would be closer or farther away from the earth, while also explaining the retrograde motions of the other five planets for the same reason.

Additional epicycles were added later to conform to observations, but the system eventually became so cumbersome that Aristarchus in the second century BCE and Copernicus in the sixteenth century of the CE realized that some of these anomalous appearances could be attributed to our geocentric perspective, rather than the actual motions of the planets. Thus they could be eliminated and a simpler system achieved by adopting the heliocentric theory. Thus the crucial issue was whether to accept the testimony of our senses or attribute the observable discrepancies to the misleading appearances of the planets owing to our geocentric perspective.

Having explained the structure and motions of the planets within the limits of human reason, one would suppose the discussion to be complete. However, accepting the principle that "all things that are in motion must be moved by something," that applied to naturally circular as well as unnatural motions, Aristotle was forced to explain how the first sphere of the fixed stars acquired its motion. The question did not pertain to a temporal first cause since all the motions were eternal, but to a *sufficient* explanation of the ultimate origin of motion itself. While the revolutions of the other spheres could be traced back to the revolving effect of the fixed stars transmitted successively to each of them, the logic of the argument demanded an explanation also of the origin of the motion of the fixed stars. This Aristotle attributed to a ninth sphere that accounted for the daily rotation of the entire universe in twenty-four hours from east to west. Moreover, if everything that moves requires a mover, to avoid an infinite regress the explanation had to terminate in an 'unmoved mover,' the origin of Aristotle's famous 'Prime Mover,' that existed beyond the ninth sphere in a spaceless Empyrean Heaven.

Although the concept of the Prime Mover is presented also in Book VIII of the *Physics,* the most comprehensive discussion of its nature occurs in the various books of the Metaphysics. This culminating work treats of subjects beyond physics, such as being, intellect, and thought, especially as attributed to the Prime Mover. He called this discipline 'first philosophy' which, in its search for the basic principles of explanation and of the nature of 'being qua being,' is the ultimate source of Wisdom. Thus metaphysics became known as the 'queen of the sciences,' the origin of its significance for philosophy throughout its history. It is in Book Lamda of the *Metaphysics* that Aristotle presents the concise argument for the Prime Mover.

> There is, then, something which is always moved with an unceasing motion, which is motion in a circle; and this is plain not in theory only but in fact. Therefore the first heaven [the outermost sphere of the fixed stars] must be eternal. There is therefore also something which moves it. And since that which is moved and moves is intermediate [derivative], there is something which moves without being moved, being eternal, substance, and actuality. And the object of desire and the object of thought move in this way, they move without being moved. (p. 879; brackets added)

There are in Greek, as in English, two senses of the expression "to be moved," one referring to a physical displacement and the other to a psychical or emotional reaction to something, such as a tragedy or a beloved object. So while the motion of the outermost sphere or *primum mobile* is transmitted to the other spheres by contact, to avoid an infinite regress *its* motion is attributed to its desire for or attraction to the Unmoved Mover: "The final cause [the Prime Mover], then, produces motion as being loved, but all other things move by being moved [contiguously]" (p. 879; brackets added). While this may sound strange, in Aristotle's system it is consistent with the doctrine expressed in the *De Anima* that 'appetition' or 'desire' is the source of action or motion in an organism, its being affected "movingly" one might say. In the same work he also describes thinking as an activity of the soul whereby 'contemplation' produces action, therefore it "follows that there is a justification for regarding these two as the sources of movement, i.e., appetite and . . . thought . . ." (p. 598).

When we remember that the heavens are not composed of physical matter but of a finer kind of substance, the aither, which in the *Metaphysics* is described as divine and the celestial bodies as gods, it is not as difficult to understand the rationale behind Aristotle's thinking. For example, in Book Lambda he commends his "forefathers in the most remote ages" for having the "inspired" belief that "these [celestial] bodies are gods and that the divine encloses the whole of nature" (p. 884; brackets added). Although in the same book he adopts the view that there are a plurality of unmoved movers corresponding to the total number of planetary spheres, because the Prime Mover is the ultimate cause our discussion will be confined to it.

The Prime Mover represents the culmination of the principles discussed previously. As the source of motion by being the object of desire it is the *efficient* cause; as pure actuality devoid of all substance and potentiality it is the *formal* cause; and as the self-sufficient being who contemplates its inner thoughts it is the *final* cause. So the Prime Mover is not a creative or personal God in the Christian sense, but a kind of Ideal Philosopher that reflects on its own thoughts. As he says in Book Epsilon of the *Metaphysics*: "Therefore it must be of itself that the divine thought thinks (since it is the most excellent of things), and its thinking is a thinking on thinking" (p. 885). Despite the difference between a personal creator God and a contemplative eternal God, one can appreciate that Thomas Aquinas' transformation of Aristotle's Prime Mover into the Christian God was not so difficult. The creation of a brilliant mind, Aristotle's cosmology brings together in a coherent system his extensive knowledge of astronomy with the primitive animistic conception that only life or soul can be the ultimate source of motion.

The reason for such a lengthy description of Aristotle's account of scientific explanation and astronomical theory is that its later destruction in the seventeenth century is the only example in history of the complete replacement of an entire worldview, clearly illustrating how a deeply entrenched cognitive system when confronted with anomalies, disconfirming evidence, refuting arguments, and corrected presuppositions can be discarded and replaced. In my opinion, this is exactly the situation regarding the current religious worldview despite some claims that it is hardwired and therefore inviolable. The fact that it has been possible to outgrow the mythical antecedents of the religious worldview and to replace Aristotle's cosmology by the modern scientific worldview belies these claims.

Though neurophysiological research has shown that John Locke's view that at birth our consciousness is a "blank slate" is false and that Kant was correct in attributing to cognition innate endowments, he too was mistaken in thinking that these inherent "categories of the mind" unalterably determine and therefore limit how we think. The radical reorganization of our cognitive outlook brought about by the scientific revolution of the sixteenth and seventeenth centuries preceding Kant refutes this. Since then, though there have been other major changes in our ways of conceiving the universe and human existence, such as the origin of species, the discovery of extragalactic universes, the age of the universe, relativity theory, quantum mechanics, and the realization that consciousness is a product of our brains, rather than an independent soul, these have been additions to or revisions of the worldview of science, not a rejection or replacement of it.

Nonetheless, despite the complete rejection of Aristotle's cosmological system about two millennia later, we should not minimize his tremendous contributions to Western thought that were the background for the scientific revolution of the sixteenth and seventeenth centuries. There are four main reasons for his conception of scientific inquiry and cosmological system being so influential: (1) his constant assertion that sensory observations are the foundations of knowledge, superior to dialectical arguments; (2) the reasonableness of his common sense presuppositions; (3) his remarkable ability to clarify theoretical problems and offer credible solutions within the limits of what was then known; and (4) the magisterial comprehensiveness of his system which surpassed all others from the thirteenth to the seventeenth centuries. It therefore took a convergence of a number of developments to bring about its downfall.

Primary was the realization that sensory evidence, as important as it is, cannot go unchallenged. Although the geocentric point of view was supported by such ordinary experiences as not feeling the movement of the earth and seeing the rising and setting of the sun, it nevertheless created certain astronomical problems that could be eliminated by viewing the planetary orbits from a nongeocentric perspective. Tycho Brahe's more exact naked eye astronomical observations supplied Kepler with the evidence that enabled him to formulate his elliptical laws of planetary motion that provided an exact explanation of the variations in the distance and brightness of the planets. Then Galileo's telescopic discoveries (as well as the invention of the microscope at the turn of the seventeenth century) reinforced the realization that direct sensory observations were not only misleading, but limited and thus had to be supplemented by invented instruments.

In addition, Aristotle's physical laws based on common sense inferences, such as falling objects accelerate proportional to their weights, was corrected by Galileo's incline plane experiments demonstrating that they fall proportional to the squares of the times, while his inadequate explanation of projectile motion inspired

Buridan and Oresme in the fourteenth century to devise an alternative theory in terms of inertial force, showing that revised explanations and experimental inquiry are required to correct direct observations by probing nature to discover more basic causes and laws. These crucial additions are reflected in Newton's declaration that "in Natural Philosophy, the Investigation . . . consists in making Experiments and Observations and in drawing general Conclusions from them by Induction, and admitting of no Objections against the Conclusions but such as are taken from Experiment, or other certain Truths."[7]

Moreover, in contrast to Aristotle's primary reliance on syllogistic deduction owing to his generic or 'organismic' explanations, experimental inquiry required mathematics to express the magnitudes, ratios, and formulae embedded in the laws of nature. Yet from our discussion it also is obvious that most of the new discoveries and explanations were based on the critical examination and rejection of Aristotle's previous interpretations, so even if Galileo can be called the father of modern science because of his novel telescopic and experimental discoveries and revised laws and conceptual changes, at least Aristotle's empirical orientation provided the essential background. Thus his contributions proved as crucial for the development of scientific inquiry, as a first approximation, as the later use of instrumentation and experiments were to discovering new evidence, causes, and mathematical correlations to correct and supplement direct observation.

NOTES

1. The quotations from Aristotle's writings can be found in Richard McKeon, *The Basic Works of Aristotle* (New York: Random House, 1968). So unless otherwise indicated the page references are to this work.

2. W. D. Ross, *Aristotle*, fifth ed. rev. (London: Methuen and Co., 1949), pp. 3–4.

3. Werner Jaeger, *Aristotle*, trans. by Richard Robinson, 2nd ed. (Oxford: At The Clarendon Press, 1948), p. 316.

4. Galileo Galilei, *Dialogues Concerning the Two Chief New Sciences*, trans. by Stillman Drake, 2nd ed. (Berkeley: University of California Press, 1970), p. 321.

5. Charles Darwin to William Ogle, on the occasion of the publication of the trans. of Aristotle's *The Parts of Animals,* 1882 in *The Life and Letters of Charles Darwin*, Vol. II, ed. by Francis Darwin (New York: 1896), p. 427. Quoted from John Herman Randall, *Aristotle* (New York: Columbia University Press, 1960), p. 219.

6. Cf. Arthur Koestler, *The Sleepwalkers* (New York: Grosset's Universal Library, 1959), pp. 288.

7. Sir Isaac Newton, *Opticks*, Book III, sec. V, Query 31, based on the 4th ed. London, 1730, prepared by Duane H. D. Roller (New York: Dover Publications, 1952), p. 404.

Chapter 4

HELLENISTIC SCIENCE IN ALEXANDRIA

THE FOUNDING OF ALEXANDRIA

We turn now in this rendition of the growth of science to a new era, called "the Hellenistic Age," and to a different location, Alexandria in Egypt. Exhausted by the Peloponnesian wars and the devastating effects of the plagues, the decline of Athens and loss of its empire enabled the Macedonian King, Phillip II, to defeat Thebes and Athens in 338 BCE in the battle of Chaironeia, placing Athens under Macedonian rule. Following Philip's death two years later he was succeeded by his son, Alexander the Great, who then was just twenty years of age. Despite his youth, Alexander resumed his father's preparations to invade Persia and in twelve years of brilliant tactical campaigns succeeded in conquering their extensive empire, including destroying the magnificent palace at Persepolis. Then, tragically, in 323 Alexander died at age thirty-three, probably from an illness contracted while he was preparing a voyage by sea around Arabia. Except for Napoleon, never in history has another person achieved such conquests at such a young age. His death and that of his former tutor Aristotle a year later marked the end of the Hellenic period and the beginning of the Hellenistic age.

His only son, Alexander Aegus, killed at the age of twelve, was born posthumously to Alexander's widow, Roxana, a Bactrian princess. Thus not having an heir before he died, Alexander left his signet ring as a token of inheritance to his Macedonian general Perdiccus. But the bitter rivalry among his commanders led to his vast empire being divided into three kingdoms under the rule of three dynasties: (1) Greece and Macedonia governed by the Antigonids, (2) Western Asia by the Seleucids, and (3) Egypt by the Ptolemies. As it was the Ptolemies who transformed the city chosen by Alexander to be his namesake into a great metropolis replacing Memphis as the capital of Egypt, along with its becoming the leading center of scientific and humanistic research during the Hellenistic period, our discussion will be confined to it. While other Greek cities, such as Pergamum, Rhodes, and Antioch also were outstanding centers of learning during this period, it was Alexandria which attracted and produced the scholars that made the greatest contributions to the advance of science in the succeeding three centuries.

It is a tribute to the enduring Greek intellectual tradition that a Macedonian general and his son, even if persuaded by their choice of directors, would be motivated not only to establish Alexandria as the capital of Egypt, but also to make it an outstanding research center. For it was the first two Ptolemies, Ptolemy I Soter (the savior) and his son Ptolemy II Philadelphus (brother-loving), who founded and funded the famous centers of research in Alexandria, the Museum and the Library. But the two well-known Peripatetics initially responsible for their organization and development were Demetrius of Philemon, who was invited by Ptolemy Soter to be the first head of the Museum and later director of the Library, and Strato of Lampsacus who, after having been tutor to the crown prince, was selected by Ptolemy Philadelphus the second head of the Museum.

Both Demetrius and Strato had been students of Theophrastus following his replacement of Aristotle as head of the Lyceum when the latter (a Macedonian) found it expedient to leave Athens when a rebellion broke out against Macedonian rule after the death of

Alexander. The close tie between Athens and Alexandria is further evident in Demetrius having been governor of Athens before being selected head of the Museum, while Strato returned to Athens after directing the Museum in Alexandria to lead the Lyceum for eighteen years following Theophrastus' death. When governor of Athens Demetrius had convinced his teacher Theophrastus to expand the Lyceum by adding a large lecture hall, accommodations for teachers and students, and a library to house the collection of Aristotle's extensive works. This enlarged institution was called the Museum to honor the Muses, hence the origin of its name. It was this Museum in Athens that became the model for the greater Museum in Alexandria. Thus the latter's success in surpassing Athens as the main research center from the third to the first centuries BCE was due to its close connection with the Lyceum in Athens.

Following the example of Aristotle's Lyceum, it was largely Strato who decided that the kind of research pursued in the Museum would include the more empirical sciences of physics, geology, anatomy, and medicine, along with astronomy and mathematics, while research in Plato's Academy focused more exclusively on the abstract disciplines of astronomy and mathematics. Receiving free room and board along with a stipend, perhaps for the first time in history scholars were state supported to devote their lives exclusively to the pursuit of knowledge, explaining why the Museum in Alexandria contributed so much to the progress of science. As Sarton states:

> It was because of its creation and because of the enlightened patronage which enabled it to function without hindrance that the third century witnessed such an astounding renaissance. The fellows of the Museum were permitted to undertake and to continue their investigations in complete freedom. As far as can be known, collective research was now organized for the first time, and it was organized without political or religious directives, without purpose other than the search for truth.[1]

Like the Museum, the Library was built and endowed by Ptolemy Soter and greatly expanded by his son who appointed Demetrius its first director. Eventually becoming the most famous library in antiquity, said to have contained more than 700,000 scrolls and the prototype of all subsequent libraries, the Ptolemies went to extraordinary lengths to acquire its superlative collection of papyri. As Sarton continues:

> The Kings of Egypt were so eager to enrich their library that they employed high-handed methods for that purpose. Ptolemaios III Evergetes . . . ordered that all travelers reaching Alexandria from abroad should surrender their books. If these books were not in the Library, they were kept, while copies on cheap papyrus were given to the owners. He asked the librarian of Athens to lend him the state copies of Aeschylus, Sophocles, and Euripides in order to have transcripts made of them, paying as a guarantee of return the sum of fifteen talents; then he decided to keep them, considering that they were worth more than the money he had deposited and he returned copies instead of the originals. (pp. 143–44)

As is true of modern libraries (before the Internet), the Library contained the collective repository of human culture and history that scholars could consult when pursuing their own physical, mathematical, historical, or literary research. Without such records each generation would have had to start anew except for what could be remembered and transmitted orally. This is why the destruction of the Library, reputedly during Caesar's siege of Alexandria in 48 BCE, with its huge collection of classical works, was such a tragedy (as was true of the recent sacking of the museum and library in Baghdad) because so much of what it contained is lost forever. But the Library, which was devoted mainly to humanistic works, was not merely the repository of books.

In addition to collecting and cataloguing the rolls of papyri, the librarians were philologists involved in authenticating and editing them. It was owing to one of the later librarians, Aristophanes of Byzantium, that a system of punctuation was developed and then with the help of Aristarchus of Samothrace (not the astronomer from Samos) rules of grammar were created by differentiating the various parts of speech. It seems incredible that all the early Greek epics, tragedies, histories, philosophies, and mathematical treatises were written without any explicit knowledge of grammar but only by an assimilated usage, yet that seems to be the case. As Sarton again states:

> It is marvelous that all of the intricate beauties of the Greek language, a very complex grammar as well as a rich and well-integrated vocabulary, were created to a large extent unconsciously. The main creators of Greek literature did not know grammar, but the Alexandrian philologists extracted grammar from their writings . . . where it was implicitly contained. (p. 155)

Having described the origins of the Museum and Library and the nature of the research activities in both, we turn now to an account of some of the more outstanding scholars associated with them. Unlike my previous attempts to reconstruct the thought processes of the philosophers in order to analyze and critique their presuppositions and reasoning, this will not be possible in the case of the mathematicians and scientists to be cited. The reason is that their treatises normally do not disclose the reasoning that led to their conclusions or discoveries, but mainly the outcomes themselves, with the exception of Archimedes' *Method*. This is especially true of mathematics which usually is mysteriously esoteric and opaque to the uninitiated. While everyone can observe that the elements have different properties, that some objects fall and others rise, that the planets vary in their orbital motions, and that animals have an anatomical structure, looking at a mathematical theorem or proof does not convey a similar recognition or understanding. Replicating the three famous mathematical problems of antiquity, the determination of conic sections or the area and volume of a sphere, the trisection of an angle or duplication of the cube, and the quadrature (squaring) of the circle, requires a specialized mathematical skill and learning.

Therefore, rather than a critical examination of the thought processes of the scholars, this chapter will summarize some of their more outstanding achievements with the intent of evaluating their contributions to the advance of mathematics and science, along with correcting some misconceptions in the process. As an example of the latter, it is often claimed that what distinguishes classical from modern science is that while the former is qualitative, descriptive, and nonmathematical, the latter is experimental, explanatory, and quantitative, as if the distinction were absolute. However, even in our previous discussion of the Presocratics we have seen that this is not true, but especially with the advance of science in the Hellenistic period we find clear examples of applied mathematics and of (simple) experiments to test hypotheses. The rise of science is a gradual process that is, unlike evolution, governed by intelligent design.

As claimed previously, becoming aware of the possibility of organizing empirical data mathematically and of using experiments to discover new phenomena and test hypotheses depended upon first realizing the limitations and deceptiveness of ordinary observations and commonsense generalizations and explanations. Thus based on the previous investigations of statics and hydrostatics by Aristotle, Strato and Archimedes could make new discoveries in those fields. Given the concentric orbits of Eudoxus, succeeding Hellenistic astronomers such as Heraclides, Apollonius, and Hipparchus could introduce modifications to make the concentric model conform more to exact astronomical observations. Similarly, based on the research in the medical school of Hippocrates, Herophilus and Erasistratus could advance the knowledge of anatomy and physiology, all of which illustrates how science progresses by more acute observations, clearer discrimination of problems, refining the conceptual interpretations and theoretical explanations, and devising new methodologies and instruments to expedite the process.

EUCLID

Turning now to the contributions themselves, considering the crucial role today of mathematics as the "language of science" it is surprising how little is known about the personal lives of the ancient Greek mathematicians, such as Pythagoras, Theaetetus, Archytas, Eudoxus, and Euclid, whose investigations were so significant that they were cited by the founders of modern classical physics as their predecessors. For example, the little we do know about Euclid is derived mainly from the fifth-century CE philosopher and mathematician Proclus, who provided the following description:

> . . . Euclid . . . put together the *Elements*, collecting many of Eudoxus' Theorems, perfecting many of Theaetetus', and also bringing to irrefragable demonstration the things which were only somewhat loosely proved by his predecessors. This man lived in the time of the first Ptolemy. For Archimedes, who came immediately after the first (Ptolemy), makes mention of Euclid. . . . He is then younger than the pupils of Plato, but older than Eratosthenes and Archimedes, the latter having been contemporaries, as Eratosthenes somewhere says.[2]

This tells us that he lived at the time of Ptolemy Soter while the reference to Eudoxus and Theaetetus suggests that he may have visited or studied at Plato's Academy. Another statement by the mathematician Pappus who lived in the last half of the third century BCE informs us that Apollonius of Perge "spent a very long time with the pupils of Euclid at Alexandria," which indicates that he established a school at Alexandria in the latter half of the third century BCE. But more important than his biography is the fact that we have extant the *Elements* which, as Struik asserts, has been one of the most influential books of all time.

> The *Elements* form, next to the Bible, probably the most reproduced and studied book in the history of the Western World. More

> than a thousand editions have appeared since the invention of printing, and before that time manuscript copies dominated much of the teaching of geometry. Most of our school geometry is taken, often literally, from eight or nine of the thirteen books; and the Euclidean tradition still weighs heavily on our elementary instruction. For the professional mathematician these books have always had an inescapable fascination (even though their pupils often sighed) and their logical structure has influenced scientific thinking perhaps more than any other text in the world.[3]

This assertion is confirmed by the impact that the *Elements* had on two of the greatest scientists, Newton and Einstein, the former having modeled his *Principia Mathematica* on the *Elements* and Einstein reporting in his "Autobiographical Notes" the effect that reading a book on Euclidean plane geometry had on him as a boy of twelve: "Here were assertions . . . which—though by no means evident—could nevertheless be proved with such certainty that any doubt appeared to be out of the question. This lucidity and certainty made an indescribable impression upon me."[4]

The thirteen books of the *Elements* deal with such diverse subjects as plane geometry, Eudoxus' theory of proportions, the Pythagorean theorem and the golden section, number theory, Theaetetus' theory of irrational numbers, quadratic irrationals and their quadratic roots, and solid geometry plus the five Platonic solids. As Euclid himself acknowledged, many of the theorems in the *Elements* had been discovered before him and some of the books are expanded versions of topics of earlier mathematicians.

Even if partially derivative, the *Elements* still consists of many new proofs, extended inquiries, and a deductive system that eclipsed all previous books on geometry. But its greatest innovation lies in the use of the axiomatic method illustrated in Book I. Beginning as books on mathematics do today with definitions, postulates, and axioms, he illustrated how theorems were deduced from these premises, although even here he was not original. As Sarton points out:

"The most amazing part . . . is Euclid's choice of postulates. Aristotle was, of course, Euclid's teacher in such matters; he had devoted much attention to mathematical principles, had shown the unavoidability of postulates and the need of reducing them to a minimum; yet the choice of postulates was Euclid's" (p. 39). Here again the legacy of Aristotle is apparent. Even though in the Lyceum there was more emphasis placed on empirical inquiry than on mathematical studies, it was Aristotle's *Organon* containing his books on logic that especially explored and described the structure of axiomatic systems.

Euclid's fifth postulate concerning parallel lines is particularly well known because it was the critique of this postulate in the nineteenth century by Gauss, Lobachevsky, Bolyai, and Riemann that led to the development of various non-Euclidean geometries. It was Riemann's geometry of curved space that Einstein used in 1915 in his general theory of relativity to describe his new conception of a curved space-time continuum to replace Newton's Euclidean absolute linear space and flow of time.

Although best known for his *Elements,* Euclid wrote other works: a book on plane geometry called *Data* that still survives, as well as books on "Conics," "Surface-Loci," and several additional books on geometry, all of which have been lost. In addition, he wrote a work on applied mathematics titled *Phaenomena* intended to be used in astronomical calculations and perhaps a work on music, referred to as *Elements of Music,* which also have been lost. Probably these were burnt in the fire that destroyed much of the collection in the Library of Alexandria, but at least we have the *Elements* that has been called "the greatest mathematical text-book of all time."

ARCHIMEDES

This Hellenistic mathematician, almost as famous as Euclid, is described by Heath as "perhaps the greatest mathematical genius the world has ever seen." A citizen of Syracuse, the capital of Sicily,

Archimedes was a close friend of the ruler of Syracuse, King Hieron and his son Galen. His father, Pheidias, was an astronomer. After an illustrious life devoted to mathematics, he was killed in 212 BCE at the end of the siege and sack of Syracuse by the Roman General Marcus Marcellus at age seventy-five, which would place his birth in 287. Thus his life spanned much of the third century BCE. There are numerous accounts of how he died, but most converge on his being so engrossed in a mathematical diagram he had drawn in the sand that he had not noticed that the city had fallen and ignoring a Roman soldier's command, the soldier struck him down with his sword. It is said that Marcellus, who had been prevented from taking the city by assault due to Archimedes' marvelous war machines forcing him to undertake a long siege, wanted very much to meet this mechanical genius and was distraught at his death.

These terrifying war machines are described by Plutarch in his *Life of Marcellus*.

> When, therefore, the Romans assaulted them by sea and land, the Syracusans were stricken dumb with terror; they thought that nothing could withstand so furious an onset by such forces. But Archimedes began to ply his engines, and shot against the land forces of the assailants all sorts of missiles and immense masses of stones, which came down with incredible din and speed; nothing whatever could ward off their weight, but they knocked down in heaps those who stood in their way, and threw their ranks into confusion. At the same time huge beams were suddenly projected over the ships from the walls, which sank some of them with great weights plunging down from on high; others were seized at the prow by iron claws, or beaks like the beaks of cranes, drawn straight up into the air, and then plunged stern foremost into the depths, or were turned round and round by means of enginery within the city and dashed upon the steep cliffs that jutted out beneath the wall of the city, with great destruction of the fighting men on board, who perished in the wrecks. Frequently, too, a ship would be lifted out of the water into mid-air, whirled hither and

> thither as it hung there, a dreadful spectacle, until it . . . would fall empty upon the walls or slip away from the clutch that had held it.[5]

In addition to these ingenious war machines. he invented an endless screw, a water-screw for irrigating fields and pumping water out of mines, compound pulleys, and an orrery (a model of the solar system) displaying the orbital periods of the moon, sun, and the planets. Cicero, who saw it about two centuries later, said that it was so exact in depicting the motions of the sun and moon that it even showed their eclipses. There are numerous anecdotes about Archimedes, such as his famous boast to King Hieron that "give me a place to stand on, and I can move the earth," whereupon the King asked for a comparable demonstration. To accommodate, Archimedes attached a compound pulley to the bow of a three mast ship loaded with passengers and freight and, grasping the end of the pulley, "he drew the ship along smoothly and safely as if she were moving through the sea."

Yet despite the spectacular consequences of his inventions and practical value of his law of specific weights, he dismissed such accomplishments as having no significance. As Plutarch again relates in his *Life of Marcellus*:

> . . . Archimedes possessed such a lofty spirit, so profound a soul, and such a wealth of scientific theory, that although his inventions had won for him a name and fame for superhuman sagacity, he would not consent to leave behind him any treatise on this subject, but regarding the work of an engineer and every art that ministers to the needs of life as ignoble and vulgar, he devoted his earnest efforts only to those studies the subtlety and charm of which are not affected by the claims of [practical] necessity.
>
> These studies, he thought, are not to be compared with any others; in them the subject matter vies with the demonstration, the former supplying grandeur and beauty, the latter precision and surpassing power. (Cohen and Drabkin, p. 317: brackets added)

In confirmation of his devotion to mathematics he asked that after his death there be placed on his tomb a drawing of a sphere circumscribed within a cylinder together with the ratio of the cylinder to the sphere. Finding the tomb in a ruined condition in 75 BCE when he was quaestor of Sicily, Cicero had it restored.

According to Diodorus, Archimedes spent considerable time in Alexandria when he was young, probably studying at the school attributed to Euclid, after which he returned to Syracuse where he devoted the rest of his life entirely to mathematical research carrying on a lively correspondence with a number of mathematicians. Some of these were friends he had met at Alexandria, such as Conan of Samos to whom he sent many of his discourses and Eratosthenes of Cyrene who received a copy of the *Method*.

The *Works* of Archimedes, edited by Heath, shows the range of his mathematical genius. Among the most famous works, some having the salutation "Archimedes to Dositheus greeting[s]," are two books, one "On the Sphere and Cylinder" demonstrating that "the surface of any sphere is four times as great as a circle in it" and that a cylinder with its base equal to the circumference of a sphere and its height equal to the diameter will be 1½ times in volume; the second "On Conoids and Spheroids" used a method anticipating integral calculus to calculate the volumes and areas of segments of various geometrical figures. In addition were books "On the Equilibrium of Planes" describing the principles governing equilibrium and levers with theorems in statics.

As his invention of the orrery indicated, he was also involved in astronomical investigations, such as calculating the distances of the planets. In the *Sand-reckoner* he devised an apparatus for measuring the diameter of the sun and proved that numbers greater than the amount of sand filling the universe could be calculated. Other works include a treatise on "The Quadrature of the Parabola;" two books "On Floating Bodies" demonstrating the "Archimedean Principle" that objects displace their weight when immersed in water; a "Book of Lemmas;" the famous "Cattle-Problem" to find the "number of bulls and cows of each of four colors or to find eight unknown qualities;" and the *Method,* a palimpsest discovered in 1906 by Heilberg that distinguishes between the discovery and the proof of theorems. As Heath states, it is also significant because in contrast to the usual mathematical treatises, which disguised the methodology, in the *Method* "we have a sort of lifting of the veil, a glimpse of the interior of Archimedes' workshop [or mind] as it were."[6]

The *Works* contain a profusion of intricate geometrical diagrams used in his calculations and proofs because lacking algebraic notations and functions all his demonstrations had to be illustrated by such diagrams. He combined two methods in his demonstrations: (1) the method of reductio ad absurdum in which a premise contrary to what one wants to prove is assumed with the expectation of deriving a contradiction thereby proving the truth of the intended premise, and (2) the method of approximation whereby areas or volumes are approached by increasing or decreasing geometrical figures to approximate desired magnitudes. Given these astonishing works and his lofty ideals we can only concur with Pappus that he was "a wonderful man, a man so richly endowed that his name will be celebrated forever by all mankind. . . ."

ARISTARCHUS

Known as "the ancient Copernicus," Aristarchus' amazing adoption of the heliocentric hypothesis is a superlative illustration of the superiority of Greek speculative reasoning over biblical revelation. As indicated previously, on purely a priori grounds the Pythagoreans were the first we know of to attribute a revolving motion to the earth, along with the counter-earth and other planets, around a central fire or hearth. Then, as Aëtius asserts, "Heraclides and [the Pythagorean] Ecphantus make the earth move, not in the sense of translation, but by way of turning on an axle, like a wheel, from west to east, about its own centre." Heraclides further observed that Mercury and Venus revolve around the sun, but having approached a heliocentric theory he shied away from the full version, instead claiming (as later Tycho Brahe will) that the sun with its two satellites, along with the other planets, revolve around the stationary earth in the center of the universe. It was Aristarchus who, despite the overwhelming terrestrial evidence, accepted the tremendous challenge of transporting the earth into another revolving planet.

In his only remaining work *On the Sizes and Distances of the Sun and the Moon*, using the method of triangulation by measuring the angular positions of the sun and the moon when the latter is exactly at half moon, he calculated the sizes and distances of the sun and the moon relative to the earth or to each other: for example, "*The distance of the sun from the earth is greater than eighteen times, but less than twenty times, the distance of the moon from the earth.*"[7] Although there is no mention of the heliocentric theory in that early work, it may have been his measurement of the larger size of the sun that led him later to assign to it the central position in the universe. It is most regrettable that the work in which Aristarchus presented his heliocentric hypothesis has been lost, along with the reasoning that led him to his conclusion.

We also do not know the exact dates of his birth and death, only the approximate extent of his life from about the last decade of the fourth century to the latter third of the third century BCE. This can be inferred from his recorded observation of the summer solstice in 281–280 and the fact that the publication of the work just mentioned preceded Archimedes' *The Sand-reckoner* written before 216. Apparently he was about twenty-five years older than Archimedes but younger than Strato of Lampsacus under whom he studied. There is confirmation of Aristarchus' avowal of the heliocentric view by Plutarch who relates "that Cleanthes [a contemporary Stoic] held that Aristarchus ought to be indicted for the impiety of 'putting the Hearth of the Universe in motion'" (p. 304; brackets added), recalling the Catholic Inquisition's indictment of Galileo. It is to Archimedes that we are indebted for the only extant statement of

Aristarchus' position that occurs in the *Works of Archimedes with the Method of Archimedes* cited above.

> Now you [King Gelon] are aware that "universe" is the name given by most astronomers to the sphere whose centre is the centre of the earth. . . . But Aristarchus of Samos brought out a book consisting of some hypotheses, in which the premises lead to the result that the universe is many times greater than that now so-called. His hypotheses are that the fixed stars and the sun remain unmoved, that the earth revolves about the sun in the circumference of a circle, the sun lying in the middle of the orbit, and that the sphere of the fixed stars, situated about the same centre as the sun, is so great that the circle in which he supposes the earth to revolve bears such a proportion to the distance of the fixed stars as the centre of the sphere bears to its surface. (pp. 221–22: brackets added)

This was an incredible hypothesis for the time because it not only ignored all the terrestrial objections, it presupposed a universe enormously larger than anyone had ever conceived. For it to be possible for the earth to orbit the sun without there being any evidence of a displacement of the fixed stars, which he states were "unmoved," the relative size of the earth's orbit to that of the fixed stars must be as a point to the latter's enormous circumference. That is, for there to be no stellar parallax (a displacement of the fixed stars as seen from various positions of the earth as it traverses its orbit), Aristarchus realized that the size of the earth's orbit should be as a point to the stars' sphere. Unable to make this imaginative leap eighteen centuries later, even Tycho Brahe adopted Heraclides' partial heliocentric model rather than Copernicus' complete heliocentric system. Moreover, since Aristarchus claimed that the fixed stars were stationary he must have held Heraclides' view that the apparent diurnal rotation of the entire universe from east to west was due to the earth's rotation in the opposite direction, attributing two motions to the earth similar to Copernicus.

APOLLONIUS OF PERGA AND HIPPARCHUS

But while Kepler and Galileo supported the position of Copernicus, there was no one to advance the theory of Aristarchus even though Eudoxus' system of twenty-seven homocentric spheres could not explain the observed variations in size and brightness of the orbiting planets, nor the apparent retrograde motions of the five planets excluding the sun and the moon. While Heraclides had emended Eudoxus' system of earth centered spheres by placing the sun at the center of the orbits of Mercury and Venus, he still retained the earth at the center of the universe. The subsequent innovations of planets revolving on epicycles with their centers on deferents to account for retrograde motions along with eccentric orbits to explain the variations in sizes and brightness apparently seemed less radical than Aristarchus' more drastic revision.

Later Apollonius of Perga, a famed astronomer and mathematician (whose treatise on conic sections was used by Kepler to formulate his first two planetary laws), in contrast to Aristarchus adopted the innovations of epicycles and eccentrics to accommodate the discrepant astronomical observations. The system of epicycles could explain retrograde motions while eccentric planetary orbits, having their centers located slightly off center from the earth, could account for the planets appearing closer or farther away in their circular trajectories around the earth, thus explaining their variations in size and brightness. This system also was adopted by the second century BCE astronomer Hipparchus who was both an outstanding mathematician and one of the greatest astronomers of antiquity. He was the founder of the branch of mathematics known as trigonometry, indispensable for calculations in astronomy today as in the past.

CLAUDIUS PTOLEMY, HERO OF ALEXANDRIA, AND GALEN

While the Hellenistic age officially terminated at the end of the first century BCE, its legacy extended through the second century CE, with three Greek scientists marking the finale of the great legacy of ancient Greek mathematical and scientific research. Though there will be an abbreviated regeneration of this tradition by the Arabs from the ninth through the thirteenth centuries, its sustained renewal will await the emergence of modern classical science from the sixteenth to the seventeenth centuries.

Continuing our discussion of astronomy with one of the three finalists, Claudius Ptolemy, the celebrated Greco-Egyptian mathematician, astronomer, and geographer who, living in the second century CE and making his observations in Alexandria, was the last great astronomer of antiquity. Although credited with a number of discoveries of his own, his reputation is based mainly on his collating and systematizing the extensive research in geography and astronomy of the scientists at Alexandria. While acknowledging Hipparchus as his greatest predecessor, it was his magisterial *Almagest* (the Arabic name of his treatise) that became the authoritative astronomical work replacing all previous treatises until Copernicus' system became accepted in the seventeenth century. As such it was one of the most influential scientific treatises of all time. Even the name has an interesting history as related by Charles Singer:

> The Greeks called it the *megalê mathematiē syntaxis*, i.e. "great mathematical composition," but also used the superlative form *megistê syntaxis.* This was transliterated by the Arabic writers as al-magistê which passed into medieval Latin as *Almagestum*.[8]

The second major example of the continuation of Hellenistic scholarship is Hero or Heron of Alexandria who, although his exact dates are unknown, probably lived in the first century CE. He is particularly important because of the large number of his treatises that have survived, including *Pneumatics*, *Mechanics*, *On the Dioptra*, and *On the Construction of Automata*, showing the range of his scientific interests, along with a number of works in mathematics. Also he created several ingenious contrivances, some of which were merely playful while others had considerable scientific value, such as an instrument for measuring heights and distances. But from the point of view of this study, it is his description of an experiment that is particularly significant because it illustrates how experimental inquiry was beginning to be used in the Hellenistic period.

The following example taken from his *Pneumatics,* and though simplistic compared to modern experiments, nonetheless explicitly describes an experiment to illustrate his argument and support his theoretical hypothesis and interpretations. Aristotle had claimed that because "nature abhors a vacuum" the terrestrial realm is completely filled with the four elements, fire, air, earth, and water, and that the celestial realm is entirely composed of the aither. In contrast, Hero believed that all substances are composed of 'invisible particles' separated by 'minute vacua' that explains their compressibility and elasticity.

> They, then, who assert that there is absolutely no vacuum may invent many arguments on this subject, and perhaps seem to discourse most plausibly though they offer no tangible proof. If, however, it be shown by an appeal to sensible phenomena that . . . a vacuum exists . . . but scattered in minute portions; and that by compression bodies fill up these scattered vacua, those who bring forward such plausible arguments [against the vacuum] . . . will no longer be able to make good their ground.[9] (brackets added)

Hero's own counterargument is advanced in three stages: first by showing that a supposedly empty vessel is actually filled with air which prevents water from entering it at the same time.

> This may be seen from the following experiment. Let the vessel which seems to be empty be inverted, and, being carefully kept upright, pressed down into water; the water will not enter it even though it be entirely immersed: so that it is manifest that the air, being matter, and having itself filled all the space in the vessel, does not allow the water to enter. Now if we bore the bottom of the vessel, the water will enter through the mouth, but the air will escape through the hole. . . . Hence it must be assumed that air is matter. (p. 249)

Second, having shown that to prevent the water from entering the vessel the air must be material, he describes what gives it its materiality. "Now the air, as those who have treated of physics are agreed, is composed of particles minute and light, and for the most part invisible." Then using the analogy of horn shavings and sponges, he argues that it is the void spaces in the air that accounts for the compressibility of substances.

> The particles of the air are in contact with each other, yet they do not fit closely in every part, but void spaces are left between them, as in the sands on the sea shore: the grains of sand must be imagined to correspond to the particles of air, and the air between the grains of sand to the void spaces between the particles of air. Hence, when any force is applied to it, the air is compressed, and, contrary to its nature, falls into the vacant spaces from the pressure exerted on its particles: but when the force is withdrawn, the air returns again to its former position from the elasticity of its particles, as is the case with horn shavings and sponges, which, when compressed and set free again, return to the same position and exhibit the same bulk. (p. 250)

Third, he concludes his proof by asserting that "a vacuum exists also naturally, but scattered in minute portions" among the insensible particles "and that by compression bodies fill up these scattered vacua. . . ." While this is a simple example as compared to modern experiments, it explicitly illustrates the essential aspects of the experimental method, except for the lack of a mathematical correlation or law. He used several experimental examples to illustrate what he intends to explain; he describes the essential property of the explanatory element, the materiality of the air; and finally he introduces the hypothesis of invisible particles to explain why air is material and what it is about the particles, their 'scattered vacua,' that accounts for the compressibility. The enormous potential of experimental inquiry was gestating but would not show real signs of life until about a millennium and a half later.

Before turning to the third scholar culminating the Hellenistic scientific tradition, the renowned physician Galen, something should be said about the outstanding Greek contributions to the advance of medical inquiries and practice. Of all the sciences, medicine is the most empirical because it is based on directly observable illnesses, symptoms, and injuries. Although the causes themselves are not directly disclosed nor are the anatomical structures and functions of the various organs and physiological systems hidden in the organism, they can be exposed due to injuries and by dissection of human corpses and animals. Thus while speculation and theories still played a large role in early medical investigations, they were more grounded in empirical evidence than most other areas of scientific inquiry—as evidenced by their influence on Aristotle. Furthermore, they were especially amenable to experimentation since physicians could prescribe remedies, treatments, and operations whose effects could be directly ascertained.

Thus there arose throughout ancient Greece numerous schools of medicine, themost famous of which was that of Hippocrates of Cos, already mentioned. Other outstanding representatives of this tradition were Herophilos of Chalcedon born in the last third of the fourth century BCE and Erasistratos also born on the Isle of Cos at the end of the fourth century. As J. Beaujeu states:

> At about 300 B.C., medicine, like all the other sciences was given a sudden impetus in Alexandria, the capital and leading scientific

> centre of the Hellenistic world. . . . Two of the greatest physicians of antiquity, Herophilos and Erasistratos, came to Alexandria to establish schools. Both men . . . did a great deal to develop anatomy and physiology, mainly by their systematic dissections. In fact, even in Galen's day (second century A.D.) physicians still came to Alexandria to study their method.[10]

But the most famous of the Hellenistic physicians is Galen who was born in Pergamum in the second century CE and studied in the well-known medical centers at Smyrna, Corinth, and Alexandria. After his studies he returned to Pergamum becoming physician to the College of Gladiators whence he could observe the terrible injuries that contributed to his outstanding knowledge of anatomy and practice of surgery. His reputation led to his being called to Rome by Marcus Aurelius to serve as personal physician to Commodious, the heir-apparent. Remaining in Rome for twenty years where he is credited with writing over 500 works on such subjects as anatomy, physiology, etiology, diagnosis, dietetics, and pharmacology, he returned to Pergamum where he died about 200 CE.

Although his support of the humoral theory of temperament proved incorrect, he was famous for his empirical approach to medicine, especially his dissection of animals, corpses, and even apes because of their close resemblance to humans. In his tremendously influential *Therapeutic* he endeavored to trace disorders to the malfunctions of various organs which stimulated research into their etiologies and morphologies. His use of experiments to enhance his inquiries and test his hypotheses is clearly described in the following procedure.

> Now the method of demonstration is as follows. One has to divide the Peritoneum in front of the ureters [urethras], then secure these with ligatures, and . . . having bandaged up the animal, let him go (for he will not continue to urinate). After this one loosens the external bandages and shows the bladder empty and the ureters quite full and distended—almost on the point of rupturing; on

> removing the ligature from them one plainly sees the bladder becoming filled with urine.[11]

He also describes experiments to demonstrate the function of the heart and the different composition of the blood in the veins and arteries connected to different organs. But what is especially impressive is his discovery, by dissecting various nerves connected to the brain, that they were the cause of the paralyses of different parts of the body.

> Dissection has taught us that the nerves which go to the parts of the face come from the brain itself. Hence when one of these parts is paralyzed along with the whole body, you may be sure that the seat of the paralysis is in the brain itself; but when these facial parts are unaffected, the paralysis originates in the [spinal] cord. . . . Since apoplexy [a stroke] at once destroys all psychic functions, it is clear that the brain itself is affected. . . . Therefore one who knows, by dissection, the origin of the nerves that pass to each part will be better able to cure each part deprived of sensation and movement. . . .[12] (brackets added)

According to Beaujeu, his explanation of the functions of the spinal cord in particular was not surpassed until the beginning of the nineteenth century, and "his correlation of injuries to various parts of the spinal cord and the cranial and cervical nerves with paralysis of given organs, or given parts of the body, are rightly considered among the most brilliant medical observations of all times."[13]

What I find especially impressive about the above investigation is his discovery that the brain not only is the crucial organ causing certain paralyses of the body, but that apoplexy or strokes "destroy physic functions." When one considers that Aristotle attributed consciousness to the heart and that the brain played practically no role in the explanation of consciousness or cognitive processes in the thinking of such modern philosophers as Descartes, Locke, and Kant, who explained cognitive structures and functions solely in terms of the mind or the soul without even mentioning the brain, one can appreciate how astute and advanced Galen was in his understanding of the nervous system, particularly the brain. Even today many religionists attribute consciousness or cognition to the soul, rather than the brain, despite all the neurological evidence to the contrary and the explanatory vacuity of the soul.

What makes this even more surprising is the fact that Galen's remarkable capacity for objective observations, controlled experimentation, and disciplined reasoning was not impeded by his deep conviction that nature is teleological with God as the source of its purposeful organization. Today he would be called a believer in "intelligent design." Yet that religious convictions did not interfere with his sound scientific inquiry is somewhat explainable by the fact that his research was in the biological and medical sciences where ascertaining the purposeful functions of organs and systems plays a vital role.

His extensive writings were in such demand that they eventually were translated into Latin, Syriac, Arabic, and Hebrew dominating biological and medical thought throughout the Middle Ages until the beginning of modern science in the sixteenth and seventeenth centuries. Like Ptolemy in astronomy, his reputation was so authoritative that it precluded criticism. When Andreas Vesalius, a leading professor at the University of Padua that had become the major center of anatomical and physiological research in the world, announced in 1543 (the year when Copernicus' heliocentric theory was published) that he had discovered errors in Galen's theories, he aroused a storm of protest. It was not until William Harvey, a student at Padua, in 1628 demonstrated the circulation of the blood and the function of the heart as a pump that the authority of Galen finally receded.

Though three Alexandrian mathematicians, Diophantus, Pappas, and Theon of Alexandria continued the development of algebra, geometry, and trigonometry in the third century CE, Ptolemy and Galen represent the apex and coda of the incredible contributions of the ancient Greeks to scientific inquiry. One can even ask whether the advent of science and technology, which have so transformed civilization, would have occurred had it not been for the unprecedented creative genius of the Greeks. One also wonders what gave rise to this tremendous creativity and why it occurred when and where it did? Was it nature or nurture or both that enticed the Greeks to demystify the world making it more intelligible?

Moreover, which brings us to the next phase of our study, it explains why the founders of modern classical science cite ancient Greek scholars as their precursor: Copernicus citing the Pythagorean Philolaus and the Hellenistic astronomer Heraclides as having declared that the earth moves, while Aristarchus is known as the "Copernicus of antiquity"; Apollonius' research into the mathematics of conic sections aided Kepler's replacement of the venerable circular planetary orbits with ellipses and formulation of his first two astronomical laws; Aristotle's law of falling bodies was the motivation for Galileo's alleged Tower of Pisa experiments refuting his law while he claimed Archimedes and Strato as his mentors; Galen's research and theories were the background for the experiments of Vesalius and Fabricius; Erasistratus' rejection of the humoral theory and discovery of the connection of arteries and veins anticipated Harvey; and Eudoxus' mathematical method of exhaustion and Archimedes' method of integration antedated the discovery of differential and integral calculus by Leibniz and Newton. Thus while the stage was set for the creation of modern classical science, neither Rome nor Christianity provided the talent nor cultural background for its enactment. It would take over a millennium and a half for the cognitive connections and transitions with their Greek predecessors to be reestablished in the modern era.

NOTES

1. George Sarton, *A History of Science*, Vol. II (Cambridge: Harvard University Press, 1959), p. 34. Future references to this work will be indicated by his name and page number.

2. Proclus on Euclid, I. Quoted from Sir Thomas L. Heath, *A History of Greek Mathematics*, Vol. I (Oxford: At The Clarendon Press, 1921), p. 354.

3. Dirk J. Struik, *A Concise History of Mathematics*, 3d. rev. ed. (New York: Dover Publishing 1967), pp. 50–51.

4. Albert Einstein, "Autobiographical Notes," Paul A. Schilpp ed., Vol. VII, *Albert Einstein: Philosopher-Scientist* (Evanston: The Library of Living Philosophers, Inc., 1949), p. 9.

5. Plutarch, *Life of Marcellus,* pp. 14–17, trans. by Bernadotte Perrin (London, 1917). Quoted from M. R. Cohen and I. E. Drabkin, *A Source Book in Greek Science* (Cambridge: Harvard University Press, 1948), p. 316.

6. Quoted from the "Introductory Notes" to *The Method of Archimedes* in *The Works of Archimedes With The Method of Archimedes*, ed. by T. L. Heath (New York: Dover Publications, 1953), p. 7.

7. Sir Thomas Heath, *Aristarchus of Samos: The Ancient Copernicus* (Oxford: At the Clarendon Press, 1966), p. 377. The following citations in the text are to this work unless otherwise indicated.

8. Charles Singer, *A Short History* of *Scientific Ideas to 1900* (New York: Oxford University Press, 1959), p. 90.

9. Morris R. Cohen and I. E. Drabkin, op. cit., pp. 249–51. Each of the quotations is within the pages cited.

10. J. Beaujeu, "Medicine," Part II, chap. 5, *History of Science: Ancient and Medieval Science*, ed. by René Taton, trans. by A. J. Pomerans (New York: Basic Books, 1963), p. 341.

11. Galen, *On the Natural Faculties*, I, 13, trans. by A. J. Brock (London, 1916) in Morris R. Cohen and I. E. Drabkin, op. cit., p. 481.

12. Galen, *On the Parts Affected*, III, 4, trans. by C. G. Kuhn in Morris R. Cohen and I. E. Drabkin, op. cit., pp. 497–98.

13. J. Beaujeu, J., Part II, ch. 5, René Taton, *History of Science: Ancient and Medieval Science*, op. cit., p. 360.

Chapter 5

THE DECLINE OF SCIENTIFIC INQUIRY

THE ROMANS

As gifted as they were in conquering and administering an empire and in disciplines such as jurisprudence, engineering, architecture, and literature, the Romans did not produce a single outstanding scientist or mathematician who was not Græco-Roman. Though Rome subsidized research in the Museum and Library in Alexandria when the eastern Mediterranean became her protectorate after 200 BCE and Greek learning was diffused among the ruling class, the Romans never produced a notable scholar in physics, astronomy, or mathematics, seemingly content to leave that to the Greeks. Their curriculum included enough science and mathematics to train engineers and technicians, but not to create scholars in the more theoretical disciplines. As Beaujeu states,

> . . . Roman engineers perfected methods of building roads, bridges, aqueducts, canals, vaults and steps, and of manufacturing glass and metal articles. . . . But it remains a fact that . . . they . . . tended to leave science to the Greeks, and lacked the ability to bring to mathematics the intellectual acumen in which they so excelled when it came to jurisprudence. In short there is no such thing as Roman science. . . .[1]

There are a number of apparent reasons for this deficiency. First, if the Persian empire and the military city-state of the Spartans can be taken as exemplars, the kind of training, discipline, and skills required for such outstanding military regimes seems to be incompatible with the aptitude and motivation for abstract theoretical inquiries, which often have no immediate practical consequences. It is not just that military training and warfare themselves are not suited to such reflective thinking, but that the demands and goals of such militaristic societies as a whole are not conducive to those pursuits. Although the Romans produced outstanding writers such as Cicero, Horace, Virgil, Seneca, and Epictetus, along with historians or biographers such as Pliny the Elder and Plutarch, the empirical disciplines in which they excelled, such as civil engineering, geography, medicine, and calendar reform, apparently reflected their practical or militaristic needs which excluded theoretical scientific curiosity.

Second, rather than adopting the idealized ethical system of Plato or even that of Aristotle, the Romans were attracted to the Greek systems of Stoicism and Epicureanism whose goal was not to attain knowledge of the good or virtue for its own sake, but to develop philosophies as guides for how best to live in the world. In a civilization fraught with so much cruelty, civil strife, political intrigue, moral excesses, and the usual vicissitudes of daily life—Pompey, Caesar, and Cicero were murdered while Mark Anthony, Cleopatra, Cassius, and Brutus committed suicide—the need for a compensating philosophy was preeminent. Thus Stoicism was the philosophy of Cicero, Seneca, Epictetus, and the philosopher Emperor Marcus Aurelius, along with prominent military leaders, civil servants, and businessmen. Cicero declared that owing to Lucretius' poem *On Nature*, which presented Epicureanism in verse, that philosophy "seized the whole of Italy."

In our previous description of the contributions of the Academy, Lyceum, and Museum to the development of scientific and mathematical thought, no mention was made of these other schools of philosophy, situated in their Stoa and Garden, that arose in Athens at the end of the fourth century BCE. Unlike the academic pursuits of the previous institutions, these philosophies developed personal ethical systems and worldviews to mitigate the adverse conditions of life. In Athens they were so successful, according to A. A. Long, that "Platonists and Peripatetics never exercised a monopoly in Greek philosophy, and they were soon outshone in the extent of their influence by the new Stoic and Epicurean schools."[2] As Lucretius wrote in his poem *On Nature*, "do you doubt that the power over men's fears and cares lies altogether in philosophy, especially since our whole life is a struggle in the dark?"[3]

One might wonder why the progress of the ancient Greeks in understanding and explaining natural phenomena was not seen as providing whatever "power" was available to allay man's "fears and cares." My guess is that the answer is twofold. First, the development of natural philosophy had not attained a sufficient consensus regarding the correct methodology and theoretical framework to be used in investigating natural phenomena. While today there is general agreement about the functions of observation, controlled experimentation, conceptual interpretation, theory construction, and the role of mathematics in scientific inquiry, that was not true at that time. Moreover, the impressive advances in physics, astronomy, chemistry, biology, microbiology, genetics, neurophysiology, pharmacology, and medicine since the latter nineteenth and especially the twentieth century have confirmed their approximate truth. Furthermore, the extraordinary success in applying the atomic-molecular framework to nearly all areas of science has reinforced this confidence. Though there had been important discoveries and explanations by the end of the third century CE, there was nothing like the confirmed consilience of knowledge that exists today.

Second, although Archimedes and Hero had used their theoretical knowledge to develop ingenious pulleys, water screws, and various machines while Galen had made considerable advances in understanding human anatomy and neurophysiology, this was insufficient for believing that such knowledge could be used to improve human conditions. Furthermore, while there had been some recognition by Leucippus and Democritus that an understanding of nature and human existence required discovering the underlying atomic causes of observable phenomena, this was too elementary to provide any useful knowledge. This would await developments in the atomic-molecular theory in the eighteenth century showing that the Greek 'Periodic Chart,' consisting of fire, earth, air, and water, was inadequate because these presumed irreducible substances were found to be composed of more basic elements and forces.

The discovery of these elements, such as oxygen, hydrogen, nitrogen, sodium, and chlorine, along with their atomic properties and exact combining ratios in molecular substances like water, acids, and salt, awaited the industrial revolution and later developments in physics and chemistry in the eighteenth and nineteenth centuries. Then advances in the medical sciences, such as the recognition of the roles of bacteria and viruses in causing infections and transmitting diseases along with the discovery of vaccines and antibodies used in preventive medicine, contributed greatly to the "power" we now have to reduce many of our "fears and cares." The more we learn about the natures and interactions of things, the more control we can exert over them. Not having achieved such control, the Romans instead embraced Stoicism and Epicureanism to bring tranquility or happiness to their lives.

STOICISM

Zeno, who founded Stoicism, was born in Cyprus in 333/332 BCE, but at the age of twenty-two settled in Athens where, in his early thirties, he established a school in a garden at one edge of the Agora, the market place and civic center of ancient Athens. Acquiring its name from the painted colonnade or Stoa that bordered the porch where Zeno strolled as he lectured, the school looked onto the altars, temples, and public buildings across the Panathenaic Way to the monumental Acropolis aloft in the distance. Its history is divided into three periods: the early Stoa directed by Zeno who died about 263 BCE and his notable successors Cleanthes, Chrysippus, and Antipater; the Middle Stoa directed by Panaetius and Posidonius; and the Late Stoa that included such illustrious authors as Cicero, Seneca, and Epictetus, along with the Emperor Marcus Aurelius. Zeno was held in such esteem by the Athenians that he was awarded a golden crown while he lived and upon his death a tomb was built in his honor and funeral reliefs carved in the Academy and Lyceum. Chrysippus, who headed the Stoa from about 232 BCE until his death near the end of the century, was a distinguished scholar whose extensive writings were a major source of Stoic doctrine.

In addition to the previous reasons, Stoicism became popular in the Graeco-Roman world because the educated citizens were losing faith in the anthropomorphic gods of the state religion and the numerous oriental mystery cults, turning to philosophy for ethical guidance and solace. Three of the greatest sources of mental anguish are remorse over choices whose outcomes are deeply regretted; tragic events over which we have no control but devastate our lives; and an acute awareness of the cruel injustices that inflict individuals and societies. In each case the anguish depends upon thinking and wishing that things could have happened differently, but if one believes that everything is inevitable and occurs for the best when considered from a larger perspective, then presumably much of this anguish can be offset. As Marcus Aurelius wrote: "Nothing is harmful to the part which is advantageous to the whole. For the whole contains nothing which is not advantageous to itself. . . . As long as I remember that I am part of such a whole I shall be well content with all that happens."[4] The difficulty is in believing that all the horrors of life are somehow advantageous to the whole of existence.

To this end the Stoics developed their predestined cosmology, theory of knowledge, and ethical principles.[5] Since human fate is dependent on the destiny of the cosmos or divine providence, which for the Stoics came to the same thing, man's highest aspiration is in recognizing the harmony of the two. The Stoics apparently found in Heraclitus' cosmology the theoretical system best suited for this. As in his diffuse cosmic system, the Stoics melded an inert matter, *pneuma*, with creative fire, soul, Logos, laws of nature, and divine providence, into their cosmology.

Adopting Aristotle's conception of prime matter as an indeterminate material substrate for all that exists, they considered this a 'passive substance' without qualitative properties that gives the universe its three-dimensional spherical shape. Pervading this inert matter is an 'active agent' providing its fiery, creative character. This '*pneuma*' or vital aspect literally means 'breath,' so that while acting as a force it is more like an animate, vitalistic agency. Believing that nurture and growth are due to a 'vital heat,' this germinating *pneuma* is further described as a 'creative' or 'artistic fire.' Composed also of air, the various combinations of air and fire constituting the *pneuma* somehow created stones or metals, plants, animals with souls, and human beings with minds.

As in Heraclitus, there is a cosmic cycle that when directed inward produces unity and when directed outward results in diversity of quantity and quality. In a process of eternal recurrence, each cycle begins with the creative fire generating the four elements, fire, earth, air, and water, that constitute the natural world and terminate eventually in a conflagration which destroys them, after which the process is renewed. As a cosmology this system is as incredible as that of Heraclitus, illustrating how fanciful a theory of the universe becomes when speculatively forced into a preconceived ethical (or religious) framework, rather than based on empirical evidence.

Again borrowing from Heraclitus, the Stoics further endowed the universe with an inherent *Logos* consisting of rational laws that completely determine all eventualities. However, unlike the mechanistic necessity of the Atomists, the deterministic laws imposed by this *Logos* produce the best of all possible worlds. Because of this providential necessity, the cosmos can be equated with divine providence or Zeus. Natural disasters such as plagues, famines, hurricanes, earthquakes, droughts, or untimely deaths are not intended by the Divine Cosmos to be harmful to humans, but are inevitable consequences of this being the best possible universe, the same rationale used by Christians. Apart from considering this providential necessity as a "higher destiny" that determines the fate of human beings, it is never defined except as producing the best of all possible worlds.

It is this notion of a providentially perfect universe that is the crucial *compensating* conception in Stoicism—as in Christianity. Given the proper understanding of the Cosmos, because the *Logos* that controls the Cosmos exists also in human beings as their rationally motivating faculty human destiny can be equated with divine providence. Thus the Stoics devoted much effort to describing what constitutes knowledge and how it can be attained, the most admirable of their rather obscure philosophy. They analyzed the structure and role of the components of propositions to explain their intentional (conceptual) meanings and referential (denotative) designations, claiming as logicians do today that the validity of truth-functional statements is determined by the truth-values of their components. I believe their conception of truth is what today would be called "justified true belief." Diogenes Laertius, a younger contemporary of Zeno, claimed that true knowledge is "a disposition in the acceptance of impressions which cannot be shaken by argument."[6]

Thus the final justification of knowledge is a secure defense based on a system of irrefutably true statements that correspond to the necessary causal connections inherent in the Cosmos. When the systematic implications in our discourse about the world represent necessary causal cosmic connections, then a harmony exists between our inner *Logos* and the *Logos* governing the world. Thus knowledge and truth are justified when the *Logos* in man, which is the source of 'articulate thought' or 'rational discourse,' coincides with the *Logos* that is the rational structure of the universe. It is this concordance that should obviate any affective opposition, rejection, or sorrow however painful or tormenting events would otherwise seem.

Having equated the 'is' with the 'ought,' the Stoics maintained that whatever occurs is as it ought to be. Voluntary action does not mean freedom of choice, but the wisdom to act in accordance with the preordained laws of nature. Since whatever has value is what either agrees with nature or consorts with nature, the highest human value is "acting in accordance with nature." Furthermore, the Stoics claimed that the 'primary impulse' of all living creatures instilled by the *Logos* is for self-preservation, which, therefore, is the basic motivation in ethics. Accepting one's fate and acting according to the laws of nature thus entails 'enhancement and self realization.' When such action becomes spontaneous or natural, as in the sage or wise man, that person has attained '*eudemonia*,' a state of happiness or well-being wherein all fears, cares, wants, or sufferings have been disdained and dismissed.

So what does all this come to? Not much, in my opinion. It is surprising that the intelligent Roman literati that we mentioned subscribed to it. Just as the stoic cosmology did not have a semblance of reality, neither did their ethics. There are three flaws in their view. (1) The most obvious is that if everything were predestined by the *Logos*, why should any effort be required to change or improve anything, including one's state of mind? The fact that human beings are irrefutably confronted with choices that do make a difference refutes the underlying conception of predestination (as it does in similar religious views). Even if the laws or processes of nature were absolutely determined (which is not supported by quantum mechanics nor the contingencies of evolution), the fact that humans are not just inanimate physical objects but conscious beings who can think, reflect, feel moral obligations, and imagine alternatives differentiates them from natural processes which are passive recipients of their causes.

(2) Their presupposition that human reason is strong enough to subdue and subjugate the fears, cares, pains, and anguish that are caused by external events is sheer delusion. Hume's claim that "reason is the slave of the passions," while exaggerated, is closer to

the truth. Though resignation induced by reason can alleviate some distresses, it certainly cannot eliminate all. One would have to be an emotionless zombie or robot to go through life indifferent to all the agony, frustrations, and suffering, as well as the pleasures, joys, and satisfactions that living as a human being involves.

(3) Furthermore, the belief that one should ignore all the misery, disappointments, and catastrophes because they are necessary for this to be the best possible world is incredible. Who is to say, and on what grounds, that this is the best of all possible worlds? That argument is no more convincing than the religious argument that the world was created by intelligent design or a loving God, despite all the discrepancies. What we know about human history, which Hegel aptly described as a "slaughter bench," as well as the horrendous natural occurrences whose devastating effects often depend on the slightest contingencies, belies such a naive view of things. When Leibniz proposed that this is "the best of all possible worlds" in the seventeenth century he was deservedly ridiculed by Voltaire. It is no wonder that Epicureanism became much more popular in Rome than Stoicism.

EPICUREANISM

The belief that human well-being or misfortunes are in the hands of the gods and that the soul has an afterlife in which there will be a final accounting for one's behavior was prevalent in Greek and Roman religions, as it was throughout antiquity and the Medieval Period. As in the Middle Ages, when such beliefs were held more literally and strictly than today, they could be a dreadful source of anxiety not only for oneself, but also for deceased loved ones. For Epicurus, these awful fears were seen as the primary obstacles to attaining a tranquil life, even more than the heartbreaks of life itself.

Born in 341 BCE on the island of Samos, after having attracted followers in Mytilene and later in Lampsacus he created a place of refuge in Athens in 307/6 when he was thirty-four years old, about the same age as Zeno when he established the Stoa. Consisting of a house and a garden, the residence resembled the Stoa in not being a teaching or research institute like the Academy and Lyceum, but a place where Epicurus' followers, which included women and slaves, could listen to his uplifting philosophy and read and discuss his numerous works. Unlike Stoicism, which advocated a virtuous participation in public office, Epicurus extolled a 'quiet life of tranquility' undisturbed by political ambition, avarice, or threatening religious beliefs. To his followers he was considered a "savior" and a "dispenser of light."

Diogenes Laertius describes him as a prolific writer exceeding all others in "the number of his works," but unfortunately only fragments from charred papyri and a few letters remain. We know of his doctrines mainly from Diogenes Laertius, Cicero, Seneca, Plutarch, and particularly the inspired expository poem, *On Nature*, by Lucretius who was an ardent admirer of Epicurus, as the following encomium declares.

> You, who first was able to lift so bright a light from amid such darkness, and to illuminate the blessing of life—you I follow, O glory of the Greek race, and in the tracks you once trod I now firmly place my footsteps, not so much desiring to vie with you as yearning to imitate you because of my love. . . .[7]

Instead of the convoluted cosmology of Heraclitus, Epicurus adopted the atomism of Leucippus and Democritus as the cosmological foundation of his ethical philosophy. Having discussed ancient atomism previously, I shall just mention several innovations introduced by Epicurus, as presented by Lucretius, to justify atomism: the concepts of 'minima' and the 'swerve.' His justification of atomism is especially cogent, pointing out that many of our sensations or observations, such as feeling heat and cold, observing the drying of wet clothes, the wearing away of stone pavements, or the disintegration of substances, do not visually disclose their components or processes suggesting that nature "carries on her affairs by means of invisible particles" (pp. 13–14: Bk. I, 299–329). Unless these particles existed we would have to conclude that natural processes violate Parmenides' maxim that "nothing can come to be from nothing nor cease to be."

Another argument introduced by the Epicureans to demonstrate the existence of the atoms shows a remarkable similarity to modern scientific thinking. The British botanist Robert Brown demonstrated in 1827 that pollen grains dissolved in water and observed under a microscope display a random motion. Excluding external causes, he attributed this erratic movement to the impact of unobservable atomic or molecular particles composing the water. Although at the time this did not persuade scientists of the existence of these particles, this changed after Einstein published, among six famous papers in 1905, one on "Brownian Motion" describing and predicting the exact kinetic motion of the molecules producing the irregular movement of the suspended particles. When his predictions were later confirmed experimentally by Jean Perrin this convinced many physicists of the existence of atomic-molecular particles. Citing the example first used by Democritus and later emended by Robert Brown and Einstein in terms of pollen grains dissolved in water, Lucretius wrote that in seeking evidence to support atomism

> you would do well to observe these motes which you see dancing in the sunbeams: this dancing indicates that beneath it there are hidden motions of matter which are invisible. You will see that many motes, struck by unseen blows, change their courses and are forced to move . . . in every direction. Truly this change in the direction of all the motes is caused by the atoms. (pp. 13–14: Bk. I, 299–29)

The second innovation is the prescient view that the atoms as defined by Leucippus and Democritus were not solid entities whose primary properties of size, shape, and weight must be assumed, but that they are determined by the number and arrangement of inherent 'parts' or 'minima.' Though these minima are elementary constituents of the atoms they are inseparable from them (as is true of quarks), leaving the atoms indivisible. As Lucretius states, these tiny parts within the atom

> which our senses cannot discern, surely are without parts and are of the least possible magnitude. . . . Other similar [minima] in. . . close-packed formation, make up completely the nature of the atom. Since these least parts cannot exist by themselves, they must cling together in the atom, from which they can in no way be separated. (p. 23: Bk. I, 600–10; brackets added)

This conception of minima foreshadows Murray Gell-Mann's theory of quarks whose fractional charges and strong interaction were posited to explain the properties of the hadronic subatomic particles, the baryons and mesons, from which they too cannot be separated. It is amazing how the thinking of these early philosophers could at times approach that of modern scientists.

Because the deterministic universe of the Atomists did not allow for free will, Epicurus introduced a third innovation, a cosmological 'swerve' to explain the accidental collisions of atoms and to allow for the indeterminacy or flexibility required for free will. As described by Lucretius, when the atoms are moving through the void, "at utterly unfixed times and places, they swerve a little from their course, just enough so that you can say that the direction is altered. If the atoms did not have this swerve . . . no collisions would occur. . . . Thus nature would not have created anything" (p. 47: Bk. II, 216–24). This swerve also permits free will, for if the atoms "did not swerve and cause a certain beginning of action that may break the chains of Fate and prevent cause from following cause through infinite time, whence [would] come this power of free will for living creatures throughout this earth?" (p. 48: Bk. II, 252–56; brackets added).

Having adopted Democritus' atomic cosmology though allowing for free will, Epicurus argued that there was nothing to fear from the existence of the gods nor that our souls would have to suffer in the afterlife. Given the universality of the belief in the gods, along with their appearance in dreams and mystical visions, again like Democritus he accepted this as evidence of their existence. But adhering to Democritus' atomic framework he claimed the gods were composed of very fine atoms having a kind of pallid, quasi-human form who enjoyed an eternal state of uninterrupted blissfulness. Existing in the ethereal spaces separating individual worlds (again following Democritus) they are completely indifferent to human affairs and have no influence over natural events. As Lucretius states, "gods in their whole being and from their own power enjoy unending life in perfect peace, far removed and separated from our experience."

Having disposed of the fear of the gods, Epicurus also eliminated the fear of retribution in the afterlife. Like the gods, souls are corporeal composed of breath-like minute atomic particles consisting of air, warmth, and wind, plus a fourth particle whose motions "produce sensations and the thoughts that stir in the mind" (p. 86: Bk. III, 238–39). Though unnamed because of its allusiveness, this fourth particle is the finest, smoothest, and most mobile of atoms to explain the "swiftness of thought." The unique mixture of these atoms determines individual personalities. Just as today there is no explanation of how chemical-electrical neuronal discharges cause sensations, conscious awareness, and thought, this was true of the Epicureans who offered no suggestion as to how these experiences are produced by atomic motions.

But their main objective being to *deny* the existence of the soul after death, they offer several arguments to show that it could not exist apart from the body. First, we have no memory of its preexisting the body; second, the mind seems to originate at the birth of the infant and develop along with it; third and primary, the soul atoms are too fine and tenuous to survive the body. Dispersed when the body dies, this prevents any sensations or conscious awareness that could be the source of pain or suffering after death. Thus the goal of eliminating those religious beliefs that could hinder attaining tranquility and well-being had been achieved. The following verse (originally in Greek) scratched on a wall in Herculeum expresses this:

> There is nothing to fear in God
> There is nothing to feel in death;
> What is good is easily procured
> What is bad is easily endured.[8]

Having disposed of the religious impediments to achieving a peaceful life, we turn to the basis of his ethics which is to maximize pleasure and eliminate pain. As he wrote in a letter to Menoeceus:

> We say that pleasure is the starting-point and the end of living blissfully. For we recognize pleasure as a good which is primary and innate. We begin every act of choice and avoidance from pleasure, and it is to pleasure that we return using our experience of pleasure as the criterion of every good thing.[9]

But as modern ethicists, such as John Dewey and G. E. Moore stressed, the fact that pleasures are initially desir*ed*, does not imply that all pleasures are desir*able*. While defrauding someone to obtain money, misrepresenting one's past accomplishments to gain a higher position, or having an affair while pretending to be faithfully married may bring temporary pleasures, yet they are not actually desirable because they are dishonest and usually end in painful or unpleasant consequences. Epicurus too was aware of the difference, so rather than proposing an unrestrained pursuit of pleasure, he advocated *the avoidance of pain as the primary good.* Thus simple, condonable, wholesome pleasures are preferable to the more exciting, immoral, and hazardous ones. Somewhat naively, he assumed that a serene life devoid of painful encounters was the natural preference of human beings. Thus he taught that foresight, good judgment, and self-control were essential ethical principles for attaining a tranquil life. Though Epicureanism has acquired the connotation of following an excessive pursuit of pleasure, this is quite contrary to Epicurus' philosophy, as he explicitly indicates.

> When we say that pleasure is the goal we do not mean the pleasures of the dissipated . . . but freedom from pain in the body and from disturbance in the mind. For it is not drinking and continuous parties nor sexual pleasures nor the enjoying of fish and other delicacies of a wealthy table which produce the pleasant life, but sober reasoning which searches out the causes of every act of choice and refusal and which banishes the opinions that give rise to the greatest mental confusion. (Ibid., p. 65)

Though in general superior to Stoicism in my opinion, Epicureanism had a similar overconfident belief in human rationality being the controlling factor in human nature and thus able to corral human desires, fears, and drives. While there was ample evidence in Greece and Rome of the irrationality, aggression, and cruelty in human beings, the philosophers seemed to assume that this was due primarily to ignorance, so that if taught the proper philosophy they would change their lives. Today, however, due to Freud's theory of the unconscious producing irrational fantasies and undesirable behavior, awareness of the deforming effects of an underprivileged or abusive upbringing, discovery of chemical imbalances and neurophysiological disorders causing chronic depression, paranoia, and schizophrenia, and inherited genetic predispositions to antisocial behavior, we are more aware that rationality, in many people at least, is not the dominating factor controlling their behavior.

While philosophers or scientists might be persuaded of Epicurus' arguments disproving the immortality of the soul, most people dread the thought of their own demise and long for reunion with loved ones in an afterlife, however implausible this might be. Moreover, even the ideal of a tranquil existence devoid of risks, adventure, or some distress can seem boring to those who are driven to seek a life of excitement, riches, prestige, and power whatever the risks, as history clearly indicates. Think of the tyrants, rapists, serial killers, religious fanatics, exploiters, and terrorists who have ravaged human beings, along with all the tribal, dynastic, religious, and regional wars. History is replete with the struggle to wrest privilege and power from kings, popes, nobles, aristocrats, landowners, and autocrats who usually were indifferent to the exploitation and misery of the lower classes.

Today the financial compensation of the chief executives of industries who, even when their companies lose money, are rewarded with outrageous termination bonuses is a disgrace, while teachers who have the enormous responsibility of educating our children are grossly underappreciated and underpaid. This, of course, is because CEOs have the power to compensate themselves. Another example is the recent corrupting influence of highly paid lobbyists who, instead of leaving governmental decisions to the elected officials, attempt to influence them to enact legislation promoting their particular special interests, whether they are in the best interests of the country or not. Integrity, honesty, and compassion for others are easy prey to selfishness, avarice, and egotism. History has been a continuous struggle to free governmental control, the courts, and economic systems from oppressors to make them more equable to the general public. Advocating withdrawal from civil, political, or military activities to live a quiet life free from conflict, as the Epicureans recommend, might have its personal advantages, but it would do nothing to improve the general welfare of people. Thus neither Stoicism nor Epicureanism proved to be a panacea for attaining an untroubled, serene life, any more than religions are.

NOTES

1. J. Beaujeu, "Hellenistic and Roman Science," René Taton, ed., *History of Science: Ancient and Medieval Science,* ed. by René Taton, trans. by A. J. Pomerans (New York: Basic Books, 1963), pp. 268–69.

2. A. A. Long, *Hellenistic Philosophy*, 2nd ed. (Berkeley: University of California Press, 1986), p. 9.

3. Lucretius, *On Nature*, trans. by Russell M. Greer (New York: Bobbs Merrill, 1965), p. 42.

4. Marcus Aurelius, *Meditations*, trans. by A. S. L. Grube (The Library of Liberal Arts: New York: Bobbs-Merrill, 1963), iv, p. 23. Quoted from A. A. Long, op. cit., p. 413.

5. For a fuller account of Stoicism see Richard H. Schlagel, *From Myth to Modern Mind*, Vol. I, *Theogony through Ptolemy* (New York: Peter Lang, 1995), pp., 406–16. Also see A. A. Long, op. cit., chap. 4.

6. Diogenes Laertius, *Lives of Eminent Philosophers*, 2 vols. ed. by H. S. Long (Oxford: Oxford University Press, 1964). Quoted from A. A. Long, op. cit., p. 129.

7. Lucretius, op. cit., p. 79: Bk. III, 1–5. The immediately following citations are to this work.

8. W. Hyde, *Paganism to Christianity in the Roman Empire* (Philadelphia, 1946), f.n. p. 30.

9. Epicurus, "Letter of Menoeceus," pp. 128–29. Quoted from A. A. Long, op. cit., p. 62.

Chapter 6

CHRISTIANITY AND THE REIGN OF FAITH

BACKGROUND

We now leave the classical Greco-Roman civilization with its philosophical and secular knowledge to enter the allegorical world of Christianity with its alleged revelations, inerrant scripture, epistles, and faith that dominated Western culture for a millennium, from the fourth to the thirteenth centuries CE. Representing the nadir of scientific activity, its spell began to be lifted when the scientific writings of the ancient Greeks and medieval Muslims began to reappear in the West in the twelfth and thirteenth centuries as a result mainly of the Arabic conquest of Spain. According to current statistics, church attendance has considerably diminished in Western Europe though it remains high in the United States, South America, and other parts of the world. As this whole age could not be covered in a single chapter, or even a book, the focus will be on the origins of Christianity and the reasons for its triumphal acceptance in pagan Rome despite several centuries of barbaric persecutions.[1]

Among all the paradoxes enveloping the history of Christianity, none compares to the fact that its originating figure, Jesus, who perhaps had the greatest impact on human history of any other mortal, was the least likely to accede to the role. All scholars can say with any confidence about the *historical* Jesus is that he came from Nazareth; that Mary and Joseph were his parents; that there is some mention of his having four brothers and two sisters (refuting the proposed doctrine of "Mary's virginity in perpetuity"); that like his father he was trained as a carpenter with little schooling; that he attracted followers in Galilee, Decapolis, Samaria, and Judea; and that he was crucified in Jerusalem during the administration of the Roman Procurator Pontius Pilate at the request of the Pharisees whom he had attacked, about the year 33 CE. Thus what is actually known about Jesus is in marked contrast to what most Christians believe about him.

Yet despite his simple family background and little formal education, in the gospels he is presented as having an extensive knowledge of Judaism, as well as a gift for speaking in parables and metaphors drawn from everyday life, which explains both his tremendous influence and the kind of followers he initially attracted, such as fishermen. Gospels Mark (6:1–3) and John (7:15) describe the amazement at his learning given his lack of schooling: "How is it that this man has learning, when he has never studied?" This could mean that Jesus was illiterate (as was true of Mohammad) which is what one would expect at this period from his background as the son of a carpenter.

But it was the portrayal of his having empathy for the suffering and the oppressed, commending the "meek," the "mournful," the "poor in spirit," and those "thirsting after righteousness" whose plight was generally ignored, that accounts for much of the success of his preaching, along with the angry opposition to his teaching by the Pharisees and the Sadducees that led to his crucifixion. He is shown renouncing the rich, the privileged, and the powerful while extolling the virtues of poverty, charity, and humility. Nothing like that had been preached in the synagogues nor heard in pagan temples. His reputed gospel of "loving thy neighbor as thyself" and "doing unto others as you would have them do unto you" raised the moral consciousness of Christians beyond the teachings of the schools of philosophy discussed in the last chapter. He is described as speaking with such conviction and authority of "our father who art in heaven," "the kingdom of God," and "everlasting life" that countless people have been persuaded of their truth. Following the commandments of God, believing in divine love, and having faith in eternal salvation have been the enduring legacy of Christianity.

Asserting that he was the successor of Moses and thus fulfilling the Old Testament prophesy of the coming Messiah, it is also written that he declared himself the son of God whose mission was to bring the Kingdom of God to mankind and hence was the inspiration also for the Gospels of the New Testament. This conception of a savior preaching a transcendental ethics who was crucified and then resurrected to save mankind was in marked contrast to Græco-Roman philosophies, though not completely unknown in other religious cults. According to the Gospels Jesus foresaw his betrayal by Judas, Peter's denial of knowing him, and his crucifixion that he considered to be a necessary sacrifice to redeem mankind.

He allegedly foretold his resurrection on the third day, which was confirmed by Mary Magdalene, and also predicted a day of final judgment when the righteous and the faithful will enter the Kingdom of God, whereas the wicked and the faithless will be damned in hell and the world will pass away. The "rebirth in Christ" has been a powerful idea attracting people to Christianity throughout the ages. A vividly compelling worldview, it still resembles previous myths with its dramaturgical, magical features which account for much of its attraction, but which has little resemblance to the world as we know it today.

THE CRUCIAL PROSELYTIZING OF SAINT PAUL

While these are the core Christian doctrines that have been handed down to us, believing them assumes they represent an accurate account of his sermons, teachings, and mission. Thus the crucial question is how much of it should be believed, both in terms of the *authenticity of what Jesus actually preached* and *whether what he claimed about his mission could be true.* It must be remembered that nothing *written* about him seems to have come from anyone who knew Jesus personally nor heard him speak. This was due to the fact that neither he nor his immediate converts left any written accounts, but also because the expectation of his second coming and end of the world made any historical record seem unnecessary. Though he was perhaps thirty-three years old when he was crucified, only a scant sixty days of his life are known with any reliability.

While there were a few Christian communities outside of Palestine when Jesus was crucified, it was Saul or Paul, as he was known after his conversion to Christianity in 34 on the road to Damascus, who was primarily responsible for spreading the gospel in the Jewish communities of the Diaspora. It has been estimated that his travels to the various cities mentioned in his letters, such as Antioch, Corinth, Athens, Philippi, and Ephesus, covered 10,000 miles. He had not known Jesus personally and seldom quotes him directly, but due to his close association with the disciples Peter, Stephen, James, John, Timothy, Philip, and Barnabas he was very familiar with his teachings. His identification with Jesus was such that he considered himself as another disciple. Paul's emphasis on Christ as the incarnation of God and his atonement for man's sins by his sacrificial crucifixion impressed these doctrines on the Christian converts.

It is said that without the learned and compelling proselytizing and epistles of Paul, who has been called the "second founder of Christianity," it would have been unlikely that the original Jewish cult would have outgrown its initial parochialism. Paul's insistence that the new religion not be confined to Jews but open to Gentiles, claiming that all people are "brothers in Christ," was a major reason for Christianity's expansion.

This was enhanced also by his preaching that in their personal relations women should be treated as equals despite the customary hierarchy of men above women, especially among the Greeks and

Hebrews. As his numerous references to women, such as Phaebe, Prisca, Mary, Apphia, Euodea, and Syntyche in his letters indicate, he accepted them as "fellow workers" with Phaebe even becoming a deaconess. (Romans 16:1–2) Finally, his making conversion to Christianity easier by not demanding adherence to Jewish law nor requiring circumcision also contributed to the growth of the emerging Christian communities.

Although he disparaged his speaking ability, his letters instructing these communities and addressing their questions and problems were crucial in Christianity's survival. While the "Book of Acts" (written later and attributed to the same author who wrote Luke, along with the "Acts of the Gospels") has provided additional knowledge of the beginnings of Christianity and Paul's travels, it was his epistles composed between 50 and 62 CE that were among the first written interpretation of Jesus' mission. Unlike Jesus' first disciples who were illiterate, Paul knew Hebrew and Greek, along with Latin, and enriched his exposition of Jesus' teachings with his extensive knowledge of Jewish and Roman religions, as well as Platonism and Gnosticism. Frederick B. Artz has given a vivid description of his forceful personality and influence.

> Paul shows himself a man of deep feeling and burning conviction and a magnificent stylist. There is no evidence that the Epistles were written for publication; but they so impressed those who received them that they were later collected and published. Paul had, as few men in history, that quality Aristotle called "greatness of soul"; he possessed enormous energy, an iron will, and a genius for leadership and organization. He was convinced that Jesus was a divine being, who by his death and resurrection saves those who are united with him by a mystic faith—"salvation by faith," as the doctrine was to be known in history. This mystic union with the divine one, the Christ, replaced the observance of the old Jewish law. It was the result of moral living, of worship, of prayer, of works of charity, and of the rites of baptism and the Lord's Supper.[2]

GROWTH OF CHRISTIANITY DESPITE ROMAN PERSECUTIONS

It is known that Paul and Peter passed their last days in Rome where they were martyred in the terrible persecutions ordered by Nero. According to Catholic doctrine, Jesus gave Simon the name Peter saying "on this rock *(Petros)* I will build my Church." Then claiming to have discovered his remains preserved in a shrine under the high altar of the Church of St. Peter, the Catholic Church has maintained this confirms the apostolic succession. The persecutions that led to their martyrdom was begun by Nero following the fire in 64 CE that he is believed to have ignited, but to divert accusations from himself attributed to the Christians. According to the Roman historian Tacitus:

> To put an end to the rumors he shifted the charge on to others. First those who acknowledged themselves of the [Christian] persuasion were arrested; and upon their testimony a vast number were condemned. . . . Their death was turned into a diversion. They were clothed in the dress of wild beasts, and torn to pieces by wild dogs; they were fastened to crosses, or set up to be burned, so as to serve the purpose of lamps when daylight failed. Nero gave his own garden for this spectacle.[3]

The persecutions continued under the reigns of Dalmatian, Marcus Aurelius, Deices, Valerian, and Diocletian. Moreover, many ordinary Romans were so opposed to the Christians' beliefs that rejected the Roman gods and by rumors of sacrificial killings, perhaps based on reports of drinking the blood and eating the flesh of Jesus during the sacrament of communion, that they relished the spectacle of their mutilations by attacks of gladiators and wild animals in the infamous Coliseum. As again described by Hibbert:

> Their deaths in the amphitheatre accordingly became one of the fiercest thrills that the shows there could afford. Christians were

> eaten by half-starved lions, burned alive before images of the sun-god, shot down by flights of arrows, hacked to death with axes and swords. In the reign of Diocletian alone . . . there were probably as many as three thousand martyrs. Yet their religion could not be suppressed; and while those arrested were torn to pieces in the Colosseum, the survivors were joined by convert after convert until by the time of Diocletian's death there were perhaps thirty thousand Christians in Rome, meeting together for worship. . . . (p. 67)

There has been much speculation about the reasons for the successful survival and increasing population of Christians during the first four centuries despite the extensive persecutions, some even attributing it to divine intervention, but recent studies have found natural explanations. While a fuller account can be found in my *The Vanquished Gods* referred to earlier, I shall discuss several of the reasons here.[4] Along with Paul's removing obstacles to conversion and opening Christianity to the Gentiles, recent demographic studies of ancient Rome have provided further explanations. One cause can be attributed to the Roman preference for male children that led to female infanticide by killing or exposing female babies to the elements leaving fewer survivors, while this was forbidden by the teachings of Christianity. In addition, Roman women frequently had abortions to prevent unwanted births, especially from premarital conceptions or adultery, and practiced primitive birth control methods using drugs, intrauterine devices, and alternative methods of sexual gratification to vaginal intercourse to reduce family size.

Lessening the number of Roman births because these intrusive methods often resulted in sterility or death of the mother, they too contributed to the decline in Roman population. The hostility between Roman husbands and wives was another factor, especially since the wives were betrothed at a very early age, often just after puberty, to older men they hardly knew. Because Christian women married at a later age and were freer to select their husbands, their marriages proved more congenial and fruitful. Also, the larger proportion of males to females in Roman society and the gentler, more chaste and virtuous behavior of Christian wives led to their husbands being more apt to convert to Christianity than their wives adopting the Roman pagan religion. And since the wives had greater responsibility for raising the children, they had greater influence on their children's choice of religion. Alarmed by the decline in Roman population, the emperors provided various inducements to increase marriages and the size of families, but none of these proved especially effective in stemming the decrease in population.

An additional reason for the disproportionate population increases is offered by Rodney Stark in his description of the rise of Christianity. During the first four centuries two devastating epidemics occurred. The first, perhaps produced by an outbreak of cholera in 165 that lasted for fifteen years, killed from "a quarter to a third" of the Roman population. During the second, perhaps due to an outbreak of measles in 251, "five thousand people a day were reported to have died in the city of Rome alone."[5] Stark thinks that the decrease in birthrate described previously, along with these ravaging epidemics, was a major cause of the decline of the population of the empire. But since the Christians were no more immune to these deadly epidemics than the Romans, it has been questioned how this could have increased the population of the Christians?

Stark believes this can be explained by the different cultural responses to these tragedies. Without any understanding of how these diseases were physically caused or transmitted, the Romans were entirely mystified as to their origin and how to treat them, attributing them to the displeasure, punishment, or caprice of the gods. But without knowing what they had done to deserve this and finding their usual rites and exhortations ineffective, they were left despondent and in disarray. Furthermore, fearful of their own death they tended to abandon the sick leaving them to die. In contrast, the Christians because of their religious beliefs viewed these disasters as having some divine meaning, perhaps to test their righteousness and faith or to punish their sins.

Moreover, their Christian ethics proclaiming that one should "love one's neighbor as oneself" led them to have greater compassion and caring for the ill. And since they believed in an afterlife wherein they would receive a reward for their good deeds in this life, they were less afraid of death than the Romans and more willing to sacrifice themselves for others. Dionysus, Bishop of Alexandria, has left the following account.

> Most of our brother Christians showed unbounded love and loyalty, never sparing themselves and thinking only of one another. Heedless of danger, they took charge of the sick, attending to their every need and ministering to them in Christ, and with them departed this life serenely happy. . . .[6]

He goes on to describe this unselfish concern with the contrasting indifferent behavior of the Romans to their fellow citizens.

> The heathen behaved in the very opposite way. At the first onset of the disease, they pushed the sufferers away and fled from their dearest, throwing them into the roads before they were dead and treated unburied corpses as dirt, hoping thereby to avert the spread and contagion of the fatal disease, but do what they might, they found it difficult to escape. (p. 82)

While this difference in caring for the sick did not reflect a greater ability to cure the disease, it apparently produced a higher survival rate. According to William McNeill: "[w]hen all normal services break down, quite elementary nursing will greatly reduce mortality. Simple provisions of food and water, for instance, will allow persons who are temporarily too weak to cope for themselves to recover instead of perishing miserably."[7] If caring for the sick could reduce the fatality rate by two-thirds as medical experts now believe, Stark calculates that in "a city having 10,000 inhabitants" in 160 when the initial epidemic began and the ratio of Christians to Gentiles was 1 to 249, the ratio could have increased to 1 in 134 by 170, a very significant growth in a single decade. But just the demographic changes alone would have been startling. According to Stark, because the Roman population was declining while the Christian population was increasing, the percentage of the Christian to the Roman population would have increased from 0.0017 percent in 40 AD to 56.5 percent by 350, an increase that might appear miraculous but was just a normal exponential growth rate (cf. Stark, p. 89).

After over two centuries of repression and persecution, Gallienus (reigned 253–268) was the first emperor to issue an edict of toleration permitting Christians freedom of worship and restoration of their property. This was followed by forty years of relative peace and the rapid growth just described enabling Christians to build splendid churches and intermarry with the pagans. Then in the early fourth century Galerius, who later would become Augustus, seeing in Christianity a threat to Roman religion and rule, urged Emperor Diocletian to reinstate the confiscations and persecutions. In February 303 a decree was passed to destroy all Christian churches, burn their books, reclaim their property, disband their congregations, exclude them from all public offices, and execute all who were found in Christian congregations. This persecution lasted for at least eight years bringing death to nearly 1500 Christians and relentless suffering to many more.

In 311 Galerius, facing a terminal illness, seeing the futility of his policy of repression, and implored by his wife to cease the killings, issued his own edict of toleration, again making Christianity a lawful religion. Having met the test of the Galerian and Diocletian persecutions, despite the many who renounced their faith under threat of torture and death, Christianity continued to expand. As Will Durant states in his inimitable style:

> There is no greater drama in human record than the sight of . . . Christians, scorned or oppressed by a succession of emperors, bearing all trials with a fierce tenacity, multiplying quietly, building order while their enemies generated chaos, fighting the

> sword with the word, brutality with hope, and at last defeating the strongest state that history has known. Caesar and Christ met in the arena, and Christ had won.[8]

CONSTANTINE'S TRANSFORMATION OF CHRISTIANITY

But it was Constantine who, following the battle of the Mulvian Bridge, opened the gateway to Christianity becoming the state religion after his transfer of the capital of the empire to Byzantium. Helen, his strong-willed mother, had converted to Christianity some years earlier and though Constantine had not yet committed, her influence is undoubted. As Durant recounts the events according to the historians Eusebius and Lactantius, on the afternoon of October 27, 312, before the battle with Maxentius,

> Constantine saw a flaming cross in the sky, with the Greek words *en toutoi nika*—"in this sign conquer." Early the next morning . . . [he] dreamed that a voice commanded him to have his soldiers mark upon their shields the letter X with a line drawn through it and curled around the top—the symbol of Christ. On arising he obeyed, and then advanced into the forefront of battle behind a standard . . . carrying the initials of Christ interwoven with a cross. As Maxentius displayed the Mithraic-Aurelian banner of the Unconquerable Sun, Constantine cast his lot with the Christians, who were numerous in his army, and made the engagement a turning point in the history of religion. (p. 654)

Following his defeat of Maxentius, who died in the Tiber River during the battle, Constantine made a triumphal entrance into Rome as its conqueror.

Then in 313 he and his temporary co-ruler Licinius issued the "Edict of Milan," reaffirming Galerius' declaration of toleration for all religions, along with restoring church property. Initially he was careful not to offend the pagans who still outnumbered the Christians even though the latter were steadily becoming more influential, but after Byzantium was made the capital of the empire in 330 Christianity became the official religion. In succeeding centuries, beginning with the Council of Nicea called by Constantine in 325 to define the doctrine of the Trinity, Christian theology began to be formulated under the guidance of the Catholic Church.

THE GOSPELS' AUTHENTICITY AND THE QUEST FOR THE HISTORICAL JESUS

In our previous discussion of the origins of Christianity we saw how Paul had declared some of its essential doctrines, such as the divinity of Christ and brotherly love in his epistles, and how the two "Acts" had recorded the travels of Paul and the Apostles and described the early history of the Christian communities. But it is the four Gospels of the New Testament that are the foundation of Christian doctrine.[9] These consist of the three "Synoptic Gospels," Mark, Matthew, and Luke, so named because they present a "common view" of the life and teachings of Jesus despite some discrepancies. John, in contrast, shows a marked independent influence setting it apart from the other three.

Although Christians are led to believe that the Gospels were written by the persons whose names they bear who were witnesses to Jesus' mission, critical scholars of the New Testament have rejected this, concluding that the actual authors are unknown. As G. A. Wells states: "It was what the Toronto theologian F. W. Beare calls 'second-century guesses' that gave the four canonical gospels the names by which we now know them; for they were originally 'anonymous documents' of whose authors 'nothing is known.'"[10] Yet following tradition I shall refer to them by their usual names.

Critical scholarship of the New Testament extends as far back as the German scholar Hermann Samuel Reimarus (1694–1768) who was among the first to distinguish between the historical Jesus and the Jesus portrayed in the Gospels. Since then there have been numerous critical studies throughout the centuries attempting to delineate the authentic Jesus from the one handed down to us. In the nineteenth century the name most famous for the quest for the genuine Jesus was David Friedrich Strauss whose 1,400 pages of scholarly exegesis published in 1835, *Life of Jesus Critically Examined*, attempted to separate the "mythical" from the actual "historical" Jesus. The twentieth century included the following critical studies that I have consulted or read: Albert Schweitzer, *The Quest for the Historical Jesus* (1906); G. A. Wells, *Who Was Jesus: A Critique of the New Testament Record* (1987); R. P. & A. T. Hanson, *The Bible Without Illusions* (1989); Burton L. Mack, *Who Wrote the New Testament?* with the provocative subtitle, *The Making of the Christian Myth* (1995); John Dominic Crossan, *Jesus: A Revolutionary Biography* (1995); and John Shelby Sprong, *Liberating the Gospels* (1996). As I suppose one might have expected, rather than being admired these critical studies usually aroused considerable hostility and retaliation.

But the publication that probably received the most critical acclaim, along with condemnation, is the study by the "Jesus Seminar" published under the title *The Five Gospels: The Search for the Authentic Words of Jesus* (1993). Called the "Scholar's Version" because of its fresh translation of the New Testament, it was the product of an initial collaboration in 1985 of thirty scholars who worked free of any ecclesiastical or religious constraints. After comparing the words attributed to Jesus in the surviving texts with those current in the various ancient languages in which they were preserved, the scholars concluded that "[e]ighty-two percent of the words ascribed to Jesus in the gospels were not actually spoken by him. . . ."[11] This by itself does not invalidate the conception and teachings of Jesus presented in the Gospels because the same meanings and representation can be conveyed in different terminology, but it does call in question the inerrancy attributed to the Gospels by Christian fundamentalists.

Even more striking is the personal and scholarly odyssey of Bart

D. Ehrman described in his recent book, *Misquoting Jesus: The Story Behind Who Changed the Bible and Why*.[12] He describes himself as previously being a Christian evangelical who believed that because those who wrote the Bible were inspired by God, it must be infallible. Later, finding several errors in Mark and a number of discrepancies among the Gospels, he decided to devote his life to critical biblical scholarship, learning Greek and Hebrew to be able to read the original manuscripts. Then, to his dismay, when studying the surviving Greek manuscripts of the New Testament to compare the accuracy of the translations, he found that there were innumerable transcriptions, but none that could be identified as the inspired originals. As he writes:

> It is one thing to say that the originals were inspired, but the reality is that we don't have the originals. . . . Moreover, the vast majority of Christians for the entire history of the church have not had access to the originals . . . [relying on] copies made many *centuries* later. And these copies all differ from one another, in many thousands of places. . . . (p. 10; brackets added)

So rather than discovering that the New Testament was infallible because it was inspired by God, he found that it had many authors that accounts for the conflicting versions, from which he drew the following conclusion.

> The Bible began to appear to me as a very human book. Just as human scribes had copied, and changed, the texts of scripture, so too had human authors originally *written* the texts of scripture. This was a human book from beginning to end. It was written by different human authors at different times and in different places to address different needs. Many of these authors no doubt felt they were inspired by God to say what they did, but they had their own perspectives, their own beliefs, their own views, their own needs, their own desires, their own understandings, their own theologies. . . . In all these ways they differed from one another. (pp. 11–12)

Considering that the Gospels are the main source of Christian belief and that they were written decades after Jesus died by unknown authors who were not witnesses to his life or teachings and that no originals are extant, this would account for the emendations, embellishments, and variations in interpretation. It would also support Ehrman's conclusion that they were *human* creations, not the direct word of God as usually believed by evangelicals and fundamentalists. We know from Paul's Epistles and the Acts of the Apostles that Christian communities had been established in the Levant, Asia Minor, Greece, Macedonia, and even Rome before the Gospels appeared. Furthermore, Apostles such as Peter, James, John, and Barnabas visited these communities and undoubtedly compared their accounts of Jesus with some members who also had heard Jesus preach. Since some, like Paul, were well educated, they must have been anxious to relate what they knew about the mission and sayings of Jesus to share with others and use in sermons on the Sabbath. Thus stories about the crucial events in Jesus' ministry and reports of his sayings coalesced into an "oral tradition" that the authors of the Gospels, who probably came from these communities, must have relied on when writing them.

Although Matthew was placed before Mark when the New Testament canon was first agreed upon in 185 and retained that position until the early decades of the twentieth century, a line by line comparison of the texts (a method similar to that used by the Jesus Seminar) of the four Gospels revealed extensive repetitions of Mark in Matthew and Luke. Thus the consensus now is that Mark is the oldest having been written in Greek around 70 based on the oral tradition, almost two generations after Jesus' death. Matthew appeared fifteen or twenty years later followed by John between years 95 and 110. Luke may have been written about a decade later, around 120, as a two volume work that included a sequel called the Acts of the Apostles.[13] While there was agreement that Jesus was the *Christos* (Greek meaning the Messiah) predicted by the prophets of the Old Testament who would perform miracles and be the "sacrificial lamb" bringing "the good news" of the Kingdom of God, there were considerable other differences owing to the various times, backgrounds, and motivations of the authors, as Ehrman discovered.

As indicated previously, Matthew and Luke borrowed extensively from Mark, but where they diverge from Mark they show considerable agreement. Since there is no evidence that they collaborated, it has been conjectured that the similarity can be attributed to their having access to the same independent source. A German scholar hypothesized that there must have existed a source document which he designated as *Quelle,* the German word for source. Now referred to as the "sayings Gospel Q," it has taken on a reality of its own. But because no existing manuscript has been found, its contents have been inferred from the passages in Matthew and Luke that are similar. The belief that the content of the four Gospels can be traced to additional sources has been vindicated by the discovery of a trove of ancient manuscripts during the nineteenth and twentieth centuries.

For example, in 1844 the Codex Sinaiticus (named for where it was located) was accidentally found in St. Catherine's monastery in the Sinai peninsula that also contained New Testament scripts from the early fourth century CE. Later in the century a copy of the Greek Bible, again from the fourth century that had been lost in the vaults of the Vatican library, was unexpectedly recovered. Then in the early 1930s the Chester Beatty papyri dating from the first half of the third century was unearthed. In 1945 the marvelous Nag Hammadi collection was discovered in Egypt consisting of a "fourth century CE repository of Coptic Gospels," "Christian Gnostic texts," a "complete copy of the Gospel of Thomas lost to view for centuries," along with the text of "the Sacred Book of James and the Dialogue of the Savoir."[14] Then the extraordinary Dead Sea or Qumran Scrolls, dating from the first century CE, were "accidentally stumbled upon" by a shepherd in the late 1940s, many from numerous caves in "the rugged stone hills on the western shore of the Dead Sea. . . ."[15]

Just recently the discovery in Egypt of a leather-bound papyrus codex, a 300 CE Coptic copy of the Gospel of Judas written a century earlier, has aroused considerable interest and controversy. Challenging the traditional Gospel account of Judas as the one who revealed Jesus to the authorities, it portrays Judas as a confident of Jesus who asked Judas to betray him so he would be killed freeing his spirit from his body. This view of Judas' action conforms to the Gnostic conception that the body was merely an external wrapping covering the inner spirit of Jesus that was sent to the earth by the Lord to redeem mankind. But in contradicting the role of Judas presented in the Gospels and thus challenging their authority, one can understand why it has raised a commotion.

The discovery of these fascinating ancient texts has contributed significantly to the scholarly debate over the contrast between the historical Jesus and the risen Christ portrayed in the Gospels. Grudgingly, it is being conceded that the discrepancies among the Gospels preclude their being literally or inerrantly true: that though the authors of the synoptic Gospels may have been guided by a similar oral tradition in portraying Jesus' life and mission, their individual renderings were based largely on how Jesus was viewed in their own communities and, according to Bishop Sprong, the need to revise the liturgy of the synagogue to accommodate the new Christian rites. Because it is claimed in the Gospels that Jesus was killed at the urging of the Pharisees and the Sadducees who he harshly criticized, later Christian converts tended to place as much ethnic distance between Christianity and Judaism as possible. But this overlooks the fact that the converts being the first to welcome the Apostles were overwhelmingly Jewish.

Considering the evidence, the authors of the Gospels were not concerned to write historically accurate accounts of Jesus, but to relate his mission and doctrines to the prophets and events of the Old Testament.[16] That the New Testament makes frequent references to Jesus having been descended from Abraham, Moses, and David, along with being the anticipated Messiah, confirms this. He was, after all, drawing on his Jewish heritage and preaching mainly to

Jews. Considering all the later reinterpretations and additions to Christian doctrines to fit theological and liturgical needs, it is not surprising that this was also true of the life of Jesus. As Bishop Sprong frankly states:

> I do not today regard the details of the gospel tradition as possessing literal truth in any primary way. I do not believe that the Gospels offer us either reliable eyewitness memory or realistic objective history. I do believe that the Gospels are Jewish attempts to interpret in a Jewish way the life of a Jewish man in whom the transcendence of God was believed to have been experienced in a fresh and powerful encounter. I do believe that the God met in Jesus is real, and that by approaching the scriptures through a Jewish lens, saving reality can be illuminated and . . . can still be entered. (p. 20)

While admitting that the "details" of the Gospels are not historically or literally true, Sprong nonetheless believes that Jesus' "God-encountering experience" (his alternate phrasing) is valid. But how does one distinguish the "unreliable details" from the essential experience? Can one consistently defend the authenticity of Jesus' claim reported in the Gospels that he was the son of God while also denying the "realistic objective history" of the very documents on which this claim is based? Is not our understanding of the actual Jesus only as valid as the scriptural evidence, considering that there is practically no other historical testimony? If we were to accept Sprong's argument wouldn't reading Homer's *Iliad* allow one to conclude that the dramatic "encounter" with Achilles as portrayed in the *Iliad* is a justification for believing that it is a true representation of the historical Achilles, rather than Homer's literary construction? "God encountering experiences" cannot be assumed to be authenticating considering how many nonexistent deities have been worshipfully 'encountered' in the past!

These arguments that what most Christians believe about Jesus is what tradition has handed down to us, not the actual Jesus, is reinforced by the discrepancies found in the Gospels' accounts of him.[17] If it is true that Mark is the primary source for Matthew and Luke, then many of the cherished beliefs about Jesus derived from the other Gospels will have to be questioned or rejected. For example, there is no mention in Mark of the virgin birth of Jesus nor of his being born in a manger in Bethlehem. In Mark (1:10–11) Jesus is simply described as coming from Nazareth to be baptized by John following which "the heavens opened and the Spirit descended upon him like a dove, and a voice came from heaven, 'thou art my beloved Son; with thee I am well pleased'"—a delightful description, but hardly credible in any literal sense.

In contrast, Matthew (1:18–22) relates that when Mary was "betrothed to Joseph, before they came together she was found to be with child of the Holy Spirit"; and her husband Joseph, "being a just man and unwilling to put her to shame, resolved to divorce her quietly." But before he could do this, "an angel of the Lord appeared to him in a dream, saying, 'Joseph . . . do not fear to take Mary your wife, for that which is conceived in her is of the Holy Spirit; she will bear a son, and you shall call his name Jesus. . . .'" Matthew continues with the account that after Jesus was born, "wise men from the East seeing his star in the East," came to Bethlehem "to worship him" bringing gifts of "gold and frankincense and myrrh" (2:1–11). When the wise men departed "an angel of the lord" again appeared to Joseph in another dream and said: "Rise, take the child and his mother, and flee to Egypt and remain there till I tell you; for Herod is about to search for the child, to destroy him." Joseph did as he was bidden, taking the mother and child to Egypt "to fulfill what the Lord had spoken by the prophet, 'Out of Egypt have I called my son'" (2:13–14).

Having been "tricked," Herod in a "furious rage" directed that "all the male children in Bethlehem and in all that region who were two years old or under" be killed (2:16). After Herod died and following several additional angelic warnings and dreams, Joseph withdrew to the district of Galilee and the city called Nazareth, "that what was spoken by the prophets might be fulfilled: 'He shall be called a Nazarene'" (2:23). The similarity between this description of the birth of Jesus and the Old Testament account in Exodus of Moses having been saved from the Egyptian King's decree that all Hebrew male births be "cast into the Nile" by being placed in a basket in the reeds of the river and found by the Pharaoh's daughter could hardly be coincidental. But unless one can believe that angels can actually appear in dreams conveying messages to the dreamer and suppress the question as to how the Holy Spirit can impregnate a women, none of this has any credibility. As Shakespeare wrote, "dreams . . . are the children of an idle brain, Begot of nothing but vain fantasy; Which is as thin of substance as the air; And more inconstant than the wind" (*Romero and Juliet*, Act I, sec. 4). Yet at that early age dreams where often credited with a validity they hardly merited then and no longer have.

Still a different version is presented in Luke. Addressed to one Theophilus, unlike Mark and Matthew Luke tells how John the Baptist came to be born, echoing the story of how the ninety-nine year old Abraham and Sarah of the same age (even though she "had ceased to be after the manner of women") gave birth to Isaac, again supporting Sprong's thesis of how much the authors of the New Testament borrowed from the Old. Like Abraham and Sarah, the intended parents of John, Zechariah and Elizabeth, were childless with no prospect of having children "because Elizabeth was barren and both were advanced in years" (1:7). But analogous to Abraham, an angel (later disclosed as Gabriel) appeared to Zechariah in the temple declaring that "your wife Elizabeth will bear you a son, and you shall call his name John" (1:13).

Then during the sixth month of Elizabeth's pregnancy Gabriel was sent on another mission by the Lord, this time to a virgin betrothed to Joseph named Mary, declaring: "'Hail, O favored one, the Lord is with you!'" (1:28). Perplexed, Mary asked what it meant, whereupon Gabriel explained that because "you have found favor with God . . . you will conceive in your womb and bear a son, and you shall call his name Jesus" (1:30). When Mary asked "'how can this be, since I have no husband?'" Gabriel answered: "The Holy Spirit will come upon you . . . therefore the child to be born will be called holy, the Son of God'" (1:35). Acquiescing, Mary replied: "'Behold I am the handmaid of the Lord; let it be to me according to your word'" (1:38).

Mary then went to Judah to visit Elizabeth who was pregnant with John. When they met and Elizabeth "heard the greeting of Mary, the babe leaped in her womb; and Elizabeth . . . filled with the Holy Spirit . . . exclaimed with a loud cry, 'Blessed are you among women, and blessed is the fruit of your womb!'" (1:41–42). (Is it credible that after a century when this was written the stirring in Elizabeth's womb could be known?) Staying with Elizabeth and Zechariah for about three months, Mary then returned home.

When a decree was issued by Caesar Augustus that the census of the empire should be taken Joseph took Mary, "his betrothed who was with child," from Nazareth to Bethlehem "because he was of the house and lineage of David" (2: 4). There then follows those moving lines that have become so endearing at the Christmas season.

> And while they were there, the time came for her to be delivered. And she gave birth to her first-born son [implying she had additional sons who are listed later, refuting the proposed absurd doctrine that she be considered a "virgin in perpetuity"] and wrapped him in swaddling cloths and laid him in a manger, because there was no place for them in the Inn. (2: 6–7; brackets added)

Saint Louis, King of France, during his crusade in the fourteenth century allegedly purchased these swaddling clothes, along with a vial of the Virgin's milk, and had them placed in Saint Chapel. (As regards religious beliefs, there are no limits to credulity.)

The version of Jesus in the Gospel John is entirely different not mentioning his birth at all, instead presenting the Hellenistic conception of Jesus Christ as the Word or Light that accompanied God at the beginning of creation. As we know from our previous discussion of Heraclitus and the Stoics, the Greek for "word" is Logos, and just as the Logos was the instrument of creation in their philosophies, so it is in John in the following mystical statement.

> In the beginning was the Word, and the Word was with God, and the Word was God . . . all things were made through him. . . . In him was life, and the life was the light of men. . . . The true light that enlightens every man was coming into the world. He was in the world, and the world was made through him, yet the world knew him not. . . . And the Word became flesh and dwelt among us, full of grace and truth; we have beheld his glory, glory as of the only Son from the Father. (1:1–14)

How different this reads from the previous Gospels. The Gospel then identifies the author as John the Baptist. It relates how the Jews questioned John as to whom he was. Denying that he was the Christ, he initially replies that "I am the voice of one crying in the wilderness, 'Make straight the day of the Lord,' as the prophet Isaiah said" (1:23). This reference to Isaiah again illustrates Sprong's claim of the closeness of the Old and New Testaments. The Pharisees then ask him: "'why are you baptizing, if you are neither the Christ, nor Eli'jah, nor the prophet?'" John replies: "'I baptize with water; but among you stands one whom you do not know, even he who comes after me, the thong of whose sandal I am not worthy to untie'" (1: 24–27). The difference between the beginning of John and the other three Gospels could hardly be more striking!

Turning to a further example of Gospel discrepancies taken from the end of Jesus' life, rather than the beginning, pertaining to his last words along with his entombment and resurrection, we find so much divergence that it is impossible to decide which is supposed to be the true account, if any. First, there is the discrepancy in what Jesus says on the cross just before dying: Mark (15: 34) and Matthew (27: 46) declaring that "at the ninth hour Jesus cried with a loud voice, 'My

God, my God, why hast thou forsaken me?'" while according to Luke Jesus declared "'Father, into they hands I commit my spirit!'" (23: 46). In John the account is still different with Jesus in the end simply saying: "'It is finished;' and bowed his head and gave up his spirit" (19:30).

As for the differences in the account of the entombment and resurrection, according to Mark when Pilot learned that Jesus was dead he gave his body to Joseph of Arimathe'a who wrapped it in a linen shroud and laid it in a tomb which had been cut from the rocks, rolling a stone to block the entrance. Two days later Mary Magdalene, having seen where he was laid, was joined by Mary the mother of James and Salome to bring spices to the tomb to anoint him. Wondering how they would roll the large stone from the entrance, they were relieved to see it had been removed. Entering the tomb they saw a "young man sitting on the right side, dressed in a white robe, and they were amazed" (16:1–6). The young man said to them: "'Do not be amazed: you seek Jesus of Nazareth, who was crucified. He has risen, he is not here, see the place where they laid him. But go, tell his disciples and Peter that he is going before you to Galilee, there you will see him, as he told you'" (16:6–7). "Astonished and trembling," the women left the tomb too afraid to tell anyone.

There then follows a number of footnotes in my edition of the Revised Standard Version with alternate accounts, one claiming that Jesus appeared first to Mary Magdalene who then told the others that he was alive, but they did not believe her. Other footnotes state that he appeared to the eleven disciples, upbraided them for their unbelief and urged them to spread the word to the "whole creation," after which he was "taken up into heaven, and sat down at the right hand of God" (n. 9–19). These possible alternate interpretations again raise the question of which version is true, if any.

Matthew repeats the account of Joseph of Arimathe'a taking the body of Jesus wrapped in linen to the tomb and blocking the entrance with a stone, but then diverges from Mark, adding the following account. To prevent the followers of Jesus from surreptitiously going to the "sepulcher" and removing his body to fulfill his prophesy of arising in three days, Pilot directed a group of soldiers to guard the entrance of the tomb to prevent it from being removed. Again Mary Magdalene with an unidentified Mary and without Salome went to the sepulcher just as a "great earthquake" occurred and "an angel of the Lord descended from heaven and . . . rolled back the stone, and sat upon it. His appearance was like lightning, and his raiment white as snow. And for fear of him the guards trembled and became like dead men" (28:2–4). Again, how can a reasonable person believe such an incredible story?

As in Mark, the women were told not to be afraid, that Jesus had risen, and directed to go and tell the disciples what they saw. This time they were still afraid but filled with such "joy" that they ran to tell the disciples but meeting Jesus, they took hold of his feet and worshiped him. Jesus said to them: "Do not be afraid; go and tell my brethren to go to Galilee, and there they will see me" (28: 9–10). Then, to refute anyone saying that Jesus had risen, when the soldiers told what had happened the elders gave them money to "[t]ell people, 'His disciples came by night and stole him away while we were asleep'" (28:12). The Gospel concludes with the eleven disciples going to Galilee as directed and when they saw him they worshiped him; but some doubted, so Jesus said to them: "'All authority in heaven and earth has been given to me. Go therefore and make disciples of all nations, baptizing them in the name of the Father and of the Son and of the Holy Spirit, teaching them to observe all that I have commanded you, and lo, I am with you always, to the close of the age'" (28:18–20).

Luke repeats the account of Joseph of Arimathe'a taking the body of Jesus to the tomb after which "the women who followed him from Galilee" brought spices to the tomb. Again the stone had been rolled away but when they entered the tomb, instead of finding the body, they saw "two men . . . in dazzling apparel" who said: "'Why do you seek the living among the dead?' Remember how he told you . . . that the Son of man must be . . . crucified, and on the third day rise'" (24:4–7). Now for the first time the names of Mary Magdalene, Mary the mother of James [is this Jesus' mother?], and Joanna are mentioned as taking the news to the eleven disciples and others, but were not believed.

In another version, later in the day as two of the disciples were walking and talking together, Jesus "drew near and went with them. But their eyes were kept from recognizing him." Then he asked them: "'What is this conversation which you are holding . . . as you walk?'" (24:17). Surprised, Cleopas asked if he were the only person who had not heard of the "things that have happened" recently? whereupon Jesus asked "'What things?'" (24:18–19). They then recounted the events concerning Jesus until they stopped to eat. It was only when "he took the bread and blessed and broke it and gave it to them . . . [that] their eyes were opened and they recognized him and he vanished out of their sight" (24:30–31: brackets added).

Later the two rejoined the disciples recounting what had occurred and as they were doing this "Jesus himself stood among them. But they were startled and frightened, and supposed that they saw a spirit" (24:36–37). Seeing their agitation Jesus admonished them pointing to his hands and feet to reassure them that he was real, even asking for some food and eating it to convince them. He then said: "'These are my words which I spoke to you . . . that everything written about me in the law of Moses and the prophets and the psalms must be fulfilled'" (24:44), further proof of Sprong's thesis that Jesus and the authors of the Gospels saw him as fulfilling the prophecies of the Old Testament. The Gospel ends with Jesus leading them to Bethany and after blessing them, parting from them with no mention of the second coming.

John as usual is the most diverse. He begins with the common account of Joseph of Arimathe'a taking away the body from Pilot and covering it with linen, but this time Nicode'mus aided him by "bringing a mixture of myrrh and aloes, about a hundred pounds' weight" to cover his body with spices (19:39). But in this account they placed him in a new tomb in the garden (not mentioned in the other Gospels) where he had been crucified. Also, only Mary Magdalene came to the tomb early in the morning and finding the stone removed, without entering she ran to Simon Peter and another disciple, "the one whom Jesus loved," and told them that "'[t]hey have taken the Lord out of the tomb, and we do not know where they have laid him'" (20:2). A second Peter joined them and running with the other disciple to the tomb looked in, but seeing the linen cloths lying there neither entered. Then Simon Peter arrived and entered the tomb and saw the linen cloths and the napkin which had been on Jesus' head. The Gospel then states "the other disciple, who reached the tomb first, also went in, and he saw and believed; for as yet they did not know the scripture, that he must rise from the dead" (20: 8–9). But it is curious as to what the other disciple believed, that Jesus was not there or that he had risen?

Mary Magdalene remained outside weeping and "as she wept she stooped to look into the tomb; and she saw two angels in white, sitting where the body of Jesus had lain, one at the head and one at the feet" (20:11–12). (One wonders why the two disciples who had entered previously had not see them.) When they saw Mary Magdalene crying they asked, "'why are you weeping?'" to which she answered "'Because they have taken away my Lord, and I do not know where they have laid him'" (20:13). Then turning around she saw Jesus standing, but for some reason did not recognize him. He asked "'why are you weeping? Whom do you seek?'" (20: 15). Again not recognizing him she supposed him to be the gardener (they were standing in the garden by the tomb) and so replied: "'Sir, if you have carried him away, tell me where you have laid him, and I will take him away'" (20:15). Jesus then called her by name, "Mary." Finally recognizing him, she replied in Hebrew, "'Rab-bo'ni!,'" meaning Teacher. Jesus in turn said, "'Do not hold me, for I have not yet ascended to the Father; but go to my brethren and say to them I am ascending to my Father and your Father, to my God and your God'" (20:15–17). Mary Magdalene then sought the disciples and told them what she had seen and heard.

The next evening Jesus met with the disciples and said to them: "'Peace be with you'" and showed them his wounds. When they later told Thomas of this he refused to believe until he saw the wounds himself. Eight days later Thomas encountered Jesus who, having learned of Thomas' doubt, said: "'Put your finger here in my side; and see my hands, do not be faithless, but believing.'" Thomas then exclaimed: "'My Lord and my God!'" to which Jesus replied reprovingly: "'Have you believed because you have seen me? Blessed are those who have not seen and yet believe'"(20:26–29). This is a striking example of advocating that belief be superior to empirical evidence!

The Gospel then ends with an episode not mentioned in the other Gospels, that of the disciples fishing all night without catching any fish and meeting Jesus the next morning, but again not recognizing him. Directing them to cast the net to the right side of the boat, they caught a net full of large fish, 153 of them, which Jesus invited them to eat (21:11). This apparently jogged their memories so that they now recognized him for the third time since he rose from the dead. After asking Simon Peter three times whether he loved him and hearing the response, "'Yes Lord, you know that I love you,'" the Gospel ends with the puzzling reference to the disciple "whom Jesus loved, who had lain close to his breast at the supper and had said, 'Lord, who is it that is going to betray you?'" (21:20). Now seeing him, Peter asks the enigmatic question, "'Lord what about this man?'" to which Jesus replies: "'If it is my will that he remain until I come, what is that to you? Follow me!'" (21:22–23). The Gospel adds that the rumor spread that "this disciple was not to die; yet Jesus did not say to him that he was about to die," another puzzling report.

These references to a beloved disciple depicted in the fresco by Leonardo da Vinci has raised a disturbing question as to what the relation of this disciple was to Jesus, but as there are no additional references to him, it remains a mystery. What surely is not the case is that this individual is Mary Magdalene, as some recent revisionists have suggested. While I have tried to show how little we know about the historical Jesus, this does not allow the radical claim that he was married to Mary Magdalene for which there is absolutely no evidence in the Gospels. Not only were all the disciples who were at the last supper men, but the references to "the beloved" disciple are in the masculine form, as the above quotations indicate.

It is difficult to accept that an intelligent person reading the Gospels would not find them hard to believe, filled as they are with angelic visitations, the mysterious disappearances and reappearances of Jesus, the difficulties in recognizing him, along with the blatant inconsistencies. But, then, intelligence is an easy victim of the need to believe. The discrepancies, at least, are what one would expect given the diverse origins of the Gospels. The parlor game "Gossip" is an excellent illustration of how a simple sentence can be distorted or mangled when successively transmitted to ten recipients. Equally striking is the notorious differences in the testimony of nearby witnesses to the same accident or crime.

It is only because the Gospels are taught at an early age as if they were in agreement and that any questions about their authenticity are generally discouraged by church authorities, declaring they must be taken on faith, that they are commonly believed to be self-authenticating and consistent. Moreover, one wonders how many Christians have even read the Gospels carefully given the recent statistical evidence of their lack of knowledge of the Bible. In any case, considering that the Synoptic Gospels were written between about 70 and 120 years after the death of Jesus, based on an oral tradition lacking written supporting evidence in an age fraught with ignorance and superstition, with crucial episodes often involving angels in dreams, what would one expect about their authenticity?

It can always be retorted that the discrepancies are merely complementary descriptions of the same events, but in court decisions different descriptions of the same event are not accepted as complementary, but as dubious or self-refuting, especially when there are crucial omissions. Or one can reply they must be taken on faith, but since the doctrine of "justification by faith" is itself a scriptural claim, the rebuttal begs the question. Finally, one can claim that despite their differences, the immense appeal of Jesus' message as jointly conveyed in the Gospels and their enormous transforming power through nearly two millennia are themselves a vindication of its truth.

This reply is more persuasive in my opinion, but given that the history of civilization is strewn with false beliefs that also had their ardent followers, and that faith is only an *affirmation* of belief, not a *justification*, the argument seems less compelling, especially when the sources of the belief are so doubtful, contentious, and unreliable. But the crucial refutation lies in the way science has continued to disprove religious beliefs.

While both the Old and the New Testaments, along with the Koran, can be admired for the inspired moral teachings of some of the prophets, they do not contain a single *natural* fact or explanation that is of any value. It is this complete discordance between our present understanding of natural processes and the origin and nature of human beings due to the rigorous application of the scientific method, in contrast to the mythological framework underlying religious beliefs, that is the latter's undoing. As A. Weigall graphically states:

> The constricted Jesus of Christian theology does not belong to modern times. He is dated. He is the product of the early centuries A.D., when men believed in Olympus, and drenched its altars with blood. Magic plays about him like lightning. He walks upon the waters, ascends into the air, is obeyed by the tempests, turns water into wine, blasts the fig tree, multiplies loaves and fishes, raises the dead. All these marvels made him God incarnate to the thinkers of the First Century; all these marvels make him a conventional myth to those of the Twentieth.[18]

Would that the latter were true!

NEUROPHYSIOLOGICAL EXPLANATIONS OF RELIGIOUS EXPERIENCES

Before concluding this chapter, something should be said about the neurophysiological causes of religious beliefs and conversions. In the "Critique of Religious Experience," the final chapter of *The Vanquished Gods*, I discuss the crucial role in the past of dreams, visions, hallucinations, out-of-the-body experiences (OBEs), and near-death encounters (NDEs) in shaping religious beliefs. We have just seen in the four Gospels how frequently the Holy Spirit or angels such as Gabriel conveyed messages from the Lord to Jesus, Joseph, Mary, and others through the intercession of dreams. Throughout the past, dreams, trances, mystical experiences, possession, hearing inner voices, speaking in tongues, and out-of-the-body and near-death experiences were generally considered valid evidence of encounters with some form of spiritual being.

For example, the well-attested experience after the death of loved ones of hearing their voice was considered evidence of their continued existence, while "out-of-body" or "astral projections" where the dreamer soars to distant places (but wakes up in bed), encouraged the belief that the soul can leave the body and travel to distant regions. Consider how primitive man must have responded to severe cases of schizophrenia where the person hears commanding voices, sees fantastic hallucinations, and is frightened by a loss of ego or self. These incredible experiences, along with the fact that primitive societies knew almost nothing about the actual causes of natural events, perceived nature animistically as imbued with vital forces or spirits, and explained phenomena as manifestations of intentional or miraculous causes by controlling deities, explain the acceptance and pervasiveness of the mythical and religious worldview in ancient times and its continuation today among peoples with similar levels of understanding despite the advances of science.

However, for those with a scientific background, its investigations have largely replaced these primitive beliefs with more credible and effective empirical explanations, not just of natural processes but also of the bizarre experiences themselves. Today rumbling thunder is not a display of angry gods; a terrifying flash of lightning is not a bolt from Zeus; natural disasters are not punishments of the gods (as Pat Robertson claimed of the recent devastating earthquake in Haiti); psychoses are not demonic possession; nor are startling occurrences, such as eclipses, a streaking comet, or the cry of a distant animal, omens of some disaster. This naturalizing or demystifying process also applies to unusual *subjective* experiences. We no longer regard hearing voices, having out-of-the body experiences, or seeing terrifying hallucinations as evidence of supernatural occurrences.

But in ancient times, rather than interpreting them as aberrant or delusional, those who had them often were regarded as supernatural, elect, or godlike. It was not until the development of biochemistry, molecular biology, and neurophysiology, along with the discovery that psychoses, schizophrenia, nightmares, hallucinations, and such fantastic experiences as seeing UFOs or extraterrestrial creatures could be localized and attributed to malfunctions of the brain, that they have been considered abnormal, rather than supernatural. Moreover, chemical imbalances, brain injuries, lesions, aneurysms, and genetic impairments have shown how dependent ordinary cognitive processes are on our brains, with the later replacing the psyche or soul as the cause of these experiences.

In his book on *Dreams and Dreaming*, Norman MacKenzie has given an excellent account of the historical influence of the interpretation of dreams and "dream books" in ancient Assyria, Babylon, Sumeria, Mesopotamia, and Egypt, along with the effects of dreams on Roman Emperors and biblical prophets.[19] He also cites numerous writers and artists for whom dreams played a crucial role in their creations: Piranesi, Blake, Goya, Coleridege, Robert Louis Stevenson, De Quincy, Cocteau, Verlaine, and Rimbaud. According to the testimony of De Quincy, Coleridege, and Robert Louis Stevenson a number of their literary works (sometimes written after taking opium or other drugs) were the direct result of vivid, commanding, and recurrent visions.

But by far the most extraordinary example of the influence of dreams is that of Muhammad. When approaching forty years of age he became increasingly troubled by religious questions, often withdrawing to a cave at the foot of Mt. Hira where he passed many days and nights fasting, meditating, and praying. One night in 610 when he was alone in a cave, he had a dream that changed his life and that of billions of others. As told to his major biographer, Muhammad ibn Ishaq:

> Whilst I was asleep, with a coverlet of silk brocade whereon was some writing, the angel Gabriel appeared to me and said, "Read!" I said, "I do not read." He pressed me with the coverlets so tightly that Methought "twas death." Then he let me go, and said "Read!" . . . So I recited aloud, and he departed from me at last. And I awoke from my sleep, and it was as though these words were written on my heart. I went forth until . . . I heard a voice from heaven saying, "O Mohammed! Thou art the messenger of Allah, and I am Gabriel." I raised my head toward heaven to see, and lo, Gabriel in the form of a man, with feet set evenly on the rim of the sky saying "O Mohammed! thou art the messenger of Allah, and I am Gabriel."[20]

For the next twenty-three years Muhammad had a succession of visions and dream-revelations, which he would then dictate to amanuenses who transcribed them. Shortly after his death they were assembled by his primary scribe Zaid ibn Thabit who, with the aid of three Quraish scholars, some years later wrote the text of the Koran officially used today containing 114 chapters or Sutras. It was reported that often when the visions came, Muhammad "fell to the ground in a convulsion or swoon," a physical reaction similar to Saint Paul's seizure. As Durant states, it is possible "his convulsions were epileptic seizures; they were sometimes accompanied by a sound reported by him as like the ringing of a bell—a frequent occurrence in epileptic fits" (p. 164).

If another account of his visions also is true, that he was "miraculously transported in his sleep to Jerusalem [where] a winged horse, Buraq, awaited him . . . [and] flew him to heaven, and back again," after which he "found himself, the next morning, safe in his Mecca bed" (p. 166; brackets added). This would be a perfect example of an OBE or astral projection. There hardly could be a more convincing illustration of the tremendous effect on history of these extraordinary experiences.

Although illiterate, Muhammad like Jesus had developed an amazing memory enabling him to digest and recite large portions of the Bible that he had overheard read and discussed, particularly by his wife's cousin Waraqah ibn Nawfal who could read the scriptures, as well as discussions he had with Jews and Christians living in Mecca and Medina. Thus he acquired an extensive mnemonic retention of the Jewish Talmud, along with the Old and New Testaments, evident in constant references to them in the Koran. Although it may seem incredible that such a revered and influential book as the Koran could have originated in a series of dreams and visions, this is the case according to Muhammad's own testimony. What authenticates this for the Muslims is that while Muhammad admitted his revelations occurred in dreams and visions, he claimed they actually were Allah's messages conveyed to him by the angel Gabriel.

Recently there have been attempts to provide more specific neurophysiological explanations of these religious experiences. In his book, *Neurophysiological Basis of God Beliefs*, Michael A. Persinger claims, based on clinical research, that

> God experiences are the products of the human brain . . . tempered by the person's learned history . . . correlated with transient electrical instabilities within the temporal lobe of the human brain. These temporal lobe transients (TLTs) are normal changes that are precipitated by maturation, personal dilemma, grief, fatigue, and a variety of physiological conditions. Production of TLTs create an intense sense of meaningfulness, profundity, and conviction.[21]

When activated these TLTs "contain common themes of knowing, forced thinking, inner voices, familiarity [déjà view?], and sensations of uplifting movements [levitation or heavenly ascensions?]" (p. 19: brackets added). Given this evidence, I again leave to the reader to decide which of the two causes, temporal lobe projections or visitations by angels in dreams, is the more credible explanation of the biblical messages recurrently described in the Gospels and Muhammad's dictations. But just as Ehrman's careful textual research of the original sources of the New Testament convinced him that it is not a revealed but a "human book," so it seems that the Koran, rather than being dictated by Allah, was Muhammad's own creation derived from his dreams or visions which in turn were based on his intense and extensive religious discussions.

Even more bizarre symptoms are reported by temporal lobe epileptics, according to Persinger, often "characterized by the persistent theme of religiosity. Their lives are full of repeated peaks of mystical experiences and multiple conversions" (p. 19). As described by Slater and Beard and quoted by Persinger:

> Mystical delusional experiences are remarkably common. One patient said that God, or an electrical power was making him do things; that he was the Son of God. Another said he felt God working a miracle on him. Another felt that God and the Devil were fighting within him and that God was winning. . . . Hallucinations were often extremely complex and were usually full of meaning, often of a mystical type. Nearly always there were auditory hallucinations at the same time. One patient saw God, heard voices and music and received a message that he was going to heaven. Another had a vision of Christ on the cross in the sky [similar to the experience of Constantine], and heard the voice of God saying, "You will be healed, your tears have been seen."[22]

Just recently it was reported in the journal *Psychopharmacology* that scientists studying the effects of psilocybin, the active chemical in so-called magic mushrooms used for centuries by Mexicans and other South Americans to induce mystical states, "promoted a mystical experience in two-thirds of people who took it for the first time," some claiming that it was "the 'single most spiritually significant' experience of their lives."[23] Thus what once were considered inviolable mystical visions and god encounters can now be explained as illusions produced by nervous disorders or chemicals affecting the brain.

This type of explanation has been suggested for another paranormal experience in which "10 percent of survivors of heart attacks report having a near death experience," such as "feelings of transcendence, being surrounded by light or floating outside their bodies."[24] Some feel "paralyzed or hear sounds that others do not" as they gradually awaken. However, in the article Vedantam claims that new research recently published in the journal *Neurology* indicates a possible neurological explanation that attributes near-death experiences to a different sleep-wake cycle in the brain from the ordinary.

In a study comparing 55 people with near-death experiences with 55 people who had no such experiences, neurologist Kevin Nelson of the University of Kentucky found that people who reported such experiences were also more likely to report a phenomenon known as "REM intrusion," where things normally experienced during sleep carry over into wakefulness. REM is an acronym for rapid eye movement, one of the phases of sleep. As one approaches death the fear of dying apparently activates the same neuronal process that produces the transitional state between sleeping and wakefulness accompanied by "various dreamlike phenomena," often influenced by the individual's religious beliefs.

Although still at the hypothetical stage, the explanation illustrates how throughout history scientific advances discredited or demystified various beliefs and experiences. Yet none of these naturalistic explanations replacing the divine revelation of scripture with human authors nor the cause of religious experiences by erratic brain processes negates the tremendous effect Christianity has had on Western civilization. There is no denying that the divine status and miraculous capacities attributed to Jesus Christ, along with his mission and teachings, resonated with billions of people and has made the Bible the best selling book of all time. It is not unique, however, since this is also true of Buddhism, Hinduism, and the Muslim religion which implies that each arose and has to be understood within the epoch in which it appeared.

Although scientific inquiry reemerged in the West during the twelfth and thirteen centuries, the so-called Dark Ages, it still was believed that everything was created and explainable by God. Bishop Ambrose, one of the Patristic Fathers wrote as early as the fourth century CE that no explanations of phenomena are needed except the Will of God.

> To discuss the nature and position of the earth does not help us in our hope of the life to come. It is enough to know what Scripture states, "that He hung up the earth upon nothing" (Job, xxvi, 7). Why then argue whether He hung it up in air or upon the water [the views of Anaximines and Thales]. . . . Not because the earth is in the middle, as if suspended on even balance [the position of Anaximander], but because the majesty of God constrains it by the law of His will, does it endure stable upon the unstable and the void.[25] (Brackets added)

This period is called the Dark Ages because knowledge as we know it was still largely "in the dark," as the saying goes.

It is incredible that a person about whom we know so little could have initiated such a complete change in worldview that only biblical scripture had any explanatory significance. Yet as Augustine, perhaps the greatest of the Church Fathers, who lived in the fifth century states: "'Nothing is to be accepted except on the authority of Scripture, since greater is that authority than all powers of the human mind.'"[26] Perhaps at that time the mythical account of creation in Genesis, the story of Adam and Eve as the first humans, and the belief in the global flood and Noah's Ark preserving all existing species could convince Augustine, but this should not be true today.

This uncritical acceptance of religious doctrine is enshrined in the notorious statement of another early Church Father, Tertullian (d. 222), that "I believe [in Christianity] because it is absurd" (brackets added). In truth this "absurdity" is clearly illustrated in the following statement by him: "The Son of God was crucified; that is not shameful because it is shameful, and the Son of God died; that is credible because it is absurd. And he rose from the dead; that is quite certain because it is impossible."[27] Given the normal tendency toward irrationality inherent in human beings, one can only shudder at what effect this kind of thinking would have on intellectual history or the future of humankind if it had been taken seriously.

NOTES

1. This chapter is based primarily on my book, *The Vanquished Gods* (Amherst, New York: Prometheus Books), 2001.

2. Frederick B. Artz, *The Mind of the Middle Ages*, 3rd ed. rev. (New York: Alfred A. Knopf, 1962), pp. 57–58.

3. Quoted from Christopher Hibbert, *Rome: The Biography of a City* (London: Penguin Books, 1987), pp. 66–67.

4. Cf. Richard H. Schlagel, *The Vanquished Gods*, op. cit., pp. 205–14.

5. Rodney Stark, *The Rise of Christianity* (San Francisco: HarperCollins, 1997), p. 77.

6. Dionysus, "Festival Letters," quoted by Eusebius in *Ecclesiastical History* by G. A. Williamson (Middlesex: Penguin Books, 1965), 7: 22. Quoted by Stark, op. cit., p. 82.

7. W. H. McNeill, *Plague and Peoples* (New York: Doubleday, 1976), p. 108. Quoted from Stark, p. 88; brackets added.

8. Cf. Will Durant, *The Story of Civilization*, Part III, *Caesar and Christ* (New York: Simon and Schuster, 1944), p. 652. Brackets added. The next two quotations also are to this work.

9. The following discussion of the Gospels is based on Richard H. Schlagel, *The Vanquished Gods*, op. cit., pp. 144–65.

10. F. W. Beare, *The Earliest Records of Jesus* (Oxford: Blackwell, 1964), p. 13. Quoted from G. A. Wells, *Who Was Jesus? A Critique of the New Testament Record* (La Salle, IL: Open Court, 1989), p. 5.

11. Robert W. Funk, Roy W. Hoover, and The Jesus Seminar, *The Five Gospels* (New York: Macmillan, 1993), p. 5.

12. Bart D. Ehrman, *Misquoting Jesus: The Story Behind Who Changed the Bible and Why* (New York: HarperSanFrancisco, 1955). The immediately following references are to this work.

13. Cf. Burton L. Mack, *Who Wrote the New Testament? The Making of the Christian Myth* (New York: Harper Collins, 1996), chap. 6.

14. Funk, Hoover, and The Jesus Seminar, op. cit., pp. 8–9.

15. Cf. Howard Clark Kee, *Understanding the New Testament*, 5th ed. (Englewood Cliffs, New Jersey: Prentice Hall, 1993), pp. 62–63.

16. This account and the following quotation are from John Shelby Sprong, *Liberating The Gospels: Reading the Gospels with Jewish Eyes* (New York: HarperCollins Publishers, 1996), p. 20.

17. These biblical references are all from The Holy Bible: Revised Standard Version, 1952 (New York: Thomas Nelson & Sons, 1953).

18. A. Weigall, *The Paganism in Our Christianity* (New York: 1928), pp. 18–19. Quoted from Frederick Artz, op. cit., p. 459.

19. Cf. Norman McKenzie, *Dreams and Dreaming* (New York: The Vanguard Press, 1965), pp. 277–89.

20. Will Durant, *The Story of Civilization*, Part IV, *The Age of Faith* (New York: Simon and Schuster, 1950), pp. 163–64. The following two citations in the text are to this work.

21. Michael A. Persinger, *Neurophysiological Basis of God Beliefs* (New York: Praeger, 1987), p. x.

22. E. Slater and A. W. Beard, "The Schizophrenic-like Psychoses of Epilepsy," I, Psychiatric Aspects, *British Journal of Psychiatry*, 109, 1963, pp. 95–150. Quoted by Persinger, pp. 20–21.

23. *The Washington Post*, July 11, 2006, A8.

24. Shankar Vedantam, "Near-Death Experiences Linked to Sleep Cycles," *The Washington Post*, April 11, 2006, A 2. Until otherwise indicated, the following quotations are from this article.

25. Quoted from H. O. Taylor, *The Medieval Mind*, Vol. I, 4th ed. (Cambridge: Harvard University Press, 1962), p, 73.

26. W. W. Hyde, *Paganism to Christianity in the Roman Empire* (Philadelphia: University of Pennsylvania Press, 1946). Quoted from Artz, op. cit., p. 82

27. F. Lear, "Medieval Attitude Toward History," *Rice Institute Pamphlets,* 1933. Quoted from Artz, op. cit., p. 76.

Chapter 7

SCHOLASTIC RENEWAL OF EMPIRICAL INQUIRY

THE INTERVENING DARK AGES

It took nearly a millennium for the biblical spell of Christianity to begin to wane with the realization that the Bible was not the font of all knowledge and wisdom. As if aroused from a mesmerizing book, following the scriptural immersion during the Dark Ages the scholastics toward the end of the medieval period began to awaken to the world around them, largely due to the reintroduction of Hellenistic philosophy and Arabic science from the twelfth to the fourteenth centuries as a result of the Arabic conquest of Spain. With the exception of a few thinkers such as Simplicius and John Philoponus in the sixth century, scientific inquiry had ceased in the West from the fourth century CE with the demise of the Greco-Roman civilization and its replacement by the Byzantine Empire until 1453 when it was conquered by the Ottoman Sultan Mehmet II. Although the Byzantine culture is noted for the splendor of its mosaics and church architecture, its intellectual endeavors were so bound by the worldview of Christianity that there was little interest in scientific inquiry. As Lord Norwich graphically characterizes this fixation in his book *Byzantium*, "[t]he Byzantines were mystics, for whom Christ, his Mother and the Saints were as real as members of their own families."[1]

So engrossed with formulating such Christian doctrines as the immaculate conception of Anne, the mother of Mary, who in turn was said as a virgin to have given birth to Jesus by the intervention of the Holy Spirit; the adoption of the Athanasian consubstantial definition of the trinity in which the Father, the Son, and the Holy Spirit are of the same substance though existing separately; the sacrament of the Eucharist in which the bread and wine retain their usual appearance despite being miraculously transformed into the flesh and blood of Christ; the existence of the soul surviving the body in the afterlife; the reconciliation of freewill with the omniscience and omnipotence of God; and the question of whether salvation depended upon God's grace, good works, or both, they had no interest in inquiries about natural phenomena.

Not amenable to empirical or rational solutions, these creeds also were too esoteric to be decided by the sayings of Jesus who mainly spoke in parables and country metaphors, nor by the Gospels. As Charles Freeman states in his study of the rise of Christianity: "when one puts together the Gospels, the letters of Paul, the Book of Revelation and the Old Testament, there is no sense of a coherent 'axiomatic' base on which to build theological truth."[2] Lacking such a base, these alleged 'theological truths' were decided by whichever faction of the church was the strongest and then accepted on faith.

Although usually claimed to be justified by revealed scripture, following considerable dissention they were decided by church councils or papal decrees reinforced by the threat of excommunication. For example, the Council of Nicaea was called by Constantine in 325 to decide the correct formulation of the Nicene Creed, but the final adoption awaited Theodosius' Council in 381. The initial presentation of a scientific hypothesis or theory also can be quite contentious, but sooner or later after a number of experimental tests the issue is resolved objectively. But nothing like this occurs in theological disputes, which accounts for their authoritarian and dogmatic nature in contrast to the empirical confirmation and emendation of scientific theories.

Because there are no discernable referents to terms such as 'Gabriel,' 'heaven,' 'immaculate conception,' 'immortal soul,' 'Satan,' 'God,' just as there are none to 'Zeus,' 'celestial orbs,' 'the Demiurge,' or the 'Prime Mover,' these concepts are entirely verbal. Though given a defined or intentional meaning, they do not have the referential or extensional meaning of the theoretical terms of science which, even when designating unobservable entities, have indirect empirical evidence and testable consequences. In contrast, religious concepts can be given any interpretation consistent with the conceptual framework depending on one's intent. But like earlier mythologies, when the conceptual framework in which they are embedded comes to be seen as fictional, they can be discarded. This was what became of the pagan religions, as well as the Genesis account of creation which was once universally believed by Hebrews and Christians, but has been largely discarded because it has no evidential support or explanatory value. This is difficult for some people to acknowledge, but so was heliocentricism, glaciers as the causes of the earth's topology, Darwin's theory of evolution, the age of the universe, and the brain as the source of conscious processes, but given the compelling evidence these theories eventually have to be accepted.

THE ARAB LEGACY OF SCIENTIFIC INQUIRY

The renewal of scientific inquiry, based on the earlier achievements of the Hellenic and Hellenistic philosophers and mathematicians, occurred in the ninth century owing to the emergence of Islam as a world power. By the time of his death in 632 Muhammad had conquered and converted to Muslim the entire Arabian peninsula. Then due to the military and colonizing prowess of succeeding caliphs, by the middle of the eighth century the Arabs had extended their domain in the Middle East from the Mediterranean to as far as the Indus River and the Punjab region of India, along with all of North Africa and Spain up to the Pyrenees. Known as the Baghdad Caliphate, it encompassed about three times the extent of the Byzantine Empire under Justinian and a much greater area than that conquered by Alexander the Great.

In contrast to the Church Fathers quoted earlier and the Christian tradition in general, Muhammad believed that because God's creation, in addition to his revelations, displays his power and wisdom he encouraged studying nature to enhance one's understanding of and faith in God. Thus "he besought his disciples to seek knowledge from the cradle to the grave, no matter if their search took them as far afield as China, for 'he who travels in search of knowledge, travels along Allah's path to paradise.'"[3] Initially this injunction took the form of translating the philosophical, scientific, and mathematical legacy of the ancient Greeks. The closing of Plato's Academy and Aristotle's Lyceum in 529 by the Christian Emperor Justinian and the persecution of the heretical Nestorian Church by the Byzantines, forcing the Nestorians to emigrate to Jundishapur in Persia, led to the latter becoming a major scientific center where the works of Greek scholars were translated into Persian and Syriac.

Then the founding of Baghdad in 762 by the Abbasid Caliph al-Mansur and subsequent flourishing of the Abbasid Dynasty brought about the greatest period of conquest, splendor, and prosperity in Islamic history. Caliph al-Mamun later established the Baghdad Academy of Science, attracting scholars from India, Persia, Syria, and Egypt, along with prominent Jews, that eclipsed Jundishapur as the major scientific center in the Middle East. Finally, the conquest of Spain by the Amayyad Caliphate led to Cordoba becoming another influential center of learning and major transmission of Arabic texts to the West because of it affinity to the rest of Europe.

While at first Arabic scholarship was confined to translations, this eventuated in critical commentaries on the works being translated. As the Arabic language did not include the theoretical concepts nor the technical terms of the original texts, the scholars had to replicate the reasoning and devise meanings for the specialized terms being translated. This in turn led to the detection of mistakes, new discoveries, correction of astronomical tables, and revised explanations. Thus from the ninth through the twelfth centuries, owing to their extensive conquests and the adoption of the Arabic language and culture by the conquered peoples, "Arabic became *the* international scientific language; any scientific text of importance was written in Arabic and read all over the cultured world" (p. 396). It was the dissemination of these works following their translation into Latin from the Andalusian region of Spain that initiated the revival of scholarship in the West from the twelfth to the fifteenth centuries leading to the construction of the methodology and theories of modern classical science. While there is insufficient space to list all the contributors and their discussions, some indication of the most outstanding should be mentioned.

PARTICULAR ARAB CONTRIBUTIONS

The first half of the ninth century includes the works of al-Kindi and al-Khwarizmi. The former, called the Arabic "philosopher king," taught in Baghdad and is known for his treatises on meteorology, specific weights, geometrical optics, and physiology, along with a critical work on alchemy which was a precursor of chemistry as astrology was of astronomy. In addition to translating portions of Plotinus' *Ennaeds* and a work by Proclus, he wrote the influential *Theory of Aristotle*, an early source of the Neoplatonic interpretation of the Stagirite. Al-Khwarizmi, who was influenced by the Hindus, was the most famous Arabic mathematician from whom we acquired the word 'algorithm' and whose treatise on *al-jabr* (*Algebra*) gave to the West that important branch of mathematics so essential in classical and modern science for the formulation of functional equations as scientific laws. He derived from the Hindus the so-called "Arabic numbers" which made calculations much simpler than with Roman numerals, along with the place notation of numbers, such as 2009.

The second half of the ninth century included al-Battani and al-Razi, the latter known to the West as Rhazes. Considered the greatest of the Arabic astronomers, al-Battani worked mainly in the observatory at Baghdad. After studying the Arabic version of Ptolemy's work, renamed the *Almagest*, he derived more accurate measurements of the obliquity and the precession of the equinoxes, along with more precise tables for the orbital motions of the sun and the moon. Rhazes was the greatest of the Arabic alchemists performing experiments in a well equipped laboratory where he attempted to explain the transmutation of substances and metals based on combining ratios of the Empedoclean four elements, fire, earth, air, and water, plus their defining qualities hot, cold, wet, and dry. Though paving the way for chemistry, alchemy still remained a pseudoscience based on false assumptions and fictitious chemical processes.

The most prominent scholar of the first half of the tenth century was al-Farabi who worked in Aleppo and Damascus. Known as the "second Aristotle," he translated the latter's *Organon* into Arabic which promoted the use of Aristotelian logic in deductive scientific explanations. His medical investigations led him to interpret Aristotle's material cause as the imperceptible ingredients of a medication and his formal cause as the active power or agency arising from the material cause that is responsible for the healing effects. In the second half of the tenth century Abu l-Wafa stands out as one of the greatest mathematicians, following al-Khwarizmi, who pursued his research mainly in Baghdad writing commentaries on Euclid, Diophantos, and al-Khwarizmi and contributing greatly to the advance of trigonometry.

Three outstanding scholars worked in the first half of the eleventh century. Al-Biruni, although a Persian, epitomized the interdisciplinary nature of much of Muslim research having made significant contributions to medicine, astronomy, mathematics, physics, geography, and history in what is considered the Golden

Age of Islamic scholarship. Following the procedure of Archimedes, he determined the specific weights of many precious stones and metals, while his *Chronology of Ancient Nations* is a valued contribution to ancient history. In one area of investigation, optics, the Arabs even surpassed the Greeks.

Ibn al-Hatham or Alhazen, as he is known in the West, who was born in Babylon but taught in Cairo, opposed Ptolemy's optical theory that vision is produced by the eyes sending visual rays to the object which then projects an image to the observer. Instead, in *The Treasury of Optics* he presented a more plausible Aristotelian explanation in which the form of the object is conveyed to the eye directly where it is transmuted by the lens into the visual image. He also discussed the nature, propagation, and reflection of light, along with optical illusions. According to Charles Singer, his investigation of reflecting mirrors in *On the Burning-sphere* represents "real scientific advance, and exhibits a profound and accurate conception of the nature of focusing, magnifying, and inversion of the image, and of formation of rings and colours by experiment . . . far beyond anything of the kind produced by the Greeks."[4]

An equally famous figure of the early eleventh century was the Persian scholar Ibn Sina or Avicenna, as he is known in the West. In addition to his outstanding contributions to philosophy and medical investigations he is known for his research in astronomy and physics. His *Canon of Medicine*, one of the most influential medical works of the Middle Ages, was known for systemizing the biological works of Aristotle, the anatomical and physiological investigations of Galen, and the medical advances of the Arabs in general. Even more renown as a philosopher, his best-known philosophical work, *Healing*, interpreted Aristotle as a Neoplatonist describing his successive intelligences and planetary spheres as emanations of the Prime Mover, along with claiming that Aristotle's intellectual part of the soul was immortal, discarding its personal identity after death when it returns to the Divine Intellect.

The last of the great Muslim scholars to be mentioned was not from the Islamic Middle East but from Spain. Born in Cordoba of Spanish parents in the twelfth century, his Arabic name was Ibn Rushd but he is commonly known as Averroës. Like Avicenna, he wrote a series of commentaries on Aristotle from a Neoplatonic perspective earning him the title "the commentator," as Aristotle was known as "the philosopher." Rejecting the Genesis account of creation incorporated into the Qur'an, he argued that the cosmos is eternally renewed by a continuous series of divine acts (foreshadowing Descartes). While only the Prime Mover is uncaused and eternal, owing to God's constant creation the universe too is everlasting. He divided knowledge into three categories: (1) the highest truths attainable by philosophical reflection; (2) lesser truths accessible to the educated class; and (3) partial truths available to the masses via allegorical symbolism, a classification applicable today if one substituted science for philosophical reflection. Severely criticized by orthodox Muslims, his books were burned by royal decree.

In addition to these accomplishments, the Muslims made detailed studies in botany, zoology, geology, geography, and musicology. They also improved scientific instruments like the astrolabe used in astronomical measurements and the balance in mechanics and chemistry, along with developing more exact water clocks for marking the hours of prayer. Given these outstanding scientific and mathematical achievements for half a millennium or so, while such activities in Western Europe and Byzantium lay dormant, this raises the question of what brought about the decline of Arabic scholarly research beginning in the thirteenth century and its eventually being overtaken later by the West owing to the Arabic reintroduction of the ancient works of the Greeks.

The first reason was the conquest of Baghdad, along with much of the Middle East, in the middle of the eleventh century by the Seljuk Turks who had recently been converted to Islam. Extremely conservative and repressive, they disdained any investigations not directly supportive of orthodox Muslim beliefs, hence opposed scientific inquiry. Second was the capture of Cordoba in the twelfth century by the Spanish King of Castile, Ferdinand III, terminating scientific inquiry in what had become the most flourishing Islamic research center in the West. Third was the repressive influence of the religious views of al-Ghazali (d. 1111) under whose direction all writings leading "to loss of belief in the creator and in the origin of the world" were condemned. Although steeped in Greek and Muslim thought having studied philosophy, theology, and law, like Muhammad, when nearing forty years of age he had a religious conversion which attracted him to the Sufic doctrine of the primacy of personal religious experience.

As Martin Luther five centuries later, al-Ghazali found the rituals and doctrines of conventional religion an impediment to inner faith based on direct experience of God. In two works, the *Destruction of Philosophy* and the *Revival of the Science of Religion*, he dismissed arguments for God based on scientific evidence or rational proofs, grounding religious belief entirely on mystical experience, prayer, and faith. He also denied that Muhammad was the final source of mystical revelations claiming their recurrence continued to enrich Muslim religion. Becoming renown for his eloquent defense of Sufism, in the following century his Sufic religious orders sprang up throughout Islam.

The fourth reason was the result of the barbaric Mongol invasion of the thirteenth century, a much more destructive incursion that devastated Islamic civilization. These conquests of the main centers of Arabic research accompanied by a rise in Islamic fundamentalism are considered the main reasons for Islam's fatal decline following five centuries of outstanding scientific achievements.

REVIVAL OF WESTERN SCIENTIFIC INQUIRY

But while this Arabic tradition of scientific research atrophied after the thirteenth century its legacy continued, the proximity to Europe of the Muslim schools at Cordoba and Salerno facilitating the transmission to the West of their trove of Greek and Arabic manuscripts when translated into Latin. The stimulus of this influx of knowledge in the twelfth and thirteenth centuries, at the time when universities were being established at Oxford and Paris, was the major reason for the revival of learning in the West. A. C. Crombie's book, *Grosseteste and Experimental Science*, describes this revival. Although he acknowledged in the Preface to the Second Impression that "some of the expressions I used about the extent of the [western] medieval contributions to the structure and methods of research of modern experimental science now seem to be exaggerated,"[5] nevertheless, his account of Grosseteste's influence on the renewal of scientific inquiry in Western Christendom is a significant scholarly contribution. The thesis of his book is that "a systematic theory of experimental science was understood and practiced by enough [medieval] philosophers for their work to produce the methodological revolution to which modern science owes its origin" (p. 9; brackets added). Science today is such a pervasive aspect of our culture that we are apt to forget the time and effort required to achieve a correct formulation of the methodology required to produced these tremendous advances and achievements. Whether Grosseteste and Oxford University where he taught were the primary originators of this conception of science

> or, as seems more probable, it was the product of a more general European response, in which Oxford took the lead, to the new ideas coming from classical Greece and medieval Islam, there is no doubt that from the time of Grosseteste the experimental science which has ever since been an essential part of the Western world began to appear in centre after centre. (p. 14–15)

What the scholastics from the twelfth to the fourteenth centuries inherited from the Muslims, in order of importance, was first, the methodology and cosmology of Aristotle; second, the philosophical system of Plato that was particularly influential in the twelfth century owing to his conceptions of the independently existing

Receptacle, the Realm of Forms, and the Demiurge, along with role of mathematics in attaining knowledge of the Forms; third, Augustine's interpretation of knowledge as divine illumination; fourth, the emanation theory of Plotinus; and fifth, the optical investigations of Ptolemy, Avicenna, and Averroës, among others. But since it was Aristotle's system especially that was the prevailing influence until its overthrow by scientific developments in the seventeenth century that led to Newton's mechanics, a review of his system would be helpful even though we have described his philosophy in chapter 3.

ARISTOTLE'S COSMOLOGY AND SCIENTIFIC KNOWLEDGE

Like the ancient Greeks in general, Aristotle's intent was to account for the permanence and order in nature despite the prevalence of change and becoming. Believing that the world existed objectively as we ordinarily perceive it, to account for the permanence of objects he endowed them with an underlying substance supporting their defining forms and accidental attributes. The designation of an object's substance is relative as his example of a bronze statue illustrates. While the bronze of the statue is its observable substance, the alloys copper and tin are its more basic substances, with these being further analyzable until one arrives at "prime matter," the indestructible, first matter. What characterizes an object is its defining form or its essential nature.

Having created syllogistic logic, this became the essential formalism of Aristotle's method of explanation in opposition to Plato's emphasis on mathematics, which he believed applied to abstract magnitudes rather than concretely existing objects. Aware that ordinary experience discloses numerous kinds of objects and regular sequences but not their actual causes, his mode of explanation began with distinguishing the defining properties of an object from those which are variable and accidental.

Thus all scientific knowledge begins with *inductions* to discover the generic forms or defining natures of objects which serve as the premises for deducing or explaining their particular properties. His general mode of explanation, based on syllogistic reasoning, consisted of deducing particular attributes from these universally true premises as described and illustrated previously.

As these syllogistic explanations consist of showing or denying that entities have the properties they have because they belong to a higher genus, it was the obvious limitation of these deductive demonstrations that gave the early scholastics such a bad reputation. Yet sensing the deficiencies in his schema, they emended it by redefining the premises as theories, asserting that scientific explanations consist of deducing phenomena from true theories. But while this was an improvement, there still was the problem of distinguishing true from false theories. They prescribed experimental tests, but there were differences of opinion as to how reliable or certain the results of the tests were.

Notwithstanding its limitations, as an attempt to provide a natural explanation of phenomena Aristotle's deductive method of proving an empirical conclusion was superior to the belief of the early Church Fathers that simply attributing something to God was a sufficient explanation. Adelard of Bath, for example, a late twelfth-century English scholar in his *Quaestiones naturales* describes an imaginary conversation with his nephew who declares that "a certain natural event could be explained only as 'a wonderful effect of the wonderful Divine Will,'" to which Adelard replies that while "it was certainly the Creator's will that it should happen, 'it is also not without a natural reason, and that this was open to human investigation'" (p. 12). He adds that this "does not detract from God," because nature is "not without system and human science should be given a hearing on those points it has covered." What he did not foresee is that when natural reason or science is let in, God is eventually forced out.

Aristotle also maintained that an adequate explanation must

include his famous four causes, the material, formal, efficient, and final described previously. Moreover, in his *Metaphysics* he considered the study of "Being qua Being" as a separate discipline. As we shall see, the development of modern science involved a transformation in the meaning of the first three causes, while the final cause and the study of "being as such," in Aristotle's sense, were eliminated altogether.

In addition to his methodology, it was primarily Aristotle's cosmology which had to be corrected and revised to construct modern classical science. His spherical cosmos consisted of the seven planetary spheres composed of an incorruptible, weightless aither conveying their celestial bodies in uniform circular motions eastward at various velocities to explain their appearances, with a total of forty-seven spheres to counteract the influence of their adjacent spheres. The eighth outermost sphere of the fixed stars produced the daily rotation of the entire cosmos westward, while the Prime Mover in the outer nonspatial Empyrean Heaven was the ultimate source of motion by being loved.

In contrast to the celestial realm, the terrestrial world consisted of the four elements, fire, air, earth, and water, that undergo change and destruction, the first two having the natural tendency to rise to the innermost heaven while the latter two fall to the center of the universe. It was to this world of change and becoming that Aristotle's method of scientific explanation particularly applied. With this in mind, we can see how the scholastics revised his methodology and cosmology to achieve a more effective method of explanation, along with a more exact understanding of change reinterpreted as due to motion and forces whose explanation was the primary aim of modern classical science.

CONTRIBUTIONS OF ROBERT GROSSETESTE

Returning to Grosseteste and the scholastics, according to Crombie, Grosseteste was the leading figure in the renewal of scientific research during the thirteenth century having founded "a distinguished school of scientific thought" at Oxford University dedicated to pursuing "the methods and implications of experimental investigation and scientific explanation" that was continued by Roger Bacon, John Pecham, William of Ockham, Duns Scotus, Thomas Bradwardine, John of Dumbleton, and others (p. 14). Albertus Magnus played a similar role in what is now Germany. By the end of the thirteenth century the developments extended to Padua and other Italian cities. During the fifteenth and sixteenth centuries they reached "Leonardo da Vinci and the Italian physiologists and mathematical physicists. And so it went to Kepler, Galileo, William Gilbert, Francis Bacon, William Harvey, Descartes, Robert Hooke, Newton, Leibniz, and the world of the seventeenth century" (p. 15).

It is this development that we shall now briefly follow beginning with the works that were available to Grosseteste to show how important this influence was.

> By the end of the twelfth century not only the *Analytics*, but also the *Physics*, *De Generatione* et *Corruptione*, *De Anima*, *Parva Naturalia*, and the first four books of the *Metaphysics* of Aristotle had been translated into Latin from the Greek, and the *De Caelo* and the first three books of the *Meteorologica* had been translated from the Arabic. During the same period Latin versions from either Greek or Arabic, and sometimes from both, were made of the *Elements*, *Optics*, and *Catoptrics* of Euclid . . . part of the *Conics* of Apollonius, the *Almagest* and *Optics* of Ptolemy . . . and numerous works by Hippocrates and Galen. (pp. 35–36)

It was such works as these, particularly Aristotle's, that were the incentive and background for Grosseteste's seminal investigations. Crombie describes the general consensus regarding the prospects for scientific knowledge at the time.

> Knowledge of changing physical things was the least certain scientific knowledge, and physics . . . the least certain science. Nev-

> ertheless . . . [i]t was possible to arrive at a probable knowledge of the causal necessities or laws, belonging to the form, according to which natural things behaved. Such laws were suspected when certain phenomena were seen to be frequently correlated. It was the function of natural science to discover and define as accurately as possible the form [or laws] that could give rise to demonstration, as the middle term of a syllogism, and so to knowledge of the causes . . . of observed effects. (p. 61; brackets added)

By reinterpreting "the form that could give rise to demonstration" as laws describing "correlations in nature," rather than merely definitions of objects in terms of their species and genus, Grosseteste brought Aristotle's method closer to that of modern scientific inquiry. In addition, while retaining Aristotle's general method of induction and deduction renamed "resolution and composition," he relied on experimental verification and falsification as the means of selecting the correct premises, another significant advance over Aristotle. He also utilized the Arabic principle of *reductio ad impossible* (or as it is known today, *reductio ad absurdum*), the method of refuting a premise or a theory by demonstrating that its implications lead to erroneous conclusions or absurdity, another notable method for evaluating hypotheses (cf. p. 84). Furthermore, he concluded that the use of these forms of reasoning and verifying criteria imply that nature is both "simple and uniform," additional crucial presuppositions of scientific inquiry because they exclude miracles and occult explanations.

Yet he was not alone in seeing the need for further methodological clarifications, Peter Abelard (1079–1142) having made the important distinction between practical knowledge based on observing the causes and effects of things, and theoretical knowledge dependent on understanding how the causes and effects are produced. As quoted by Crombie:

> For many people are practiced in action but have little scientific understanding; they have tested the healing power of medicines and are good at healing because of their experience alone, but they

> do not know much about the natural causes. For they . . . are not instructed in the theoretical. . . . Many people on the other hand have understanding but not practical ability, and these can impart knowledge to others but cannot put it into practice themselves. The man of understanding is he who has the ability to grasp and ponder the hidden causes of things . . . those from which things originate, and these are to be investigated more by reason than by sensory experiences. (pp. 29–30)

An acute observation, it is mitigated by the fact that rather than the 'hidden causes' being scientific and empirical, the "causes Abelard sought in nature were the necessary causes of metaphysics, causes that could be the start of certain demonstration . . ." (p. 30).

Grosseteste agreed that scientific explanations depended upon discovering the "hidden causes" underlying the correlations in nature because he maintained that "in natural science the inductive, resolutive method was aimed at bringing to light causes that were hidden 'from us,'" but he denied that such knowledge was certain because "it was usually impossible to reach a complete definition of the cause of an experienced effect because it was impossible to exhaust all the possibilities" (pp. 81–82). Thus he believed as Abelard that the deficiencies of scientific explanations could be surmounted by metaphysical knowledge, by a "Divine illumination . . . alone man could have certain knowledge of the essence of the real" (p. 131).

In addition to these modifications of Aristotle's system of explanation, Grosseteste rejected his belief that mathematics could tell us nothing about the natural world because it only referred to abstract quantities. "Mathematics," he claimed, "could often provide the reason for occurrences in the world of experience because although the subject that mathematics studied was abstract quantity, *mathematical entities actually existed as quantitative aspects of physical things*" (p. 91; italics added). Considering the indispensable role of mathematics in modern science, this was an astute pronouncement. Furthermore, having made a careful study of the merits of the Aristotelian and Ptolemaic astronomical systems, rather than Aristotle's reliance on concentric spheres, Grosseteste chose Ptolemy's epicycles and eccentrics for "saving the appearances" of the planetary motions, but retained Aristotle's explanation of the cause of celestial motion.

Still, as one would expect, he was not able to foresee some crucial aspects of modern science nor entirely shed prevalent misconceptions of his age. His notion of "hidden causes," for example, was not in agreement with later scientific developments which restored Leucippus and Democritus' atomic theory as the theoretical foundation of modern science. Furthermore, his conception of "divine illumination" supplementing scientific inquiry was discarded by later scientists. But in some respects his theory of the origin of the universe, called his "metaphysics of light," with a little charity could be interpreted as prescient. He seems to have been influenced by Augustine's notion of God as an infinite corporeal light, along with the Neoplatonic interpretation of Aristotle by Avicenna and Averroës. The latter reinterpreted Aristotle's celestial spheres in terms of Plotinus' theory of emanation claiming that the physical world was the last of a hierarchical series of a descending radiation of the divine light.

According to Crombie's description, Grosseteste had prime matter created first, then light which, in penetrating matter, creates its three spatial dimensions and motion.

> In the beginning of time God had created out of nothing unformed matter (*materia prima*) and light (*lux*) which, by autodiffusion [spontaneous diffusion], had produced the dimensions of space and then all subsequent beings. For this reason Grosseteste believed that the study of optics was the key to the understanding of the physical world. . . . (p. 104)

Grosseteste's account of creation, as quoted below by Crombie, though still within the theoretical framework of Aristotle's form and matter, seems to attribute three dimensionality, along with corporeality, to the diffusion of light itself.

> The first corporeal form . . . I hold to be light. For light (*lux*) of its own nature diffuses itself in all directions, so that from a point of light a sphere of light of any size may be instantaneously generated. . . . Corporeality is what necessarily follows the extension of matter in three dimensions, since each of these, that is corporeality and matter, is a substance simple in itself and lacking all dimensions. But . . . lacking matter and dimension, it was impossible for it to become extended . . . except by . . . suddenly diffusing itself in every direction. . . . And, in fact, it is light, I suggest, of which this operation is part of its nature, namely, to . . . instantaneously diffuse itself in every direction. (p. 106; brackets added)

Although there is no mention in this quotation of the prime matter or prime form, it is included later: "nothing is present . . . in every body except primitive matter (*materia prima*) and primitive form (*forma prima*) and magnitude, which necessarily follows from these two, and whatever is entailed by magnitude as such, as position and shape" (p. 107). By following Adelard's injunction that though it is God's will that things exist as they do this does not preclude trying to describe scientifically how this came about, Grosseteste made an extraordinary effort to explain how, from Aristotle's limited schema of form and matter, the "first corporeal form light" and "prime matter," which initially had no dimensions and motion, became the material and efficient causes of the physical universe.

Thus his "metaphysics of light" attempted to describe how light could infuse prime matter with its spatial dimensions and, as the first principle of motion, endow matter with movement enabling it to generate the universe. Despite its obscurity, I found this attempt to explain the origin of the universe by the radiation of light "prescient" because it foreshadows somewhat the Big Bang theory which describes the universe as originating from a singularity whose *radiation* creates the space of the expanding universe and its later contents. Though the initial creation of the Big Bang occurred nearly fourteen billion years ago, not instantaneously as Grosseteste claimed, there does appear to be some similarity.

What we have presented is just the framework of his theory because it is beyond the scope of this book to describe his description of the emanation of Aristotle's celestial orbs, the creation of the earth and generation of plants and animals, as well as his explanations of more specific occurrences like heat, the tides, and such optical phenomena as reflection and refraction, vision, colors, rainbows, thunder and so forth. His coverage is amazing, certainly justifying Crombie's high praise of him as a predominant influence in the renewal of scientific inquiry in the thirteenth century.

The point where Grosseteste diverged from later scientific developments was in his acceptance of the prevalent conception of "metaphysical knowledge" supplementing scientific explanation. At the time factual knowledge generally was divided into three categories: (1) scientific knowledge which was only probable because the premises derived from experimental facts could not provide deductions known with certainty; (2) mathematics which did provide certainty because its demonstrations were based on self-evident principles, but did not convey sufficient empirical evidence of existents; (3) and metaphysics which did provide absolute knowledge complementing scientific explanations because it was based on divine revelation of the eternal forms or laws existing in God's mind (a Platonic interpretation).

This conception of metaphysical knowledge as the illumination of divine archetypes enabled the scholastics to answer the question left by Plato as to how it was possible to acquire knowledge of an eternal reality when perceptions only disclose a transient world of imperfect things. Somewhat similar to Plato, Grosseteste drew an analogy between Plato's Good whose light illuminates the world and God whose spiritual light illuminates in humans the Divine Forms. As Grosseteste wrote,

> the highest part of the human soul, which is called intelligence . . . if it were not obscured and weighed down by the mass of the body, would itself have complete knowledge from the irradiation received from the superior light without the help of sense,

> just as it will have when the soul is drawn forth from the body, and as perhaps those people have who are free from the love and imaginings of corporeal things. (p. 73, brackets added)

At that time no one had any knowledge of the brain as the basis of consciousness and intelligence, so it was reasonable to imagine that the "intellectual soul" (Aristotle's term) as a separately existing entity could be a source of knowledge and continue to exist after death, while today there is nearly conclusive evidence that consciousness and intellectual capabilities are wholly dependent on the brain. Nor is there any justification for thinking that a divine illumination of the forms in God's mind can supplement scientific knowledge. Thus despite his many contributions to the renewal and advancement of science, Grosseteste could not free himself from preconceptions that would be discarded by modern scientists.

CONTRIBUTIONS OF LATER SCHOLASTICS

Before concluding this discussion, some recognition should be given to the contributions of later scholars following the death of Grosseteste in 1253 who also played a role in the gradual advance in scientific thinking. According to Crombie, the "writer who most thoroughly grasped, and who most elaborately developed Grosseteste's attitude to nature and theory of science was Roger Bacon" (ca. 1219–1292) (p. 139). While reaffirming most of Grosseteste's convictions about scientific method, according to Crombie he especially emphasized the importance of experimentation in acquiring new knowledge and opening up new fields of inquiry (cf. p. 142). Furthermore, he attributes greater significance to mathematics than Grosseteste stating that it "was the 'door and key of the sciences in this world'" because "all categories depend on a knowledge of quantity, concerning which mathematics treats, and therefore the whole excellence of logic depends on mathematics" (p. 143).

While he held like Grosseteste that the emanation of light creates the world, he denied Grosseteste's thesis that light is transmitted instantaneously even though this is not apparent in ordinary vision: "light is multiplied in time . . . nevertheless the multiplication does not occupy a sensible time . . . since anyone has experience that he himself does not perceive the time in which light travels from east to west"(p. 146). He made extensive investigations in optics, especially of the rainbow, and of the function of the eyes in vision. His theory of vision, according to Crombie, "was one of the most important written during the Middle Ages and it became a point of departure for seventeenth-century work" (p. 151).

Three other important scholars of the thirteenth century were John Pecham (d. 1292) and Witelo (b. ca. 1230) whose texts on optics were influential up to the seventeenth century, and John Duns Scotus (ca. 1265–1308) whose main contribution was to scientific methodology. The latter vigorously opposed Grosseteste's claim "that man could know nothing certain about the world without Divine illumination," a view he believed led to scientific skepticism (p. 168). In contrast, he argued that as the number of observed causal correlations increased, the principle of the uniformity of nature that "whatever occurs in a great many cases from some [natural] cause . . . is the effect of that cause," justifies believing it with certainty (p. 170; brackets added). This is a scientific principle relied on today, even if it does not guarantee certainty.

Guillaume de St. Cloud, also influenced by Roger Bacon, was known for his excellent astronomical observations that led to establishing in Paris an outstanding school of astronomers. His contemporary, the Franciscan astronomer Bernard de Verdun, who was more of a theoretician, "seems to have been the first medieval astronomer to decide firmly for the Ptolemaic as against the Aristotelian astronomical system" (p. 202). Even more significantly, Pietro d'Abana in his *Lucidator Astronomiae* written about 1310 "went so far as to suggest that the heavenly bodies were not borne on spheres but moved freely in space, and by the beginning of the fourteenth century the theory had been broached that the 'appearances' could be 'saved' by considering the earth instead of the heavens to be in motion" (p. 202). Nicole Oresme (ca. 1323–1382) even suggested that the planets moved with a relative motion in an infinite space, discarding the celestial orbs altogether (cf. p. 203). Thus began the dismantling of Aristotle's cosmological system and replacement by a mathematically based mechanistic one.

Just as Oresme contributed to the advance toward the Copernican revolution in astronomy, William of Ockham (ca. 1285–1349), by rejecting Aristotle's explanation that the cause of a moving object initially had to be in contact with the object, arguing instead that the light of the sun and the attraction of magnets were evidence of "action at a distance," prepared the way for Newton's theory of universal gravitation to explain planetary motion and the earth's attractive force. He also denied Aristotle's final cause, declaring that "all causes properly so-called are immediate causes" on the grounds that a final cause would have to produce its effect retroactively "when it does not exist," thus "this movement towards an end is not real but metaphorical" (p. 174). The denial of final causes was an important contribution because it directed scientists to look for actual physical causes rather than intentional causes, at least in the physical sciences.

Another major development in the early fourteenth century was the attempt to express functional relationships mathematically. It was Thomas Bradwardine, the real founder of the school of scientific thought associated with Merton College at Oxford who, in his *Tractatus Proportionum* in 1328, attempted to devise a mathematical method for describing functional relations (cf. pp. 178–79). In particular it was Aristotle's "Peripatetic law of motion," that the velocity of a moving object was proportional to the magnitude of the mover divided by the amount of the resistance (now easily stated in the equation $v = m/r$), that Bradwardine attempted to express. However, the formulation was complicated by the fact that the velocity did not occur when the resistance was greater than the power of the mover, only if there were an *excess* of power over the resistance, when "the proportion . . . was greater than 1" (p. 179). In attempting the mathematical formulation he devised a word-algebra using "letters of the alphabet instead of numbers for the variable quantities, but the operations of addition, division, &c., performed on those quantities were described in words instead of being represented by symbols as in modern algebra" (p. 178).

While allowing letters to stand for possible numbers that represent variable magnitudes was significant, he did not know how to connect the numbers by mathematical operators in equations representing the magnitudes. Nonetheless,

> his formulation of the problem [of motion] in terms of an equation . . . was an original and important contribution to mathematical physics in general and to dynamics in particular. Through him fourteenth-century natural philosophers, both in Oxford and Paris, got the beginnings of a conception of the use of mathematical functions in physics, and fifteenth-century Italy saw the beginnings of an experimental investigation not only of his dynamical function but also of the Peripatetic law of motion itself. (p. 180)

Thus the fourteenth and fifteenth centuries witnessed a marked advance in the investigation and conception of motion as requiring a different explanation from that of change. Change of place for Aristotle was just one of three kinds of changes, the other two being change of quality and quantity. As Clavelin asserts in his account of the scholastics' attempts to formulate a more adequate theory of motion prior to Galileo and Newton, it was necessary to distinguish motion as a state from alteration and change. What Aristotle called "local motion" or "change of place" was equivalent to

> alteration, growth, or diminution. Not only is local motion . . . not a state, but it cannot legitimately be studied in isolation from alteration or change in respect of quality. Hence the science of change is first and foremost the science of kinesis and to ignore this would be to risk losing sight of its true nature.[6]

Because Aristotle's account of motion did not included clear explanations of average motion, changes in motion due to acceleration and deceleration, nor instantaneous motion, this task confronted the scholastics before any progress in the explanation of motion could be made. As Clavelin states: "by 1320 Oxford physicists were treating velocity as an intensive magnitude, subject to intension and remission . . . and they went on to distinguish the quality of a motion (that is, its velocity) from its quantity (that is, the distance traversed)" (p. 65). By defining velocity as an *intensive* magnitude it was possible to treat it as the cause of the change in motion. Borrowing the concept of 'latitude' used by thirteenth century philosophers to characterize changing qualities in contrast to unchanging forms, the Mertonians used the concept to describe the intensive magnitudes in the process of change. This enabled them to speak of velocities as due to "increasing or decreasing latitudes or variations, and many of them used quantitative expressions from the outset" (p. 67).

It was John of Dumbleton (fl. 1331–1349) who may have introduced the initial terminology distinguishing between uniform and nonuniform motions. In his *Summa Logice et Philosophie Nauralis* he discussed measuring "latitudes of change" (*intensio*) in contrast to distance or time (*extensio*). Gradually a common terminology describing the various kinds of motion emerged due to the writings of William Hytesbury and Richard Swineshead (the famous "Calculator"), along with Dumbleton. According to Crombie,

> change was said to be 'uniform' when, in uniform local motion, equal distances were covered in equal intervals of time, and 'difform' when, in accelerated or retarded motion, increments of distance were added or subtracted in successive intervals of time. Such a 'difform' change was said to be 'uniformly difform' when equal increments were added or subtracted in successive intervals of time, and 'difformly difform' when unequal increments were so added or subtracted. (pp. 181–82)

Nicole Oresme would express these differences by his ingenious graphical representations and also by introducing the term '*velocitatio*' meaning a continuous increase of velocity, approximating the meaning of 'acceleration.' As Clavelin states: "Oresme's concept of *velocitatio* was a long way from the modern idea of acceleration," yet "it seems legitimate to translate *velocitatio* by acceleration (p. 69). While "it did not yet describe *the rate of change in velocity*, in the sense that classical mechanics later defined it, Oresme's *velocitatio* nevertheless referred to variations in speed as such, and hence transformed the latter into a distinct object of thought" (p. 69).

Furthermore, the Mertonians' and Oresme's treatment of speed as an intensive magnitude which changes from moment to moment implied that at any one moment the velocity must have a fixed or *instantaneous* intensity. Thus Clavelin concludes that the introduction of instantaneous velocity "must therefore be counted among the major contributions of medieval science" (p. 70). Yet being mainly confined to the mathematics of arithmetic and geometry, the scholastics could not define instantaneous velocity as the differential coefficient of space with respect to time (*ds/dt*), as the value of time approaches zero. This awaited the development of calculus by Newton and Leibniz in the seventeenth century. Still, Clavelin claims that the *calculationes* of the Mertonians and the *configurationes* of Oresme, which allowed medieval mathematicians to calculate the ratio of distance traversed to changes in speed, "represent the highlight of fourteenth-century mechanics and to provide clear proof of its great fruitfulness" (pp. 79–80). But Clavelin then asks the crucial question whether the Mertonians and Parisians

> were the first to express in clearly formulated language a great many of the ideas and principles that Galileo was to employ three centuries later and that, thanks to such concepts as *latitudo*, *velocitatio*, and *velocitas instantanea* and to such laws as that of the mean degree and that of distances, they laid the foundations of a science of motion which seventeenth-century scientists did no more than refine? (p. 84)

His answer is that those who claim this (such as Crombie) to be true are reading more sophistication and coherence into their contribution than is warranted. It might be thought that though the scholastics' mathematical treatment of motion did not rise to the level of Galileo, at least the theory of impetus introduced by Jean Buridan (1297–1358), as the forerunner of the concept of inertia, did approach that of seventeenth century mechanics. But even that is not true. It had been known since antiquity that one of the weakest aspects of Aristotle's explanation of motion was his treatment of so-called "unnatural" or "violent motion." In contrast to natural motions whose movement was inherent to them, forced, unnatural, or violent motions such as those of projectiles required an external "motive cause."

When thrown, what initially causes a stone or spear to move when released from the hand and produces the momentum to continue to move? What causes a potter's wheel to keep rotating when the propellant ceases? As with most familiar occurrences, we take their behavior for granted without knowing how it occurs. Plato had offered the explanation that when an object is propelled it creates a vacuum behind it so that the air rushing in to fill the vacuum continues its motion. The difficulty with this account is that the air acts both as a propellant and a resistant. Aristotle rejected this explanation proposing that when the hand propelled the object it compressed the air behind it whose expansion continued to move the object until its decompression, along with the natural tendency of the object to fall and the resistance of the air, causes it to come to rest. The flaw in this explanation is that in throwing a spear the hand is not behind the object to compress the air, therefore its decompression could not explain its continued motion.

In response to these objections, Buridan proposed dispensing with the air altogether claiming that the action of the projector "impressed an impetus" *in* the object producing the continued motion until overcome by the external resistance and gravity. As Buridan states quoted by Clavelin:

> Therefore, it seems to me . . . that the motor in moving a moving body impresses in it a certain *impetus* or a certain motive force . . . in the direction toward which the mover was moving. . . . And by the amount the motor moves that . . . body more swiftly, by the same amount it will impress in it a stronger impetus. It is by that impetus that the stone is moved after the projector ceases to move. (p. 92)

Thus the impetus impressed by the mover is proportional to the swiftness of the object and decreases relative to the strength of the resistance. Buridan adds that "in a dense and heavy body, other things being equal, there is more of prime matter than in a rare and light one. Hence a dense and heavy body receives more of that impetus" (p. 94). This would seem to be true because it takes more impetus to move a heavy object than a light one. But then the amount of the impetus impressed would be proportional not just to the velocity of the object, but also to what we would now call its mass, and Buridan referred to as its density and heaviness. Buridan's impressed impetus was so much simpler and plausible than Aristotle's impressed air that it gained wide acceptance.

But while his idea of impetus would seem to have preceded the modern concept of inertia, the fact that Buridan retained Aristotle's assumption that rest, not motion, was the eventual state of terrestrial objects, precluded this. Because like Aristotle Buridan believed that the *continued* motion of a propelled object after the propellant ceased had to have a *continuing* motive cause which he attributed to the 'impressed impetus,' he could not conceive of the object naturally continuing in motion if unimpeded. According to the law of inertia as stated by Newton, Buridan's *continuing* inertial force would *add* to the motion of the object, not just sustain it. Thus while he improved on Aristotle's explanation and his own account was a partial precursor of the modern concept of inertia, he did not fully anticipate the modern theory that an object in motion would continue in motion if in a vacuum or not impeded. Though he saw the weakness in Aristotle's view and partially emended it, he was not able to entirely free his thinking from Aristotle's presuppositions to arrive at a fully correct explanation, as was true also of the other Scholastics. This kind of gradual progression, however, is characteristic of modern science and not to be slighted, as indicated in Thomas Kuhn's accurate assessment:

> The centuries of scholasticism are the centuries in which the tradition of ancient science and philosophy was simultaneously reconstituted, assimilated, and tested for adequacy. As weak spots were discovered, they immediately became foci for the first effective research in the modern world. The great new scientific theories of the sixteenth and the seventeenth centuries all originate from rents torn by scholastic criticism in the fabric of Aristotelian thought. Most of those theories also embody key concepts created by scholastic science.[7]

NOTES

1. John Julius Norwich, *Byzantium: The Early Centuries* (New York: Alfred A. Knopf, 1999), p. 27.
2. Charles Freeman, *The Closing of the Western Mind: The Rise of Faith and the Fall of Reason* (New York: Alfred A. Knopf), p. 336.
3. R. Arnaldez and I. Massignon, "Arabic Science," in René Taton, ed., *History of Science: Ancient and Medieval Science* (New York: Basic Books Inc., 1963), pp. 385–86. This summary of Arabic scientific investigations is largely based on this work, as is the following quotation.
4. Charles Singer, *A Short History of Scientific Ideas to 1900* (New York: Oxford University Press, 1959), p. 153.
5. A. C. Crombie, *Robert Grosseteste and the Origins of Experimental Science* (Oxford: At The Clarendon Press, 1962; brackets added), p. v. Unless otherwise indicated, the following quotations are from this book.
6. Maurice Clavelin, *The Natural Philosophy of Galileo: Essay on the Origins and Formation of Classical Mechanics*, trans. by A. J. Pomerans (Cambridge, Mass.: The MIT Press, 1974), pp. 10–11. In the quotation I have substituted the English word 'kinesis' for the Greek word.
7. Thomas Kuhn, *The Copernican Revolution* (Cambridge: Harvard University Press, 1957), p. 122.

Chapter 8

ORIGINS OF THE CONCEPTION OF THE MODERN UNIVERSE

NICOLAS COPERNICUS

Nothing could mark the emergence of the modern conception of the universe as notably as Copernicus' heliocentric theory published as *De revolutionibus orbium coelestium* in 1543.[1] Although in the interest of simplicity Philolaus the Pythagorean and Aristarchus of Samos had previously proposed interchanging the sun and the earth and endowing the earth with two motions, an annual orbital revolution and a diurnal rotation on its axis, while as noted in the previous chapter the speculations of Pietro d'Albano and Nicole Oresme in the fourteenth century had "begun to make possible the conception of relative motion in an infinite, geometrical space," it was Copernicus' heliocentric theory that eventually prevailed, despite the continuing dominance of Aristotle's geocentric system at the time. Aristotelianism had the official backing of the Catholic Church throughout the seventeen century leading Newton to declare that the Aristotelians were one of his two main adversaries, Descartes being the other.

It has been argued that merely interchanging the sun and the earth and attributing two motions to the latter while retaining the rest of Aristotle's cosmology was not itself particularly revolutionary, but in fact Aristotle's system was so tightly integrated that any alteration tended to unravel the entire structure. How could the corruptible physical earth whose natural place was to be at rest in the center of the terrestrial or sublunar world possibly exist in the translunar realm composed of an unchanging, weightless aither conveying celestial bodies in eternal circular orbits; or how could an incorruptible celestial body like the sun whose natural motion was circular remain at rest in the center of the terrestrial world composed of the four elements? Even today Aristotle's cosmological distinction is unknowingly assumed when people talk about the "heavens and the earth" and "rising to heaven after death." But its destruction not only negated such terminology, its removal of humans from the center of the universe heralded the beginning of the gradual naturalization and secularization of human existence that has accompanied the fateful scientific effort to attain a more realistic understanding of the universe and the origin and nature of human beings in it.

This, of course, was not the intent of those who brought about the revolutionary changes. Those who contributed to this progressive naturalization were all believers in God, from Copernicus to Kepler, Galileo, Newton, Boyle, Priestly, Dalton, Lavoisier, Darwin, Louis Agassiz, and *perhaps* even more recent cosmologists such as Einstein and Stephen Hawking, and certainly Paul Davies. Following Adelard's admonition, they believed that in investigating the laws of nature they were rediscovering the laws that preexisted in God's mind before he imposed them on His creation.

Copernicus (1473–1543) was the nephew of Lucus Waczenrode who became the Bishop of Ermland in Poland, then a diocese of Prussia. Through the influence of his uncle he was elected a canon of the Cathedral of Frauenburg which, though he never became a priest, brought him a substantial prebend (stipend) for life and also an appointment as "Scholasicus of the Collegiate Church of the Holy

Cross in Breslau" that conferred another prebend, despite his never having visited Breslau nor performed any ecclesiastical dutie—such were the privileges or benefices of the Church. Although he began his studies at the University of Kracow, after attending for four years without earning a degree he left for Italy where he spent the next ten years studying in Italian universities which were the epicenter of the Renaissance.

He first enrolled in the University of Bologna in 1496 seeking a degree in canon law but took the opportunity to study with Domenico Maria da Novara, the distinguished professor of astronomy, who was known for his criticism of the Ptolemaic system because of its ad hoc complexity. In addition, he read the works of Nicholas of Cusa, George Peurbach, and Johanna Mueller, known as Regiomontanus, who also favored the heliocentric theory. Following a brief return to Ermland and a year's stay in Rome in the jubilee year 1500, he continued his studies at the University of Padua (where Galileo would later teach) and then earned the degree of Doctor of Canon Law in 1503 from the University of Ferrara. This was followed by two additional years of medical studies and though he never earned a medical degree he practiced medicine at Heibsberg Castle, the residence of his uncle.

He chose wisely to study in Italy because there he learned the leading theories of his day, since it was Italy that produced most of the founders of the impending scientific revolution: Vesalius (1514–1564), Fabricius (1537–1619), Galileo (1564–1642), and Harvey (1578–1657). After completing his formal studies about 1505 he returned to Ermland at age thirty-one, but after the death of his uncle in 1512 he settled in Frauenburg where he began his astronomical observations. Not especially enthusiastic about stargazing, he is reputed to have only made between sixty and seventy observations, about half of which were recorded in his *De revolutionibus.* Yet anyone who has perused his book must have been impressed by the numerous tables recording the astronomical data, detailed geometrical diagrams supporting his arguments, and acute critical comparisons of the competing astronomical systems.

It was at Frauenburg that he decided to write and circulate among his friends a preliminary version of his heliocentric theory titled *Commentariolus* (*Little Commentary*). Extremely fearful of the ridicule his heliocentric theory would arouse, he was hesitant to pursuer further publication. This apprehension apparently was justified considering the caustic denunciation by Martin Luther in 1639, a colleague of Rheticus at the University of Wittenberg, of the *Commentariolus.*

> People give ear to an upstart astrologer who strove to show that the earth revolves, not the heavens or the firmament, the sun and the moon. . . . This fool wishes to revise the entire science of astronomy; but sacred Scripture tells us [Joshua 10:13] that Joshua commanded the sun to stand still, and not the earth.[2]

Yet encouraged by the favorable response of several friends who read it, particularly Bishop Tiedmann Geise and Cardinal Nicolaus von Schönberg, whose letter entreating him to publish his theory was included with Osiander's Preface, along with the Dedication to Pope Paul III, in the eventual publication of *De revolutionibus*, Copernicus continued his research and the preparation of his manuscript. But the person whose influence was crucial was Rheticus, a German Lutheran who was a young professor of mathematics and astronomy at the University of Wittenberg (Hamlet's University). Learning of Copernicus' revolutionary manuscript, he took a leave of absence in the spring of 1539 to visit Copernicus at Frauenburg hoping to convince him to publish it, even offering to help prepare the manuscript for that purpose. Failing to persuade him at that time, Rheticus did succeed when he returned in the summer of 1540 staying with Copernicus to September of 1541 helping to ready the manuscript for printing.

Diligently reading the entire manuscript checking the tables and figures for accuracy, he even recopied it by hand to take to the printer at Nüremberg. Unfortunately, however, he was not able to supervise the final printing being forced to leave Wittenberg to accept a position at the University of Leipzig to avoid a scandal due to his sexual orientation. Thus the task fell to Osiander, another colleague of Rheticus, Luther, and Melanchthon at Wittenberg University, who inserted the contentious Preface "To the Reader on the Hypothesis in this Work." Although well intended, it was misleading because most people, except for a few such as Copernicus' friend Bishop Giese and Kepler, assumed it was written by Copernicus himself since Osiander did not acknowledge his authorship.

Perhaps reacting to the scathing statement made by Luther, he may have thought it prudent to claim that the heliocentric theory presented in the *De revolutionibus* was purely conjectural to facilitate calculations and predictions, not intended as a true theory.

> For it is the job of the astronomer to use painstaking and skilled observation in gathering together the history of the celestial movements and then—*since he cannot by any line of reasoning reach the true causes of those movements*—to think up or construct whatever causes or hypotheses he pleases such that, by the assumption of these causes, those same movements can be calculated from the principles of geometry for the past and for the future too. . . . For it is sufficiently clear that *this art is absolutely and profoundly ignorant of the causes of the apparent irregular movements*. And if it constructs . . . causes . . . nevertheless *it does not think them up in order to persuade anyone of their truth but only in order that they may provide a correct basis for calculation*. . . . [So] if anyone take as true that which has been constructed for another use, he will go away from this discipline a bigger fool than when he came to it.[3] (Brackets and italics added.)

How Copernicus responded to Osiander's Preface is unknown because he suffered a stroke in 1542, the year before his great book was published, and died the following year. According to tradition he received a copy only as he lay stricken on his deathbed.

In one sense Osiander's *apologia* is correct in that Copernicus' treatise was devoted mainly to devising a simpler *description* of the celestial motions, not with explaining the *cause* of the motions; yet in another sense it was misleading because it does not reflect Copernicus' declared belief that by interchanging the positions of the earth and the sun and attributing two motions to the earth he was able to derive a more integrated and harmonious account of the movement of the planets than alternative systems. As he states in his Preface and Dedication to Pope Paul III (the same pope who commissioned Michelangelo to paint the Sistine Chapel):

> And so I am unwilling to hide from Your Holiness that nothing except my knowledge that mathematicians have not agreed with one another in their researches moved me to think out a different scheme of drawing up the movements of the spheres of the world. . . . Moreover, they have not been able to discover or to infer the chief point of all, i.e., the form of the world and the certain commensurability of its parts. But they are in exactly the same fix as someone taking from different places hands, feet, head, and the other limbs—shaped very beautifully but not with reference to one body and without correspondence to one another—so that such parts made up a monster rather than a man. (p. 5)

Then, adding that he had discovered that others in the past had attributed several movements to the earth, he decided to do the same.

> Therefore I also, having found occasion, began to meditate upon the mobility of the Earth. And although the opinion seemed absurd, nevertheless because I knew that others before me had been granted the liberty of constructing whatever circles they pleased in order to demonstrate astral phenomena, I thought that I too would be readily permitted to test whether or not, by the laying down that the Earth had some movement, demonstrations less shaky than those of my predecessors could be found for the revolutions of the celestial spheres. (p. 6)

He next presents the advantages that this approach offered.

> And so, having laid down the movements which I attribute to the earth farther on in the work, I finally discovered by the help of long and numerous observations that if the movements of the other wandering stars are correlated with the circular movement of the Earth, and if the movements are computed in accordance with the revolution of each planet, *not only do all their phenomena follow from that but also this correlation binds together so closely the order and magnitudes of all the planets and of their spheres or orbital circles and the heavens themselves that nothing can be shifted around in any parts of them without disrupting the remaining parts and the universe as a whole.* (p. 6; italics added)

This clearly shows that he definitely believed his astronomical system presented a more harmoniously integrated description of the planetary motions than that of the existing alternatives! He adds that only mathematicians are competent to evaluate his conclusions and thus leaves the final judgment of its worth to "Your Holiness in particular and to that *of all other learned mathematicians*" (p. 7; italics added). In not appealing to the Pope as an ecclesiastical authority, but as a mathematician, this could have represented a turning point in the history of knowledge by claiming the autonomy of science in such matters. Yet the issue was not to be so easily decided, as the disgraceful treatments of Galileo and Giordano Bruno by the Catholic Church will illustrate.

As the previous quotations and discussion indicate, it was the disagreements among the mathematical astronomers, the uncertainties in the calculations and predictions of the movements of the sun and moon so that even computing the length of the months and the solar year was imprecise, along with the use of diverse principles and assumptions in the demonstrations resulting in a "monstrous" system, that persuaded Copernicus of the need for a more coherent theory. In particular, it was the fact that to preserve the uniform angular motion of the planets Ptolemy had to locate their center on an equant, a point slightly removed from the center of the universe, rather than from the center itself, that especially vexed Copernicus.

Furthermore, the discovery that the orbital motions of all the planets except the sun and the moon are not continuously circular, as Aristotle held, but interrupted by a loop-like pattern where they reverse their trajectory and then move forward again requiring the introduction of deferents and epicycles to depict their trajectories, was another unsettling factor. In addition, more accurate observations disclosed that Mercury and Venus revolve around the sun supporting the heliocentric theory, while the discovery that Mars varies in brightness and velocity implied that its orbit around the earth could not be exactly circular. But since the circularity of the planetary orbits was an irrevocable assumption of Copernicus, such counterevidence convinced him that the deviations must be deceptions owing to the geocentric perspective.

Furthermore, by attributing to the smaller earth a daily axial rotation from west to east he could eliminate the inequitable diurnal rotation of the entire universe from east to west while also establishing the unified direction of the movements of all the celestial bodies, a significant simplification. In addition, the magnetic attraction between the moon and the earth might even explain the tides, presaging gravitational forces. But it was his belief that by describing the motions of the planets in reference to the central sun their respective distances and orbital velocities could be more harmoniously integrated so "that nothing can be shifted around in any parts of them without disrupting the remaining parts and the universe as a whole" (p. 6), that was the decisive factor. Though he still utilized Ptolemy's deferents and epicycles to "save the appearances," by adopting the heliocentric view he was able to reduce Ptolemy's 80 circles to 34, a considerable simplification. Finally, that his theoretical revisions produced more precise calculations and accurate astronomical predictions facilitating needed calendar reform was an additional significant benefit.

But accepting the fact that our normal astronomical observations were deceptive due to the unfelt motions of the earth required a tremendous adjustment in thinking, even though it had its presidents. As Herbert Butterfield states in his classic work on *The Origins of Modern Science,*

> of all forms of mental activity, the most difficult to induce . . . is the art of handling the same bundle of data as before, but placing them in a new system of relations with one another by giving them a different framework, all of which virtually means putting on a different kind of thinking cap for the moment.[4]

As indicated previously, the heliocentric theory adopted by Copernicus retained crucial astronomical assumptions of Aristotle, such as uniform motion and circular orbits, yet its appeal over previous astronomical systems was due to the mathematical precision of its presentation and its greater harmony and simplicity owing to the elimination of some of Ptolemy's epicycles and deferents. Still, his retention of the two major assumptions of the older cosmological systems, uniform motions and circular orbits, shows how difficult it is to cast off previous conceptions. Nonetheless, declaring that the sun was far more worthy to be placed in the center of the universe than the earth was a major breakthrough even though ironically, considering his strong aversion to Ptolemy's use of the equant, his own system was not quite heliocentric since "all the movements of the skies were reckoned not from the Sun itself, but from the centre of the earth's orbit, which came somewhat to the side."

Two other defects of his system was the absence of a mathematical correlation of the orbital dimensions of the planets and their distances from the sun, and the lack of an adequate explanation of how the sun at the center of the universe caused the motion of the planets. There is no hint of a dynamic explanation in terms of physical forces and astronomical laws involving mechanistic principles; instead, Copernicus relies on the ancient Aristotelian conception that uniform circular motion is inherent to a spherical body. Even the earth acquired its two circular motions, rotational and orbital, due to its being a sphere despite its material and changeable nature. We have to await Kepler's adoption of elliptical orbits and nonuniform motion as described in his first two astronomical laws, along with his third law correlating the planets' orbital sizes and speeds with their distances from the sun due to diminishing forces radiating from the sun, before the ancient cosmology could be replaced with modern celestial mechanics. Still, it was Copernicus who formed a bridge between the ancient and modern astronomical worlds heralding a new age.

JOHANNES KEPLER

It has been said that Kepler believed that Copernicus did not exploit his astronomical innovations sufficiently because he remained too close to Ptolemy, but this certainly was not true of Kepler himself whose marvelous astronomical discoveries began the process of dismantling the ancient conception of the solar system replacing it with the modern one. Although his personal life was marked with tragedy, suffering, and financial struggles, these were offset somewhat by his extraordinary intellectual gifts. Like other geniuses, he recalled events in his childhood that foretold the direction his later life would take, such as the sight of a great comet when he was six years old and of a lunar eclipse at age nine.

He was born in the medieval town of Weil der Stadt in Swabia in 1571, eighteen years after Copernicus' death and the publication of the *De revolutionibus.* Shortly thereafter his parents moved to Leonberg in the duchy of Württemberg where, in recognition of his precociousness, he was allowed to attend the Latin rather than the inferior German school. Completing his studies, he entered the convent school of Adelberg at thirteen and two years later the seminary at the Cistercian monastery, Maulbroon. An outstanding student, he was praised not only for his intellectual ability, but also for his admirable character and his ardent religious disposition and observances.

After completing his studies at the monastery he entered the seminary at the University of Tübingen whose Spartan regime required the students to arise at 4:30 in the morning, maintain a strict religious discipline throughout the day, and pursue a rigorous course of studies. Though these courses were mainly theological, he also attended lectures on ethics, dialectics, rhetoric, Greek, Hebrew, astronomy, mathematics, and physics. Graduating at the age of twenty he pursued further studies for four years at the Faculty of Theology earning a Master's Degree in 1591, graduating second among fourteen candidates. His professors were so pleased with the quality of his work that they approved a renewal of his fellowship aid on the grounds that he "has such a superior and magnificent mind that something special may be expected of him," thus "we wish on our part, to continue . . . his stipend, as he requests. . . ."[5]

At the university he read in Greek Aristotle's *Analytica Posteriora*, *Physica*, and *Meteorologica*, finding the latter the most stimulating. Avowing that he "loved mathematics above all other studies," he was especially attracted to the writings of Pythagoras, Plato, and the Neoplatonist, Nicholas of Cusa. But it was Julius Caesar Scaliger's *Exercitationes exotericea* that especially aroused in the young scholar questions in theology, philosophy, and physics. He also studied with Magister Michael Maestlin, professor of mathematics and astronomy, who lectured on Euclid, Archimedes, and Apollonius, along with current developments in trigonometry. One of the leading astronomers of the day, he used his own textbook, *Epitome Astronomie,* in his course on astronomy taken by Kepler, although at the time Kepler did not display the enthusiasm for astronomy that he would later. It was Maestlin's lectures that introduced Kepler to the Copernican heliocentric system, though perhaps for reasons of prudence Maestlin declared his preference for Ptolemy's geocentrism.

But the same simplicity and promise of an integrated orbital structure that appealed to Copernicus also attracted Kepler who, unlike Maestlin, became an ardent supporter of heliocentricism. Yet the two men remained friends for life, the older professor helping to secure Kepler's first appointment while his favorable comments on Kepler's first book was crucial for its publication. Along with his growing interest in astronomy, like famous astronomers at the time, such as Tycho Brahe, Kepler maintained a strong interest in astrology having cast horoscopes with considerable success since a young man. Despite his interest in these subjects his major focus of studies at the time was theology. But this changed when the Faculty Senate at Tübingen recommended him for the position of professor of mathematics at the Protestant Seminary in Graz, in the province of Styria, Austria. Taken by surprise at the offer, after some hesitation he accepted and left for Graz in 1594 at age twenty-three.

Because there was little interest in either mathematics or astronomy among the seminarians, Kepler revised his teaching program to include other courses, such as rhetoric, history, and ethics. Having become the District Mathematician he was also involved in the issuing of calendars which, in those days, were like almanacs containing all kinds of useful information regarding the prediction of the weather, harvests, plagues, auspicious days for ceremonial or religious occasions, and prospects for war and peace. Although he continued his astrological forecasts he seemed ambivalent as to their authenticity, decrying "the customary rules and prophecies as horrible superstition, as 'necromantic monkeyshines,' but . . . held just as positively to the conviction of an influence of the stars on earthly events and human fate, a conviction that one cannot imagine as absent from his view of nature" (p. 58).

As important as these occupations were, they were incidental to his growing interest in astronomy and more ardent defense of the Copernican heliostatic universe. What particularly appealed to him was the crucial position and role of the sun in Copernicus' system, suggesting the possibility of correlating and explaining the relative distances and sizes of the planetary orbits due to the sun's influence. As mentioned previously, it was the fact that Copernicus' innovation raised the likelihood of asking such questions is what made it so significant. Thus Kepler began inquiring why there were just six planets, Mercury, Venus, Earth (with the Moon as its satellite), Mars, Jupiter, and Saturn, what accounts for their distances from the sun, the sizes of their orbits, and the lengths of their orbital periods. Why, for instance, does it take Saturn thirty years to complete its orbit, rather than twenty-four as one would expect since it is twice the distance of Jupiter whose orbital period is only twelve years? Unlike Copernicus, whose fame consists in defending a novel astronomical system which he accepted as given, by adopting the system and examining it more closely Kepler could ask why it existed as it did, illustrating how scientific inquiry advances.

Unlike religious explanations, which are considered final and infallible because they are based on God's revelations and creations, Kepler's questioning shows why science has progressed as it has because new discoveries and theoretical innovations raise new problems calling for new explanations. Yet this does not mean that religious devotion and scientific inquiry are necessarily incompatible; in fact, it was Kepler's belief that God would have created the universe according to the most elegant plan that also led him to ask the questions he did.

What initially guided him in his explanation was Plato's belief that the Divine Craftsman or Demiurge constructed the universe from geometrical proportions. Furthermore, as described in the *Timaeus* and discussed previously in the chapter on Plato, he assigned a three-dimensional geometrical configuration to each of the four elements. As three of these geometrical entities were constructed from similar triangles, their decompositions and reconstructions due to their interactions were explained as different configurational exchanges of these triangles.[6] The five polyhedrons or perfect geometrical solids composing the entire universe, in their order of creation, were the tetrahedron, octahedron, icosahedron, the cube, and the dodecahedron. Plato assigned the first four respectively to fire, air, water, and earth, the last to the whole cosmos.

But the real significance for Kepler in there being just five Pythagorean or Platonic solids was to explain that there had to be just six planets because there were only five perfect solids to join them, evidence that this did not occur by chance but must have been ordained by God. As described by Arthur Koestler, they had

> to be spaced in such a manner that the five solids could be exactly fitted into the intervals as an invisible skeleton or frame. And lo, they fitted! Or at least, they seemed to fit, more or less. Into the orbit, or sphere, of Saturn he inscribed a cube; and into the cube another sphere, which was that of Jupiter. Inscribed in that was the tetrahedron, and inscribed in it the sphere of Mars. Between the spheres of Mars and Earth came the dodecahedron; between Earth and Venus the icosahedron; between Venus and Mercury the octahedron. Eureka! The mystery of the universe was solved by the young Kepler, teacher at the Protestant school in Graz.[7]

Then realizing that plane geometrical figures were inadequate to represent spatial dimensions and recalling that Euclid had proven that only five polyhedrons could be constructed in a three-dimensional space, Kepler inferred that because five spatial intervals separated the six planets God must have used the five geometrical configurations to separate the six planets. As he wrote in his first book:

> It is my intention, reader, to show in this little book that the most great and good Creator, in the creation of this moving universe, and the arrangement of the heavens, looked . . . to those five regular solids, which have been so celebrated from the time of Pythagoras and Plato down to our own, and that he fitted to the nature of those solids, the number of the heavens, their proportions, and the law of their motions.[8]

Momentarily at least, he was convinced that he had discovered a priori the underlying principle correlating the three dimensions of space, the five perfect solids, and the six planets. As a Christian Platonist he believed this could not be accidental because any geometrical order of the universe must have preexisted in God's mind before creation. This also agreed with his Neoplatonic conception that it was possible for human thought to replicate Divine thought. (Einstein was a recent example of this when he referred to the thought of the "Old One" as the ultimate referent of his theories.) To complete the deduction Kepler showed how, also in the *Mysterium Cosmographicum* translated as *The Secret of the Universe*, the various orbital distances, sizes, and speeds could be derived from his principle.

> The Earth is the circle which is the measure of all. Construct a dodechahedron [a polyhedron composed of twelve pentagons] round it. The circle surrounding that will be Mars. Round Mars construct a tetrahedron [a pyramid consisting of four equilateral triangles]. The circle surrounding that will be Jupiter. Round Jupiter construct a cube. The circle surrounding that will be Saturn. Now construct an icosahedron [composed of twenty equilateral triangles] inside the Earth. The circle inscribed within will be Venus. Inside Venus inscribe an octahedron [consisting of eight equilateral triangles]. The circle inscribed within that will be Mercury. There you have an explanation of the number of Planets [and their orbital distances from the sun]. (p. 69; brackets added)

Although at first Kepler was enraptured with the results, his studies in astronomy and mathematics overrode his religious and Neoplatonic convictions, forcing him to acknowledge that a priori reasoning was inadequate when it led to results negated by observations. After declaring originally that the five Platonic solids were "perfect" representations of the orbital intervals, when he compared his derivations with those of Copernicus he discovered discrepancies. Yet his a priori conjecture did have the positive effect of convincing him that the sun, rather than the Earth's orbit, should be taken as the center of the universe. Thus he began searching for more exact mathematical correlations of the orbital dimensions, calculated from the sun based on the astronomical evidence: that Mercury's period is about eighty days, Venus' seven and a half months, Mars' two years, Jupiter's twelve years, and Saturn's thirty years. That his original false correlation led him to search for a more exact *explanation* of the orbital periods based on the centrality and influence of the sun was the crucial turning point. As Koestler states:

> It would be difficult to over-estimate the revolutionary significance of this proposal. For the first time since antiquity, an attempt was made not only to *describe* heavenly motions in geometrical terms, but to assign them a *physical cause.* We have arrived at the point where astronomy and physics meet again, after a divorce which lasted for two thousand years. This reunion . . . led to Kepler's three Laws, the pillars on which Newton built the modern universe. (p. 258; italics in original)

Here again Kepler replaced a less credible explanation for a more plausible one, rejecting either "intelligences" or "angels" as explanations of the motion of the planets, though still referring to the sun as a "soul." Instead, he adopted a more naturalistic conception explaining the decreasing motions of the planets the farther they are from the sun as due to a lessening of an influence from the sun.

> But if . . . we wish to make an even more exact approach to the truth, and to hope for any regularity in the ratios [between the distances and the velocities of the planets], one of two conclusions must be reached: either (1) the moving souls are weaker the further they are from the sun; or (2) there is a single moving soul in the center of all the spheres, that is, in the Sun, and *it impels each body more strongly in proportion to how near it is.* In the more distant ones on account of remoteness and the weakening of its power, it becomes faint so to speak. Thus, just as the source of light is in the Sun . . . so . . . the motion and the soul of the universe are assigned to that same Sun. . . . (p. 199; brackets and italics added)

While Kepler still retained the ancient concept of force as vitalistic or animistic, referred to as the "soul" or "*anima motrix*," his comparison of it to the rays of the sun that become less concentrated as they are dispersed represents an advance to a more plausible interpretation. Yet his religious beliefs were still so dominate that he described the cosmic arrangement of the Sun, the Heavens, and the Motive Force as analogous to God the Father, Christ the Son, and the Holy Spirit (cf. p. 63). When the *Mysterium Cosmographicum* appeared in the spring of 1597 Kepler sent copies to the leading astronomers, including Maestlin, Tycho Brahe, and Galileo with varying responses. But he was adamant as to its importance, stating in the Dedicatory Epistle to the second edition published twenty-five years later in 1621, that "almost every book on astronomy which I have published since that time could be referred to one or another of the important chapters set out in this little book. . . ."

The most consequential response came from Tycho Brahe who recognized the brilliance of the author and which led three years later to their unusual collaboration. In the intervening three years Kepler studied mathematics seeking to derive an exact correlation of the dimensions of the planetary orbits with their distances from the sun. To achieve this he needed accurate measurements of the eccentric orbits of the planets which at the time were in the possession of Tycho Brahe, considered the greatest living naked-eye observer, who was carefully guarding them for his own attempt to construct an accurate astronomical system. Still, Tycho had written Kepler graciously stating that he hoped that "some day" Kepler would be able to visit him.

Then, in the sort of extraordinary coincidence that tempts people to believe in a controlling destiny or fate, Tycho was offered the position of Imperial Mathematicus by Emperor Rudolph II, that he readily accepted. After settling in Benatky Castle near Prague, he wrote Kepler in December of 1599 asking him to join him in February of the new century to collaborate in their "common study." As the Protestant seminary in Graz had just been closed forcing Kepler to seek a new position, the offer from Tycho seemed to be heaven-sent since it satisfied two pressing needs, one of employment and the other access to Tycho's astronomical observations.

Despite the gratifying arrangement, it brought together two distinctly contrasting individuals. As vividly described by Koestler:

> Tycho was fifty-three, Kepler twenty-nine. Tycho was an aristocrat, Kepler a plebeian; Tycho a Croesus, Kepler a church-mouse; Tycho a Great Dane, Kepler a mangy mongrel. They were opposite in every respect but one: the irritable, choleric disposition which they shared. The result was constant friction, flared into heated quarrels, followed by half-hearted reconciliations. (p. 302)

As if that were not enough, this personal incompatibility was heightened by the resentment of Tycho's son-in-law and co-workers at the arrival of the obviously more talented research assistant. Still, accommodations were made and Kepler was given the assignment of measuring the orbit of Mars whose particular eccentricity could be the key to determining the dimensions of the orbits and their distances from the sun. Boasting that he would solve the problem in eight days, it took him eight years and nearly a nervous breakdown before he succeeded.

Then his collaboration with Tycho was suddenly terminated when the latter unexpectedly died a year and a half later. However, this turned out to be advantageous for Kepler because two days after Tycho's death he was appointed his successor as Imperial Mathematicus by Rudolph II, a position he held for eleven years until 1612 when the Emperor died. Despite his salary often being in arrears in his new position, the security it provided, along with Kepler's ability to obtain Tycho's astronomical observations after his death, made those his most productive years.

When working with Tycho's measurements of the eccentrics and mean distances of the planetary orbits from the sun, he became convinced that the solution to the mystery of their connection lay in the eccentric orbit of Mars. The closer he examined the data the more he realized that Mars' variations in brightness and velocity indicated that its orbit was not circular but egg-shaped or oval and that its speed increased at the perihelion, when it was nearest the sun, and decreased at the aphelion, when it was farthest away. Although this discovery contradicted the entrenched belief that the orbital motions of the planets were circular and uniform, it did conform to Kepler's causal explanation in the *Mysterium* that when closer to the sun the rays striking the planets would be denser and therefore produce a greater "impelling" force than at the aphelion where they would be more dispersed and have less impact. Thus given his confidence in Tycho's astronomical data and his own causal explanation of the planet's motions, he was ready to reject the venerable nonuniform motions, though not yet willing to reject circular orbits.

Even more important, his readiness to replace souls or the *anima motrix* as the cause of the orbital motions with a natural "force" (*vis*) indicated that he was moving closer to a purely physical explanation of the cause of the planetary motions, as stated.

> If for the word "soul" you substitute the word "force," you have the very same principle on which the Celestial Physics is established. . . . For once I believed that the cause which moves the planets was precisely a soul . . . [b]ut when I pondered that this moving cause grows weaker with distance, and that the Sun's light also grows thinner with distance from the Sun, from that I concluded that this force is something corporeal, that is, an emanation which a body emits, but an immaterial one. (p. 203, note 3)

The equivocation between the words 'corporeal' and 'immaterial' seems to indicate that Kepler intended to eliminate the spiritual connotation of the world 'force,' but not yet equate it with materiality. Still, this shows an exceptional ability to revise his ideas when they did not conform to objective facts. Unlike Copernicus who, as he said, remained largely captive to the previous tradition, Kepler gradually shed his earlier Hellenic and medieval interpretations for more naturalistic ones when confronting the evidence. *Thus his intellectual development recapitulated that of human history, demonstrating that such cognitive changes in individuals are possible given a certain open-mindedness and the recognition of the necessity of changing one's ideas to conform to the evidence, however counterintuitive they might first appear.*

In a book published between the first and second editions of the *Mysterium* titled *Astronomia Nova*, Kepler indicated that he realized that a single impelling force emanating from the sun would not be sufficient to explain what he now accepted as the ovoid shape of the planetary orbits, thus a second force was necessary. In 1600 William Gilbert published a book with the abbreviated title, *De Magnete,* in which he described the planets and the sun as having north and south magnetic poles at the opposite sides of their circumference.[9] After reading the book Kepler surmised that if all the planets, along with the sun, were polar magnets, then as the planets revolved in their orbits their polarity would reverse direction relative to the sun, so that as they orbited they would be either attracted or repelled depending upon the mutual orientation of their poles. Thus the two forces from the sun would explain the planets' ovoid pattern: the emanating force analogous to the sun's rays would account for the impelled motion while the magnetic forces would partially account for the ovate shape of the orbit and for the variation in the planet's speed, according to whether it was attracted or repelled by the sun. These revisions in his thinking were sufficient for him to be able to formulate his two laws of motion paving the way for the final theory.

Having concluded that the force of the sun decreases with the distance, he proposed his inverse speed law "that the force from the Sun varies inversely with the distance," an approximation to Newton's Law that it varies inversely with the *square* of the distance. Though imprecise, this law provided a complementary explanation of why the planets' speed varies as the alternating magnetic poles change its distance from the sun. This in turn enabled him to formulate his second astronomical law stating that a radius vector joining the planet to the sun sweeps out equal areas in equal times, replacing the ancient law of uniform circular motion by one in which the areas, not the speeds, vary uniformly with the time.

Although these developments were based on the assumption that the orbits were oval, an approximately elliptical shape, its designation was not geometrically precise. In his mathematic courses

Kepler had studied Apollonius' work on conic sections, so he knew that an ellipse was formed by a diagonal slice through a cone, but still did not identify the oval as an ellipse, even writing to Fabricius at one point that "if only the shape were a perfect ellipse all the answers could be found in Archimedes' and Apollonius' works."[10] This kind of trance-like groping for solutions is what led Koestler to describe the founders of modern science as "Sleepwalkers."

However, after measuring the exact "inner curve" of Mars's orbit (.00429 of the radius) Kepler finally conceded that its shape was elliptical, deriving his first astronomical law that the planetary orbits are elliptical with the sun as one of the foci. Now, finally, his two laws provided a simpler, consistent explanation that eliminated the ancient artifacts of epicycles, eccentrics, and equants. Although he had anticipated these laws in the earlier *Commentaries on the Movements of Mars*, they explicitly appeared in his *magnus opus*, the *Astronomia Nova* published in 1609 (the year Galileo began his astronomical observations) dedicated to the Emperor Rudolph II, his patron. For its presentation of the first astronomical laws ever discovered, along with its explanation of the planetary motions by physical forces, it is acclaimed as the "first modern treatise on astronomy."

He then made the momentous decision to replace his previous impelling force emanating from the sun analogous to light to a gravitational force analogous to magnetism, a naturally *attractive* force exerted by objects proportional to their masses that produces the planets orbital motion, justifying the claim that he introduced a new astronomical framework, celestial mechanics. As he states in the *Astronomia Nova*:

> Gravity is the mutual bodily tendency between cognate bodies towards unity or contact . . . so that . . . *if two stones were placed anywhere in space near to each other, and outside the reach of force of a third cognate body, then they would come together, after the manner of magnetic bodies, at an intermediate point, each approaching the other in proportion to the other's mass.*[11]

Yet since he maintained that the gravitational force varied directly with the distance, he had not yet arrived at the correct law that the gravitational attraction from the sun varies with the *square* of the distance. Earlier in his *Optics* published in 1604, he had argued that the strength of *light radiation* decreased inversely with the *square* of the distance, and though now he replaced an impelling force with an attractive one, this should have suggested a similar decrease in the strength of gravity with the distance.

However, as Koestler notes, he did offer the prescient interpretation that rather than being dispersed in space like light, gravity was more like a kind of electromagnetic or Einsteinian field, a structure within (or of) space that produces the object's motion. This force, he stated, "cannot be regarded as something which expands into the space between its source and the moveable body, but as something which the movable body receives out of the space which it occupies. . . ."[12] Incredibly, this reads like an anticipation of Faraday's later conception of magnetism as not analogous to a moving fluid, but as a field extending in or of space. It may have been this remarkable interpretation, that instead of gravity emanating in space it is more like a static field, which prevented him at this time from realizing that the gravitational force deceases with the square of the distance, not directly with the distance.

This conception of gravity enabled him to explain the cause of the tides as due to the combined gravitational forces of the Moon and the Earth (an explanation that was mistakenly rejected by Galileo). As he further states in the Introduction to the *Astronomia Nova*: "if the Earth ceased to attract the waters of the sea, the seas would rise and flow into the Moon. . . ." From these conclusions he drew the amazing inference that the cosmos was a "heavenly machine" that "runs like clockwork" according to purely physical laws, essentially the view arrived at by Newton. As he wrote in a letter to his friend Herwart regarding the *Nova*, quoted by Koestler:

> My aim is to show that the heavenly machine is not a kind of divine, live being, but a kind of clockwork . . . insofar as nearly all

> the manifold motions are caused by a most simple, magnetic, and material force, just as all motions of the clock are caused by a simple weight. And I also show how these physical causes are to be given numerical and geometrical expressions. (p. 340)

What an astonishing progression of thought! He not only denies his earlier conception of the universe as a "kind of divine being," but now explicitly refers to the force causing the planetary motion as "material," without any equivocation, and the total system as "a kind of clockwork," predating Newton. This purely mechanistic explanation was exemplified in the earlier *Optics* published in 1604 and in his later *Dioptrics* published a year after the *Nova*.

In 1612 the death of Rudolph II forced Kepler to seek a new position. After considering a number of alternatives he chose the post at Linz of Provincial Mathematicus to the new Emperor Matthias. Once settled there he returned to his scholarly writing. Despite his two new astronomical laws he still had not published the exact mathematical ratio relating the planetary periods to their distances from the sun. He returned to this task in the *Harmonice Mundi* published in 1619, at first regressing to his earlier Pythagorean vision of a musical harmony orchestrating the celestial orbs, ignoring his recent explanations in terms of ellipses, material forces such as magnetism and gravity, and a clockwork mechanism. Yet despite this regression, his unfailing mathematical intuition directed him to the correct mathematical law of gravitation that had alluded scientists throughout the ages.

By carefully comparing Tycho's measurements of the ratios of the periods of pairs of planets with the ratios of their mean distances from the sun he concluded that the squares of the former are equal to the cubes of the latter: $(p^1/p^2)^2 = (d^1/d^2)^3$ or the $(\text{planetary year})^2 \propto (\text{distance from the Sun})^3$, so that the periods of the revolutions vary with the 3/2th power of their distances. As he states in *Harmonice Mundi*: "it is certain and exact *that the ratio which exists between the periodic times of any two planets is precisely the ratio of the 3/2th power of the mean distances, i.e., of the spheres themselves. . . .*"[13] It was this ratio that enabled Newton to infer that the force diminished with the square of the distance.

In 1621 he published a text describing his latest discoveries titled *Epitome Astronomiae Copernicanae* in homage to Copernicus which succeeded Ptolemy's *Almagest* and Copernicus' *De Revolutionibus* as the greatest astronomical treatise of the past. It exceeded the *Astronomia Nova* in that his three laws are no longer restricted to the orbit of Mars but generalized to all the planets, along with the Moon and the satellites of Jupiter. (He later introduced the term 'satellite' to designate what Galileo called the "Moons of Jupiter" when he first discovered them.) At long last it was possible to predict the planets' orbital rotations with considerable accuracy, along with the ratio between their distances from the sun and their orbital periods. It remained for Galileo to confirm the results with his telescopic observations and for Newton to complete the celestial mechanics with his universal law of gravitation.

Yet there remained still one last unfulfilled project, the *Rudolphine Tables*, which for many years he had been promising to write but never found the time. John Napier had published in 1614 a well-known book, *Merifici Logarithmorum Canonis Descriptio*, that contained the logarithmic tables that had considerably simplified astronomical calculations but had not described how they had been computed. Aware of its importance, Kepler began writing in the years 1621–1622 a book that not only contained the logarithmic tables along with the rules and instructions for their use, but also a compilation of planetary data plus a star catalogue of over a thousand fixed stars, and even a voluminous listing of the cities throughout the world with their longitudes and latitudes. Published in 1627 titled the *Tabula Rudophinae* after his deceased patron, Rudolph II, it was the basis for all astronomical calculations for over a century—a fitting culmination of a brilliant mind whose achievements are often overshadowed by those of Copernicus and Galileo. Yet it was Kepler, as much as anyone, whose thinking drove the transformation from that ancient cosmos to the modern classic worldview.

At the beginning of this chapter I referred to the tragedy, suffering, and tribulations that accompanied Kepler's life which should be described, if only briefly, before concluding. It seems unfair that a life so devoted to heavenly bodies should be so star-crossed. In addition to his own ill health, before moving to Linz his adored six-year old son died and then his first wife, who must have been in her thirties died shortly after from what Kepler described as "melancholy despondency" following a long illness. So at age forty-one, having the responsibility of two young children, Kepler began searching for a wife to manage his household, as was the custom of the day. Confronted with eleven choices, he selected a wife seventeen years younger who worked in a Nobleman's house and was known for her modesty, thriftiness, and kindness, a person who would love his two children from his first marriage. Having a more stable and cheerful disposition, she brought about a happier marriage than his previous one. But they too had to endure sadness since three of their seven children died in infancy, reminding us of how precarious children's lives were then because of untreatable childhood diseases.

As if that were not enough, there then occurred another episode just as tragic and depressing. As he was finishing the *Harmonice Mundi*, his aged mother was charged with being a witch. Considering the outrageous accusation, bizarre proceedings, grotesque beliefs, and depraved behavior of the people involved, this provides another reason for being thankful that one did not live in that period. It is only because Kepler's mother steadfastly insisted upon her innocence, even when shown the execution chamber filled with terrible instruments of torture intended to frighten her into confessing, and that Kepler helped to prepare her defense with the aid of influential friends at court, that the desperate woman was saved from the rack and possibly being burned at the stake. After having been imprisoned for fourteen months, much of the time in chains, the court decided she was innocent because of her resolute denial of the charges despite the threats and terror and freed by the Duke. Undoubtedly because of the terrible ordeal, she died six months later at age seventy-three.

Following this nightmare and hating the city of Linz, having left for Ulm to oversee the publication of the *Rudolphine Tables* Kepler once again was looking for a suitable patron, finally deciding to enter the service of the Emperor's victorious General Wallenstein, mainly as his astrologer. Then dissatisfied with his new residence in the Duchy of Sagan and uncertain of his continued patronage when Wallenstein was dismissed by the Emperor, he set out for Ratisbon where the Emperor was residing to try once again to collect the 11,818 florins owed to him by the Crown. But three days after arriving he fell ill with a high fever from which he never recovered. And so on November 15, 1630, shortly before his sixtieth birthday and separated from his family, his travails were finally ended. He was buried in the cemetery of Saint Peter which later was so decimated by numerous battles in the vicinity that its exact location is unknown.

His was one of the most amazing intellectual odysseys in the history of science fulfilling the expectations of his professors that "he has such a superior and magnificent mind that something special may be expected of him." Gradually discarding his earlier Platonic, Aristotelian, and Neoplatonic explanations of the orbital motions of the planets, including circular orbits, uniform motion, anima motrix, and souls, he was able to devise the correct explanation by his three laws of planetary motion and two solar forces that constituted his clockwork system of celestial mechanics. What a magnificent achievement! Although his burial place is unidentified, he will be known as the "father of modern astronomy" as long as there is civilization.

NOTES

1. This chapter is based mainly on Richard H. Schlagel, *From Myth to Modern Mind: A Study of the Origins and Growth of Scientific Thought*, Volume II, *Copernicus through Quantum Mechanics* (New York: Peter Lang, 1996), chap. 2.

2. Andrew D. White, *A History of the Warfare of Science with Theology*

in Christendom (New York: Appleton Press, 1896), I, p. 126. Quoted from Kuhn, *The Copernican Revolution* (Cambridge: Harvard University Press, 1957), p. 191.

3. Nicolaus Copernicus, *On the Revolutions of Heavenly Spheres*, trans. by Charles Glenn Wallis (Amherst, New York: Prometheus Books, 1995), pp. 3–4. The immediately following quotations are to this work.

4. Herbert Butterfield, *The Origins of Modern Science*, rev. ed. (New York: Collier Books, 1962), p. 13. The following quotation is also from this work.

5. Max Casper, *Kepler*, trans. and ed. by C. Doris Hellmann (New York: Dover Publications, 1993), p. 44. The immediately following discussion and citations in the text are to this work (cf. pp. 44–47).

6. For illustrations of Plato's five polyhedrons see Richard H. Schlagel, *From Myth to Modern Mind: A Study of the Origins and Growth of Scientific Thought*, Vol. I, *Theogony through Ptolemy* (New York: Peter Lang Publishing Inc., 1995), p. 289. For Kepler's own drawings see J. V. field, *Kepler's Geometrical Cosmology* (Chicago: University of Chicago Press, 1988), pp. 53–58 and Appendix 4.

7. Arthur Koestler, *The Sleepwalkers* (New York: Grosset's Universal Library, 1959), pp. 250–51.

8. Johannes Kepler, *Mysterium Cosmographicum*, trans. by A. M. Duncan (New York: Abaris Books, 1981), p. 63. Unless otherwise indicated, the following references in the text are to this work.

9. Cf. William Gilbert, *De Magnete*, trans. by P. Floury Mottelay (New York: Dover Publications, 1948), chaps. 3 and 4.

10. Letter to David Fabricius, *Gesammelte Werke*, Vol. XIV, p. 409. Quoted from Koestler, *The Sleepwalkers*, op. cit., p. 330.

11. Johannes Kepler, *Astronomia Nova,* Introduction. Quoted from Koestler, op. cit., p. 337.

12. Johannes Kepler, *Astronomia Nova*, Vol. III, chap. 33. Quoted from Koestler, op. cit., p. 341.

13. Johannes Kepler, Book 3, chap. 5 of *Harmonice Mundi* in R. M. Hutchins, ed., *Great Books of the Western World*, Vol. 16 (Chicago: The University of Chicago Press), p. 1020.

Chapter 9

GALILEO'S DESTRUCTION OF ARISTOTLE'S COSMOS AND INITIATION OF THE NEW SCIENCE

INTRODUCTION

Although Kepler's three astronomical laws and two solar forces were inferred from a close examination of Tycho Brahe's observational measurements of the eccentric planetary orbits and their relative distances from the sun, that their justification was mainly mathematical and often vacillated from more primitive explanations to more sophisticated ones, made them problematic to most astronomers, particularly Galileo. Moreover, the Aristotelians were not prepared to accept elliptical orbits, nonuniform motions, and physical forces as the actual configurations and causes of the planetary motions. They were Catholics who had been trained to assent to authority as the final arbiter of truth, whether vested in the Church, sacred scripture, or Aristotle.

Thus while neither Copernicus nor Kepler at the time had brought about a fundamental change of thought or outlook, this was not true of Galileo whose telescopic observations and experimentally proven laws were more convincing and therefore impossible to ignore. Moreover, unlike Copernicus who was extremely timorous and Kepler whose writings were often obscure, Galileo had a disputatious character and wrote in Italian in a very clear and persuasive dialectical style, at times so incisive and biting that it aroused considerable controversy and notoriety. It also was an advantage that Galileo was not living in a provincial area of Germany, like Kepler, but near Venice or Florence, the former a major seaport and artisanal center while the latter was the hub of the Renaissance. That he taught at the universities of Pisa and Padua and later was a respected friend and court mathematician under the patronage of the powerful Medici family also contributed to his later eminence.

In a book covering so much history I shall try to present as clearly and concisely as possible the magnitude of the three major contributions of Galileo: (1) his complete demolition of Aristotle's cosmology with his startling telescopic disclosures and critical refutation of the underlying presuppositions and qualitative distinctions supporting Aristotle's astronomical system; (2) his conviction that so little was known then about the actual causes in nature that Aristotle's causal explanations should be replaced by the discovery of empirical correlations that could be expressed mathematically and experimentally verified; and (3) his discovery of several laws of motion that became the foundation of the new science of mechanics and the model of scientific inquiry.

BACKGROUND

He was born in 1564 near Pisa of a somewhat distinguished Florentine family who derived the paternal name Galilei from a prominent ancestral physician of the fifteenth century, Galileo Bonaiuti. Galileo's father, Vincenzio, was an accomplished Latinist and music critic who wrote a *Dialogue on Ancient and Modern Music* instilling in Galileo a love of music. Initially educated by a tutor, Galileo later attended a monastery in Vallombroso where he studied grammar, logic, and rhetoric becoming a Novice in the Order. However, intending that he renew the family's distinguished heritage in medicine, his father had him transferred to another monastery in Florence. In 1581 he enrolled as a medical student at the University of Pisa.

However, during his second year at the university he overheard Ostilio Ricci, mathematician to the Grand Duke, lecturing on Euclid to the court pages. Unable to attend his lectures that were restricted to members of the court, but having met Ricci socially, Galileo felt free when there was an opportunity to engage him in questions about mathematics. Recognizing Galileo's unusual aptitude for the subject, Ricci encouraged and guided him in his research with the result that Galileo became so avid in his studies of mathematics that he sought Ricci's help in trying to persuade his father to allow him to change his studies from medicine to mathematics. Although compelled as a medical student to study Aristotle's philosophy and Galen's medical works, he was more interested in Eudoxus' theory of proportions and Archimedes' works *On Plane Equilibrium* and *On Bodies in Water* when introduced to them by Ricci. It was these early studies that later motivated his research in statics and hydrostatics, the topics of his earliest writings in mathematical physics.

Disappointed that Galileo was becoming more involved in these pursuits than his medical studies, his father threatened to discontinue his financial support unless he concentrated on his medical career. Refusing to do this, he did persuade his father to continue his support for another year after which he began considering a teaching profession to support himself. Aware that there were more positions available teaching Aristotelianism than mathematics, in the summer of 1584 he started preparing lectures on Aristotle's philosophy. But after leaving the university in the spring of 1585 without a degree, he initially gave private lectures on mathematics at Florence and Siena.

He began his scientific publications a year later with *La Bilancetta* (*The Little Balance*), a book written in Italian proposing improvements in the Westphal balance for determining specific gravities and offering suggestions for refining Archimedes' procedure for calculating specific weights. He started writing a dialogue on motion in Latin that he never finished but was the background for his treatise, *De Motu*, published in 1590. The dialogue is indicative of his lifelong interest in various kinds of motions, such as free fall, the velocity of objects in a void, and projectile motion. The following year 1587 he unsuccessfully sought a vacated chair of mathematics at the University of Bologna, but when the chair of mathematics at the University of Pisa was available in 1589 he succeeded in obtaining a three year appointment with an annual salary of 60 florins. He was twenty-five years old at the time.

It was while teaching there that the famous controversial anecdote occurred about his dropping objects from the leaning tower of Pisa. Recall that in the sixth century Jean Philoponus had performed experiments with falling objects of different weights to disprove Aristotle's law that their falling velocity was proportional to their weights. While there is no evidence that he knew of Philoponus' experiments, toward the end of his life in 1637 Galileo recalled to Vincenzio Viviani, his student assistant who later became his biographer, an early public demonstration in Pisa to disprove Aristotle's law of free-falling objects.

According to Viviani, he described a public experiment in which heavy objects of the same material but of different sizes and weights were dropped from the Leaning Tower and carefully timed when they landed. Like Philoponus, he found that if made of the same solid material, so that the air resistance was negligible, they either landed simultaneously or the difference was insignificant. The controversy arises because there is no public record of there having been such a demonstration leading some scholars to dismiss the story as legendary, despite Galileo's later reference to it. But I agree with Drake that the account is probably true given Galileo's earlier criticisms of Aristotle, his penchant for public demonstrations, and there being no reason to fabricate the story.[1]

His research at Pisa consisted of revising and extending the *De*

Motu, including additional criticisms of Aristotle's distinction among, and explanation of, various kinds of motions. Fearing that these attacks on Aristotle might preclude his reappointment at the University of Pisa, at the end of his three-year contract and learning that the chair of mathematics at the more prestigious University of Padua was vacant, he applied for the position. Although Magini, who had won the competition for the chair at the University of Bologna also applied, this time Galileo's reputation had so improved that he obtained the position at a salary three times what he had earned at the University of Pisa.

Aside from the increase in salary, Venice as a great seaport, munitions and artisanal center, and the crossroads of the world, with the opportunity of meeting well-known scholars, distinguished visitors, and skilled artisans, had many advantages over Pisa, not to mention its incomparable location with splendid palazzos built on piles along the lagoon and the canals. Its impact on Galileo is evident from his numerous references in his books to sailing ships, compasses, sextants, cannons, pulleys, levers, and the demonstration of the effectiveness of one of his telescopes from the top of the campanile. Few scientists have been as curious about natural phenomena as Galileo whose investigations ranged from astronomy to insects, balances, magnets, various machines, inclined planes, music, and optics. Adept at using and inventing instruments, such as the telescope, microscope, calculator, quadrant, and various calibrators, in 1599 he invited Marc Antonio Mazzoleni and his family to live with him as his instrument maker.

In fact his first years at the University of Padua were devoted to mechanistic inquiries, beginning a treatise on *Mechanics* influenced by reading "the first modern treatise on mechanics . . . published in Latin by Guidobaldo del Monte in 1577 and translated into Italian in 1581" (p. 35). But of even greater importance was his increasing interest in the Copernican system, especially as the twofold motion attributed to the earth by Copernicus offered a possible mechanical explanation of the tides, which he mistakenly would offer as the strongest empirical evidence supporting the Copernican system. Knowing of the explanation held by Kepler that attributed the tides to the mutually attractive forces between the moon and the earth, he rejected it because he thought the lunar theory, involving mysterious forces, was too occult. Instead, he believed that just as the agitated motion of barges causes the various motions of the water in the lower bulkheads of ships, he thought the two motions of the earth, its axial rotation and orbital revolution, caused the motion of the tides.

His first open declaration of allegiance to the Copernican system was in a letter to Kepler thanking him for his gift of the *Mysterium Cosmographicum.* He wrote that once having adopted the Copernican position,

> I have discovered the causes [an allusion to the earth's motions?] of many physical effects [the tides?] which are perhaps inexplicable on the common [geocentric] hypothesis. I have written many . . . refutations of contrary arguments [to the Copernican system] which up to now I have preferred not to publish, intimidated by the fortune of our teacher Copernicus, who though he will be of immortal fame to some, is yet by an infinite number (for such is the multitude of fools) laughed at and rejected [recall the ridicule of Luther]. (p. 41; brackets added)

Apparently his academic performance and research was appreciated because he was reappointed at the University of Padua for four years at a substantial increased in salary. Shortly after in 1602 he completed the book on *Mechanics* that he had begun at Pisa. Among the numerous investigations was a return to his incline plane experiments begun in *De Motu* and his discovery of the law of free fall. Although he investigated the orbital sizes and speeds of the planets in relation to their distances from the sun, unlike Kepler he was unable to formulate the exact law describing the ratio. However, his experiments with pendulums led to the discovery that if the duration of their swings were uniform or isochronal, their speeds depended on their lengths, not on the weight of the bobs as was then believed. It was this discovery that suggested to him later that in a vacuum objects would fall with the same velocities irrespective of their weights.

He renewed his experiments with inclined planes in 1603 and by the following year confirmed experimentally that for a smooth round object accelerating from rest the distances traversed in equal times were as the odd numbers beginning with one. But as important as this discovery was, it did not explicitly relate the speed to the time. Yet it was apparent that if the successive distances increased according to this rule, then the cumulative distances (or speeds) would vary with the squares of the times. As he states, the "contraries of speeds are *times*—and indeed I do get the times by taking the square roots of the distances from rest, the same distances whose squares are as the speeds from rest" (p. 102).

Arriving at the odd number rule first, he realized that the *square roots* of the successive sums (of 1, 3, 5, 7, 9 &) gave the times (2, 3, 4, 5 &) which, when squared, indicate the ratios of the increases in distance, speed, or acceleration. On October 16 he wrote Paolo Sarpi that "he had found a proof for the square law, the odd-number rule, and other things he had been long asserting, if granted the assumption that *velocità* are proportional to distances from rest" (p. 100). In his final book written years later, the *Two New Sciences*, this law would be the key to his "new science of motion."

Despite some interest in astronomy, it was not until 1604 when a brilliant nova was sighted similar to the one observed by Tycho Brahe in 1572 that Galileo's curiosity was aroused. Because on Aristotle's system there could be no changes in the translunar realm, any new appearances in the skies generated intense interest among astronomers, as well as astrologers. The evidence for deciding whether the location was sublunar or translunar depended mainly on whether there was evidence of parallax: that is, whether the object cited remained stationary in position or appeared in a different region of the sky when seen from different locations on the earth's surface.

Yet the degree of the observed shift also depended on the velocity of the object and its distance from the observer. The motion of an automobile is obvious while an airplane high in the sky seems barely to move at all. The orbital motions of the planets, for example, are not directly evident but inferred from their different positions on successive nights, while the stars appear to be fixed because of their great distance even though they too revolve. Thus when a sighting of an apparently new astral phenomena appeared, it aroused considerable controversy as to whether it was located in the celestial or terrestrial worlds.

Like Tycho during the controversy over the appearance of the supernova in 1572, Galileo could find no evidence of parallax concluding that the nova existed in the translunar world. The occurrence raised so much controversy that Galileo gave three lectures attracting crowds of people "to explain the nature and application of paralactic reasoning to measurement of distances and to refute the Aristotelian theory that new stars and comets were sublunar phenomena. . . ." (p. 106) However, the Aristotelians' countered that while mathematics could be used effectively in the sublunar world to measure and explain terrestrial phenomena, the qualitative difference and remoteness of the celestial bodies precluded any mathematical inferences regarding their status. Yet the success of Copernicus, Tycho, and especially Kepler in discovering mathematical correlations in the movements of the planets convinced Galileo that the Aristotelian distinction regarding the application of mathematics was spurious. When it came to accepting astronomical calculations based on observations or rejecting them because they disagreed with Aristotle's system, Galileo now stood with the supporters of the heliocentric system.

Turning his attention to projectile motion, by 1609 he had decided that it consisted of two components, the horizontal trajectory and the vertical descent, which could be depicted geometrically as the two sides of a right angled triangle so that the hypotenuse connecting them represented the combined magnitude of the two motions. This vectorial addition of motions enabling Galileo to calculate a parabolic trajectory was a further important advance because it denied another of Aristotle's principles that two contrary motions could not be combined. In addition, although Aristotle had rejected instantaneous motion on the grounds that since all motion took some time instantaneous motion was a contradiction, Galileo's investigations led him to conclude that acceleration involved a continuous series of instantaneous increased in motion.

GALILEO'S TELESCOPIC OBSERVATIONS

There then occurred in the summer of 1609 when Galileo was forty-five years old the event that led to his astonishing astronomical observations supporting the heliocentric system, entering his place in history as one of the greatest scientists. While in Venice he was told of a Dutch instrument called a "spyglass" that magnified distant objects so they were seen as much closer. After returning to Padua he received a letter from a French nobleman in Paris, Jacques Bodovere, confirming that such instruments existed in northern Europe. As he later wrote in the *Sidereus Nuncius* (*Starry Messenger*):

> This finally caused me to apply myself totally to investigating the principles and figuring out the means by which I might arrive at the invention of a similar instrument, which I achieved shortly afterward on the basis of the science of refraction. And first I prepared a lead tube in whose ends I fitted two glasses, both plane on one side while the other side of one was spherically convex and of the other concave. Then, applying one eye to the concave glass, I saw objects satisfactorily large and close. Indeed, they appeared three times closer and nine times larger than when observed with natural vision only. Afterwards I made another . . . that showed objects more than sixty times larger.[2]

Continuing, he says "sparing no labor or expense, I progressed so far that I constructed for myself an instrument so excellent that things seen through it appear about a thousand times larger and more than thirty times closer than when observed with the natural faculty only" (pp. 37–38).

Then seizing the initiative when he heard that the "spyglass" (as it was initially called) was brought to Venice with the intention of selling it to the Venetian Republic, Galileo took his latest instrument to Venice where he demonstrated its powers, first to the public and secondly to the Venetian Senate from the highest point in St. Mark's Square, the campanile. Shortly after, in a letter accompanying a gift of the spyglass to the Doge, he shrewdly pointed out its nautical military potential in making distant vessels visible two hours before their arrival at the port for which he was rewarded a lifetime appointment at the University of Padua at a salary of 1,000 florins per year. But, as Galileo was aware, an even greater significance of the instrument lay in another use.

While he was not the first to realize the astronomical potential of the spyglass, his skill in improving its effectiveness, along with his adroit use of it, made him preeminent in the new discipline of telescopic astronomy. Having constructed a twenty-power instrument by the end of November, his first recorded lunar observations probably were made on December 1, 1609, when he began making sketches of the crescent shape of the Moon and its irregular surface. These observations were followed by his startling discovery of the "moons" of Jupiter—startling because in the Aristotelian system the nature of the celestial realm precluded any new occurrences. Concerned to establish the priority of these observations, he began writing the *Sidereus Nuncius* in January, completing it in March, and had it printed in Venice. He dedicated it to Cosimo de' Medici who was his former student and now the ruler of Tuscany, as well as christening the four moons of Jupiter the "Medicean Stars" in honor of the four Medicean Princes.

Remarkable for its open-mindedness in accepting the novel observations for what they were and his clarity in interpreting the data, the book immediately brought Galileo international fame.

Kepler, the most renowned European astronomer, replied as follows when he received a copy of Galileo's book.

> I may perhaps seem rash in accepting your claims so readily with no support of my own experience. But why should I not believe a most learned mathematician, whose very style attests the soundness of his judgment? He has no intention of practicing deception in a bid for vulgar publicity, nor does he pretend to have seen what he has not seen. Because he loves the truth, he does not hesitate to oppose even the most familiar opinions, and to bear the jeers of the crowd with equanimity. (pp. 94–95)

What was most detrimental to the Aristotelian qualitative distinction between the terrestrial and celestial worlds, which was considered inviolable by the scholastics, was Galileo's striking diagrams of the surface of the Moon with its ravines, mountainous, and steep valleys (which are clearly reproduced in Van Helden's book, pp. 44–46) and his description of the similarity of these to the Earth.

> By oft-repeated observations of them we have been led to the conclusion that we certainly see the surface of the Moon to be not smooth, even, and perfectly spherical, as the great crowd of philosophers have believed about this and other heavenly bodies, but, on the contrary, to be uneven, rough, and crowded with depressions and bulges. And it is like the face of the Earth itself, which is marked here and there with chains of mountains and depths of valleys. (p. 40)

The threat to the Aristotelian cosmology was so damaging that the Aristotelians struggled to come up with arguments to counter the observations, such as that the Moon was actually a crystalline sphere with the mountains and valleys lying beneath its transparent surface or that while the spyglass could be reliably used in the terrestrial world it distorted things when directed to the celestial realm. The Aristotelian Cremonini claimed that if such things were true Aristotle would have noted them, refusing to look through the spyglass. Yet even sympathetic astronomers had difficulty using the instrument which had to be handled very carefully, as Galileo's instructions indicated, as reported by Drake:

> . . . the instrument must be held firm, and hence it is good, to escape the shaking of the hand that arises from motion of the arteries and from breathing, to fix the tube in some stable place. The glasses should be kept clean and polished . . . or else cloud is generated there by the breath, humid or foggy air, or vapor which evaporates from the eye itself. . . . It is best that the tube be capable of being lengthened a bit. . . . It is good that the convex glass, which is the one far from the eye, should be partly covered and that the opening left should be oval in shape, since thus are objects seen much more distinctly. (p. 147)

The difficulty of using the instrument correctly was sadly evident when Galileo visited Bologna to demonstrate his spyglass to Magini and other leading astronomers, but was chagrin when they could not replicate his observations.

Yet, as he relates in the *Sidereus Nuncius*, despite this failure he was offered the positions of "Chief Mathematician of the University of Pisa and of Philosopher to the Grand Duke" of Tuscany. On July 10, 1610, a few months after his lifetime appointment to the University of Padua, he gladly accepted the latter offer from the Grand Duke because it carried no teaching obligations despite his title and would permit him to return to Florence, his original home, under the patronage of the powerful Grand Duke Cosimo II. Convinced of Galileo's astronomical discoveries, the latter informed the "Tuscan ambassadors at the courts in Prague, London, Paris, and Madrid . . . that Galileo would send them copies of his book and perhaps spyglasses as well," instructing them "to use their good offices to promote Galileo's discoveries" (p. 100). This confidence was justified by the end of the year when Galileo received confirmation of his discovery of the "satellites" of Jupiter (so named by Kepler) by Antonio Santini, Kepler, Thomas Harriot in England, and several observers in France.

By then he had added to his discoveries having observed what are now referred to as "the rings of Saturn," but what appeared to him as two protuberances or "ears" on each side of Saturn, and even more significantly the phases of Venus. As Copernicus had predicted these phases from the heliocentric perspective, while the Ptolemaic model predicted that one would observe constant crescent shapes of various sizes, Galileo realized that this offered another opportunity to verify the Copernican theory.

After further observations he wrote to Giuliano de' Medici on December 11 that he had observed Venus undergoing phases similar to those of the Moon, adding confirmation to the heliocentric system. By then his reputation was such that he was invited to Rome where the members of the Collegio Romano testified to Cardinal Bellarmine (who later will play such a crucial role at Galileo's trial) regarding the validity of his observations. In addition, he was inducted into the prestigious *Accademia dei Lyncei* where, at the initiation banquet in his honor, the name "*telescopium*" was suggested for the previous name "*occhilai*" or "spyglass."

But always evoking controversies because of his novel discoveries, he was soon in another dispute over floating bodies with the Aristotelians who defended Aristotle's explanation that what accounted for an object's floating or sinking in water was its shape, while Galileo defended Archimedes' explanation that it was the object's specific gravity, the ratio of its density to water. At the urging of the Grand Duke he presented his views in an essay that was the basis for his *Discourse on Bodies On or In Water*, which was published in three revised editions. He also became engaged in another dispute over sunspots. A Jesuit mathematician named Christopher Scheiner, writing under the pseudonym "Apelles" at the request of his Order, published a series of letters claiming that the recently observed "sun spots" were tiny stars revolving near the sun's surface resembling Jupiter's moons. Galileo replied in a letter in May 1613 that he could demonstrate mathematically that the "spots" were not stars but some kind of solar phenomena analogous to clouds on the earth.

Probably on the basis of this dispute and his explanation of the phases of Venus, in the third letter on sun spots he included one of the most explicit conceptions of the aims and limitations of scientific inquiry written up to that time, a conception opposed to the Aristotelian search for the essential nature of things, but similar to the twentieth century Positivists. Claiming that it was not possible to discover the true nature of the unobservable infrastructure of physical reality, like Galileo they asserted that scientific inquiry should be limited to discovering mathematical correlations in nature from which valid predictions of observable phenomena could be derived. As Galileo states:

> For in our speculating we either seek to penetrate the true and internal essence of natural substances or content ourselves with a knowledge of some of their properties. The former I hold to be as impossible an undertaking with regard to the closest elemental substances as with more remote celestial things. . . . all the things among which men wander remain equally unknown, and we pass by things both near and far with very little or no acquisition of knowledge. . . . But if what we wish to fix in our minds is the apprehension of some properties of things, then it seems to me that we need not despair of our ability to acquire this respecting distant bodies just as well as those close at hand—and perhaps in some cases even more precisely in the former than in the latter.[3]

Though this skeptical outlook was justified in Galileo's time and somewhat plausible in the early twentieth century when the anti-realist positivistic interpretation of science was prevalent, the subsequent discovery of subatomic particles, developments in nuclear physics, advances in pharmacology, creation of molecular biology leading to the amazing sequencing of genetic codes, and the detection of the background radiation left over from the Big Bang have definitely disproved it.

Because of this limitation of knowledge, Galileo criticized the Aristotelians for believing that all knowledge was contained in the writings of Aristotle, "as if this great book of the universe had been written to be read by nobody but Aristotle, and his eyes had been destined to see all for posterity." (p. 200) He even suggested that given Aristotle's respect for empirical evidence it is very likely that were he alive "in our age" he "would abandon" such views as "the inalterability of the sky . . ." (p. 201).

Galileo even had the courage at that time, in a famous "Letter to Castelli," of asserting that "concerning salvation and the establishment of the Faith" there was no higher authority than Holy Scripture, but regarding confirmed discoveries of astronomy or physics which conflict with Scripture, we should consider *the level of knowledge that existed when the Bible was written* and recognize the need for revision, especially as it is unlikely that "the same God who has given us our senses, reason, and intelligence wishes us to abandon their use. . ." (p. 226).

He then adds the even more controversial statement:

> Indeed . . . as we are unable to assert with certainty that all interpreters speak with divine inspiration, I should think it would be prudent if no one were permitted to oblige Scripture and compel it in a certain way to sustain as true some physical conclusions of which sense and demonstrative and necessary reasons may show the contrary. And who wants to set bounds to the human mind? (p. 226)

Assertions such as we cannot know "with certainty that all interpreters [of scripture] speak with divine inspiration" and implying that no one had the authority "to set bounds to the human mind" were seen as especially challenging to the Church's authority.

History has many illustrations of the fact that it is the nature of an authoritarian or totalitarian regime that it cannot allow any concessions without inviting its collapse and thus is defended at all costs. As Father Grienberger, a member of the Jesuit Collegio Romano that had previously honored Galileo, declared: "had not Aristotle been involved, the Jesuits would have agreed with everything, but that by order of the General of the Jesuits they could not oppose Aristotle in anything . . ." (p. 236). As an example of the kind of hostility Galileo's views were arousing, in Florence on December 21, 1614, Tommaso Caccini, a fiery young Dominican, "denounced from the pulpit of Santa Maria Novella the Galileists, and all mathematicians along with them, as practitioners of diabolical arts and enemies of true religion" (p. 238). While many of the Dominicans in Florence and Rome were appalled by Caccini's malicious attack, it was one of a series of events that eventually would lead to Galileo's trial and condemnation by the Inquisition.

EVENTS LEADING UP TO THE TRIAL

And in fact, following Caccini's tirade the cardinals of the Inquisition examined Galileo's Letter to Castelli to see if it contained offensive material, leading him to go to Rome toward the end of 1615 to clear himself of any allegations of heresy and to try to convince the Church authorities not to prohibit the Copernican theory. But in contrast to his tremendously successful visit four years earlier, this venture proved a dismal failure as described in Drake's summary of the proceedings:

> In view of all the documents I believe that though the pope probably wanted the Holy Office to proceed against Galileo personally, Cardinal Bellarmine counseled a less personal [more formal] procedure. First a technically independent panel of theologians would find against the notions that the earth moved and the sun stood still, and then Galileo would be informed of this and asked to

> abandon those views. Bellarmine had no doubt that he would agree; a decree of general scope could then be published and the matter would be resolved without alienating either the Medici or the several cardinals who remained favorable to Galileo. The finding of the panel was handed in on 24 February; on the 25th at the weekly meeting of the cardinals of the Inquisition, the pope instructed Bellarmine in their presence to inform Galileo of it and require him to abandon these opinions. If he resisted, then the Commissary of the Inquisition was to instruct Galileo in the presence of a notary and witnesses that if he did not obey he would be jailed. (p. 253)

As Bellarmine had recommended, he met with Galileo on February 26 in the presence of a notary, witnesses, and other officials. The document, though unsigned, states that Bellarmine "told Galileo of the official finding against the motion of the earth and stability of the sun," while the Commissary of the Inquisition with Bellarmine present "admonished Galileo in the name of the pope that he must not hold, defend, or teach in any way, orally or in writing, the said propositions on pain of imprisonment," to which he agreed! (p. 253). In light of the later trial and condemnation, the crucial point is that Galileo *did* agree not to "hold, defend, or teach in any way, orally or in writing," the doctrines of the motion of the earth and stability of the sun! The edict of 1616 was then issued prohibiting the publication of all books attempting to reconcile statements in the Bible with the Copernican theory. Galileo's books were not prohibited and he did not interpret the edict as banning Copernicus' book, but merely its suspension until corrected.

Anxious that these proceedings not be used to attack him, Galileo presented his case before the pope directly. Surprisingly, at the meeting "the pope assured him that he knew Galileo's integrity and sincerity, told him not to worry, said that not only he but the entire Congregation of the Holy Office knew about his unjust persecution, and added the unusual remark that so long as he, Paul V, lived, Galileo remained secure" (p. 256).

It is difficult to reconcile the pope's statements with the previous meeting with Cardinal Bellarmine, along with the severity of the unsigned document and the threat of imprisonment if he continued to "hold, *defend*, or teach, in any way, orally or in *writing*"the "motion of the earth and stability of the sun" (italics added). But as Galileo received several letters informing him that rumors were circulating that in Rome he had been "severely admonished and forced to do penance," when he showed these to Cardinal Bellarmine the latter, recognizing the danger to Galileo's reputation, "wrote out a signed statement [which Galileo perhaps thought exonerated him at the later trial] that Galileo had neither adjured nor done penance, but had merely been informed of the general edict governing all Catholics" (p. 256; brackets added).

Then in the fall of 1618 a public lecture was given by Orazio Grassi, a professor of mathematics at the Collegio Romano, later published as an essay, on three alleged comets which had just been sighted which provoked the usual controversy as to their location. While the Aristotelians maintained that such unusual appearances were fiery vapors in the sublunar world, Grassi supported Tycho Brahe's previous arguments that because they were not enlarged when seen through the telescope (as is true of remote objects), showed no signs of parallax, and were seen close to Venus, they were comets circling the sun in the translunar world.

Despite providing evidence of changes in the translunar world, because the comets were described as circling the sun in Tycho's modified geocentric system, Galileo believed a rebuttal was necessary despite the edict of 1616 prohibiting him from discussing or writing about the Copernican system. But being ill at the time, it was decided that his reply would be given by his assistant, Mario Guiducci, to the Florentine Academy, though when delivering the lectures and later in their published form, "Discourse on Comets," Guiducci clearly acknowledged that Galileo was the author.

Owing to his disdain for Tycho's reasoning and rejection of his astronomical system, Galileo's rebuttal was especially sarcastic and harsh, rejecting Grassi's arguments on the grounds that neither the lack of telescopic magnification nor the absence of parallax (which would not apply to them if they were not comets) were adequate to establish the nature and orbits of the three phenomena. More seriously, he disparaged the question so significant to the Aristotelians of whether the comets were in the terrestrial or celestial worlds, asserting that he never had assented "to the vain distinction (or rather contradiction) between the [terrestrial] elements and the heavens. . . ."[4] Furthermore, instead of limiting his criticism to Grassi he also slighted the mathematicians of the Collegio Romano, the apex of the Jesuit Seminaries, who had previously honored him.

Thus he not only succeeded in provoking Grassi into replying, he also caused the Jesuits to become his implacable foe. Grassi named his reply *Libra Astronomia* (*Astronomical Balance*), implying that he was weighing Galileo's arguments in the *Discourse*, which in turn led Galileo to name his second rebuttal *Il Saggiatore* (*The Assayer*), a further play on words indicating he was measuring Grassi's arguments with a precise scale. Though critical of Grassi's second book, Galileo confessed that he did not know the correct interpretation. Admitting, as he often did, that he could not "determine precisely the manner in which comets are produced" (p. 236), he seemed to think they were a form of "*simulacrum*" or "image" produced in diffused gases or vapors by the reflection and refraction of the sun's light, nor could he determine their exact location.

However, despite this indecisive explanation, the *Il Saggiatore* is significant because it contains Galileo's most explicit conception of his theory of knowledge and of scientific inquiry. As the ancient Atomists, he believed that sensations and perceptions are due to the organism's transforming the imperceptible physical stimuli that activate the sense organs, such as fiery atoms or light or sound waves, into sensations of heat or pain and perceptual qualities like colors and sounds. Thus Galileo may have been the first to introduce into *modern* science and philosophy the contentious dualism between an independent physical reality and the world as experienced that has been the major problem in modern philosophy—the dualism addressed by Descartes, Locke, Berkeley, Hume, and especially Kant, who drew the famous distinction between the phenomenal world of appearances and the unknown or noumenal world of things in themselves.

However, unlike Kant but like Locke, Galileo held that we could infer something about the *objective* physical properties of objects that Locke called "primary qualities," even though what we experience is embellished with sensations and sensory qualities, the *subjective* "secondary qualities." As he states in a famous passage in *Il Saggiatore*:

> Therefore I say that upon conceiving of a material or corporeal substance, I immediately feel the need to conceive simultaneously that it is bounded and has this or that shape, that it is in this place or that at any given time; that it moves or stays still. . . . I cannot separate it from these conditions [its primary qualities] by any stretch of my imagination. But that it must be white or red, bitter or sweet, noisy or silent, of sweet or foul odor [the secondary qualities], my mind feels no compulsion to understand as necessary accompaniments. . . . For that reason I think that tastes, odors, colors, and so forth are no more than mere names so far as pertains to the subject [or more correctly object] wherein they reside, and they have their habitation only in the sensorium. Thus, if the living creature (*l'animale*) were removed, all these qualities would be removed and annihilated. (p. 309; brackets added)

Because the Aristotelian view was that our perceptions of the independent physical world do not transform them into appearances that would cease to exist apart from the perceiver, but normally represent the external world as it is, even though the sensorium exists in us, this was another affront to the Aristotelians. Their position was the common sense realism which modern science has refuted. Today we accept imperceptibles such as ultraviolet rays, subatomic particles, and dark energy because of the experimental evidence. It is in this book also that Galileo presents his famous declaration about the importance of mathematics in investigating nature, replacing the Aristotelian reliance on deductive logic.

> Philosophy is written in this grand book—I mean the universe—which stands continually open to our gaze, but it cannot be understood unless one first learns to comprehend the language and interpret the characters in which it is written. It is written in the language of mathematics, and its characters are triangles, circles, and other geometrical figures, without which it is humanly impossible to understand a single word of it; without these, one is wandering about in a dark labyrinth. (pp. 183–84)

Despite his limitation of the language of mathematics to geometrical figures that does not include the notation of algebraic equations or formulas, this statement is one of the most eloquent ever written describing the crucial role of mathematics in scientific inquiry, one that will be extensively illustrated in his last book, *Dialogues Concerning Two New Sciences*, and confirmed in the development of modern science.

Although having alienated the Jesuits, the most prominent Catholics at the time, Pope Urban VIII (who had succeeded Gregory XV who had succeeded Paul V) was so delighted by the *Il Saggiatore* that "he had portions read to him at table." Moreover, he had been such an early admirer of Galileo that he even had written a poem in his honor. Thus the Lincean Academy, which had sponsored publication of the book, dedicated it to the pope and had his armorial crest of three bees printed on the cover. Owing to the pope's enthusiastic reception of the book, Galileo considered the time favorable to making another trip to Rome to garner permission to write his long intended book on the relative merits of the Aristotelian and Copernican systems, despite the injunction of 1616.

During his stay in Rome in April 1624 he had six audiences with the pope who not only granted him permission to write on the comparison of the two cosmological systems, but also "presented to him silver and gold medals and promised a pension to his son Vincenzio" (Drake, op. cit., p. 289), all of which had its ironies considering the later trial. The pope even went so far as to express the opinion (as stated by Drake) "that the Holy Church had never condemned Copernicus as heretical, and never would, holding it only to be rash, there being no fear that anyone would prove it true" (p. 291), another indication of what little confidence there was at the time that empirical investigations would ever be able to resolve these problems. There then begins the unfolding of the seemingly inadvertent tragic consequences of this meeting in connection with what Galileo had been told by Cardinal Bellarmine in 1616.

Recall that during that visit he had been instructed by the commission of the Inquisition in the presence of Bellarmine that "he must not hold, defend, or teach in any way, orally or in writing," the heliocentric position. The pope apparently was unaware of the prohibition and Galileo, either inadvertently or intentionally, did not inform him, a decision that would have tragic consequences. As Drake states, Galileo left Rome

> with assurances from Urban that he was free to write on the two systems of the world provided that he treated them impartially and did not go beyond the astronomical and mathematical arguments on both sides. But because Bellarmine had instructed Galileo in 1616 to regard the admonition given him by the Commissary of the Inquisition as having no official existence, so long as Bellarmine's own words to him were heeded, Galileo never told Urban VIII what had actually taken place on that occasion. In the end that omission . . . turned out to have been a fatal error.[5]

In any event, Galileo now believed he had the permission of the pope to write an evaluation of the two cosmological systems within certain restrictions.

Then in June his assistant Guiducci informed him of a critical treatise on Copernicus' system written by Francesco Ingoli in 1616 which provided an excuse to reply, so he wrote an extended critique titled *Reply to Ingoli*, which was a kind of prelude to his *Dialogue Concerning the Two Chief World Systems*. The *Reply* contains two items of particular interest because they illustrate the different approaches Galileo took in evaluating the Aristotelian cosmology: first, his rebuttal of Ingoli's argument that the earth, as the heaviest body surrounded by the three other terrestrial elements, belongs in the interior or "lowest" portion of the spherical universe, while the celestial realm, because of its ethereal nature, belongs to the "highest" region, marking the distinction between the sublunar and translunar worlds. Galileo dismisses this distinction as merely verbal, writing "that those ['lowest' and 'highest'] are words and names, proving nothing and having nothing to do with calling anything into existence" because they assume the validity of the distinction which is the question at issue (p. 293; brackets added).

The second item pertains to the argument that if the earth moves, then objects propelled vertically upward would not fall directly to where they were launched because during the time they were in the air the earth would have rotated causing the object to fall behind its projection site. Using an analogical example (without ever having tested it, apparently), the Aristotelians claimed that while on a *stationary* ship a heavy object dropped from the masthead would fall perpendicular to the mast, when the ship was *in motion* it would fall some distance away from the mast toward the stern at a distance dependent on the ship's velocity, forming a curved oblique trajectory rather than a straight vertical one.

Although Aristotle's argument seems reasonable enough, it does not take into account what Galileo had already inferred, that the object has two motions, the motion conveyed by the ship or the earth and its observed motion, so that the former motion being identical to the ship's or earth's can be discounted, leaving only the observed motion. Despite being somewhat counterintuitive, this explains why objects in the sky are not left behind or why we do not feel any rush of air as the earth rotates because it carries the air with it in its rotation. Although Descartes and Mersenne claimed that being convinced of the falsity of the Aristotelian account Galileo had not performed the experiment of dropping an object from the masthead, I believe him when he wrote,

> I have been twice as good a philosopher as those others because they, in saying what is the opposite of the effect, have also added the lie of their having seen this by experiment; and I have made the experiment—before which, physical reasoning had persuaded me that the effect must turn out as it indeed does. (p. 294)

Following a lapse of three years when he was in ill health and occupied with other projects, at the urging of friends he returned to the *Dialogue Concerning the Two Chief World Systems* that he completed by the end of 1629, but because of difficulties in getting it published, owing to the prohibitions, it did not appear until 1632. Undoubtedly the greatest polemical and literary work in the history of science, the *Dialogue* established Galileo with Plato as the two greatest dialecticians of all time. When following his critical arguments one can see him demolishing one by one the basic presuppositions, definitions, distinctions, and principles supporting Aristotle's cosmology, replacing it with the preliminary cosmology and theoretical framework of modern classical science.

Although it was stated as a condition of the consent by Urban VIII that Galileo could write on the two world systems "provided that he dealt with them impartially," Galileo states at the very beginning of the *Dialogue* that "*I have taken the Copernican side in the discourse, proceeding as with a pure mathematical hypothesis and striving by every artifice to represent it as superior to supposing the earth motionless—not, indeed, absolutely, but as against the arguments of some professed Peripatetics*."[6] If he thought that his favoring of the Copernican theory could be justified by claiming that he was treating it as "a pure mathematical hypothesis " and "not absolutely," and thus placate the pope, he was sadly mistaken as events will prove.

DIALOGUE CONCERNING THE TWO CHIEF WORLD SYSTEMS

The *Dialogue* takes place in four days and has three participants, two of whom were lifelong friends and benefactors of Galileo who had died before he had completed it and whom he wanted to honor: Sagredo, a Venetian nobleman and diplomat, represents the intelligent moderator in the dialogues, and Salviati, a wealthy Florentine aristocrat, who speaks for Galileo and is usually referred to as "the Academician." The third disputant is Simplicius, named for a famous sixth-century scholastic who represents the views of Aristotle and the Peripatetics. As some of the arguments have already been presented in the discussion of his previous works, I will focus on those that are new.

"The First Day" is devoted to rebutting the distinction (previously rejected in *Il Saggiatore*) between the terrestrial and celestial worlds enabling Galileo to dismiss objections to placing the earth in the translunar realm with the two motions assigned to it by Copernicus and Kepler. As argued previously, the rebuttal consists of showing that such distinctions as perfect circular versus imperfect rectilinear motions, natural as opposed to unnatural or violent motions, and the ethereal, weightless, incorruptible celestial realm in contrast to the material, dense, and corruptible terrestrial world have no basis in reality and therefore are essentially meaningless, indicating how much the intellectual climate had changed among some natural philosophers since Aristotle.

For example, to deny the distinction between perfect circular and imperfect rectilinear motions he cites the recent evidence of the rectilinear motion of meteors in the heavens and many instances of circular motion in the terrestrial world. Or to refute the qualitative contrast between the celestial and terrestrial worlds he points to his telescopic evidence of the similarity of the surface of the moon and the earth, along with the existence of large spots circulating the sun that resemble clouds on the earth, while the appearance of meteors, novas, and satellites refutes the immutability of the celestial realm. This evidence naturally does not dissuade Simplicius who replies that this way of

> philosophizing tends to subvert all natural philosophy, and to disorder and set in confusion heaven and earth and the whole universe. However, I believe the fundamental principles of the Peripatetics to be such that there is no danger of new sciences being erected upon the ruins. (p. 37)

"The Second Day" mainly defends the diurnal rotation of the Earth by refuting the terrestrial and celestial arguments against it. But first he again opposes Simplicius' argument that all knowledge can be found in Aristotle's writings if one is skillful enough: "Aristotle acquired his great authority only because of the strength of his proofs and the profundity of his arguments. . . . There is no doubt that whoever has the skill will be able to draw from his books demonstrations of all that can be known; for every single thing is in them" (p. 108). In refutation Galileo had stated, at the end of the first day's dialogue, his skepticism of how little anyone knows about the true nature of anything: "there is not a single effect in nature, even the least that exists, such that the most ingenious theorists can arrive at a complete understanding of it. This vain presumption of understanding everything can have no other basis than never understanding anything" (p. 101).

Turning to his support of the diurnal eastward rotation of the earth in contrast to the accepted daily westward rotation of the entire universe, his argument is based on the relativity of motion and the contrasting simplicity and harmony of the two positions. Just as it appears from a smoothly sailing ship that the shoreline recedes even though it is the ship which is withdrawing, so from the fact that the sun appears to rise in the east and set it in the west it cannot be inferred with certainty that the whole universe rotates daily from east to west, since the same effect could be produced by the opposite rotation of the earth from west to east. There being no way to test the correctness of either hypothesis, Galileo, like Copernicus, concludes that it depends upon which is the simpler and most harmonious explanation, as he has Salviati state:

> First, let us consider only the immense bulk of the starry sphere in contrast with the smallness of the terrestrial globe. . . . Now if we think of the velocity of motion required to make a complete rotation in a single day and night, I cannot persuade myself that anyone could be found who would think it the more reasonable and credible thing that it was the celestial sphere which did the turning, and the terrestrial globe which remained fixed. (p. 115)

Although it can be argued that what is considered "the more reasonable and credible thing" lies in the mind of the beholder, there are additional arguments.

Consider the tremendous discontinuity in having the fixed stars in the farthest sphere revolve in twenty-four hours while the *orbital speeds of the planets decrease* with their distances from the center, Saturn taking thirty years to complete its orbit. There would also be an "immense disparity between the motions of the stars, some of which would be moving very rapidly in vast circles, and others very slowly in little tiny circles, according as they are located farther or closer to the poles" (p.119). Finally, an additional blatant asymmetry would occur if the entire universe had a diurnal rotation from east to west in contrast to the common eastward revolution of the planets.

He then turns to the argument presented in *Reply to Ingoli* that falling objects or entities in the sky would be seen to be displaced in the opposite direction if the earth were moving independently beneath them. This objection is easily disposed of by Salviati who points out that if we observe the behavior of fluttering butterflies, flying birds, thrown balls, or dripping water in a ship's cabin, "so long as the motion is uniform and not fluctuating . . . [y]ou will discover not the least change in all the effects named, nor could you tell from any of them whether the ship was moving or standing still" (p. 187). Having previously demonstrated this by dropping a solid object from the masthead of a ship showing that it falls parallel to the masthead whether the ship is at rest or in uniform motion, he now offers another more subtle example to illustrate the same effect.

Suppose two cannon balls with the same elevation and force, but one facing east and the other west, were fired from the same position. If, while the two cannon balls were in flight, the earth revolved eastward shouldn't the one facing east have a shorter trajectory because while it was in motion the earth would have moved its launching place closer to where the ball fell reducing the distance, while the one moving westward would have its trajectory increased because the launching place would have receded eastward as the ball moved westward. Or as Salviati states:

> For when the ball goes toward the west, and the cannon, carried by the earth, goes east, the ball ought to strike the earth at a distance from the cannon equal to the sum of the two motions, one made by itself to the west, and the other by the gun, carried by the earth, toward the east. On the other hand, from the trip made by the ball shot toward the east it would be necessary to subtract that which was made by the cannon following it. (p. 126)

The fallacy consists in not recognizing that because the earth's motion is common to the system as a whole it cancels out. For example, a stone dropped from a high tower falls parallel to the tower despite the motion of the earth, because they both participate in the earth's motion, again as argued by Salviati.

> With respect to the earth, the tower, and ourselves, all of which keep moving with the diurnal motion along with the [falling] stone, the diurnal movement is as if it did not exist; it remains insensible, imperceptible, and without any effect whatever. All that remains observable is the motion which we lack, and that is the grazing drop to the base of the tower. (p. 171; brackets added)

These examples are especially significant in illustrating why revolutionary discoveries in science usually are counterintuitive because they do not conform to our normal expectations, instead presenting discordant explanations and implications. Because all our intuitive conclusions depend upon a background conceptual framework that initially is derived from ordinary experiences and common sense beliefs that do not match the new discoveries, as the opposition to heliocentricism, Darwinism, and the greater age of the universe illustrated, they were considered implausible or even heretical at the time they were introduced. This was especially true when they conflicted with religious beliefs which, though alleged to be derived from infallible revelations and incontestable religious experiences, actually are based on ancient scripture and theological doctrines that originated in a bygone age. Unlike scientists such as Galileo whose scientific background had conditioned him to recognize the tentativeness and limitations of all knowledge, and thus to be open to new developments, religions, along with most philosophies, have been conservative in professing certitude and finality. One of the greatest realizations bequeathed by scientific revolutions is that change is to be expected.

This difference is shown in those Peripatetics who refused to acknowledge Galileo's astronomical discoveries on the grounds that they were not mentioned in Aristotle's writings, though Aristotle himself was no doctrinaire authoritarian having stressed the importance of empirical evidence over his teacher Plato's a priori rationalistic approach to knowledge. While often critical of Aristotle's doctrines, Galileo shows his admiration by describing him as "a man of brilliant intellect . . ." (p. 321). To emphasize the importance of being flexible in one's thinking and open to new ideas Galileo has Salviati say,

> considering that everyone who followed the opinion of Copernicus had at first held the opposite, and was very well informed concerning the arguments of Aristotle and Ptolemy, and that on the other hand none of the followers of Ptolemy and Aristotle had been

> formerly of the Copernican opinion and had left that to come round to Aristotle's view . . . I commenced to believe that one who forsakes an opinion which he imbibed with his [mother's] milk and which is supported by multitudes, to take up another that has few followers and is rejected by all the schools and that truly seems to be a gigantic paradox, must of necessity be moved, not to say compelled, by the most effective arguments. (pp. 128–29; brackets added)

This is one of the passages that the Church commission cited as "offensive to the Church," stating that Galileo "represented it to be an argument for the truth that Ptolemaics become Copernicans, but not vice versa" (p. 477, note 103:7).

The Third Day's discussion is mainly concerned with refuting arguments rejecting an annual orbital revolution of the earth attributed to it by Aristarchus and Copernicus. Simplicius begins by declaring that the strongest objection to assigning an orbital revolution to the earth is that it would displace it from its natural position in the center of the universe. Salviati's initial rebuttal is that assigning a center to the universe assumes that it has a finite spherical shape, though it has not been "proved whether the universe is finite and has a shape, or whether it is infinite and unbounded" (p. 319), again illustrating Galileo's openness to every possibility. But that this alternative is dropped is probably due to the fact that raising this issue was dangerous considering that Giordano Bruno had been burned at the stake by the Catholic Church because of having espoused the view that the universe was infinite.

Salviati continues by asserting that assuming the universe to be finite and spherical, the question is whether the center of the universe coincides with the position of the earth or with the center of the celestial orbs, even if this is not the location of the earth. Simplicius concedes that it would be preferable to locate the center with the center of the orbs, but wonders why this would not be the position of the earth? Thus the discourse is set for Galileo to argue that the astronomical evidence indicates that a more harmoniously integrated planetary system can be constructed by assuming that the sun is at the center of the universe, not the earth. The primary evidence, as Kepler inferred from Tycho's data and Galileo's telescopic observations supported, can be seen in the contrasting coordinated movements of the planets in relation to either the earth or the sun: for example, assuming Mars circled the earth its distance from the earth as seen in its changing sizes varies so greatly that it could not possibly describe the circular orbit attributed to it by Aristotle.

Simplicius replies that even if this were true, it does not prove that the sun is in the center, to which Salviati offers additional conciliatory evidence:

> This is reasoned out from finding the three outer planets—Mars, Jupiter, and Saturn—always quite close to the earth when they are in opposition to the sun, and very distant when they are in conjunction with it. This approach and recession is of such moment that Mars when close looks sixty times as large as when it is most distant. Next, it is certain that Venus and Mercury must revolve around the sun, because of their never moving far away from it, and because of their being seen now beyond it and now on this side of it, as Venus' changes of shape conclusively prove. (p. 322)

He further adds that the telescopic evidence indicates that the orbits of Mercury and Venus are below the earth and around the sun, while those of Mars, Jupiter, and Saturn are above it, with only the moon circling the earth. This leads Sagredo to ask why, "[i]f this very ancient arrangement of the Pythagoreans is so well accommodated to the appearances . . . it has found so few followers in the course of centuries" (p. 327), to which Salviati gives the superb reply:

> No, Sagredo, my surprise is very different from yours. You wonder that there are so few followers of the Pythagorean opinion, whereas I am astonished that there have been any up to this day who have embraced and followed it. Nor can I ever sufficiently admire the

> outstanding acumen of those who have . . . through sheer force of intellect done such violence totheir own senses as to prefer what reason told them over that which sensible experience plainly showed them to the contrary . . . that Aristarchus and Copernicus were able to make reason so conquer sense that, in defiance of the latter, the former became mistress of their belief. (pp. 327–28)

He then takes up a number of other objections, such as the peculiar fact that in the Copernican system it is only the moon that revolves around the earth while all the other planets revolve around the sun, answering that here again new telescopic discoveries revealing the moons circulating Jupiter have provided new evidence that "removes this apparent anomaly of the Earth and Moon moving conjointly" (p. 340). He next considers two more recurrent issues, the retrograde motion of the planets and the absence of parallax. As for retrograde motion, he has a diagram showing that this is an illusion produced by the observer's undetected change in position as the earth orbits causing an *apparent* loop-like motion of the planet against the background of the fixed stars, so rather than refuting the motion of the earth it supports it.

As for the absence of parallax, the Peripatetics argued that if the movement of the earth can cause such illusory displacements as retrograde motions along with recessions and retrogressions of the planets, then the orbital revolution of the earth should show some displacement or parallax of the fixed stars which is not observed. Furthermore, if one tries to explain the absence as due to the distance of the fixed stars, then "in order for a fixed star to look as large as it does, it would actually have to be so immense in bulk as to exceed the earth's orbit—a thing . . . entirely unbelievable" (p. 372). Salviati replies in effect that not enough is known about how the stars would appear at greater distances to be able to draw any definite conclusion. He then provides a concise summary of the astronomical evidence that shows the heliocentric system to be the more probable of the two systems (a conclusion Urban VIII had forbidden him to take).

> See, then, how two simple noncontradictory motions assigned to the earth, performed in periods well suited to their sizes, and also conducted from west to east as in the case of all movable world bodies, supply adequate causes for all the visible phenomena. These phenomena can be reconciled with a fixed earth only *by renouncing all the symmetry that is seen among the speeds and sizes of moving bodies*, and attributing an *inconceivable velocity* to an enormous sphere beyond all the others, while lesser spheres move very slowly. Besides, one must make the motion of the former contrary to that of the latter, and to increase the improbability, must have the highest sphere transport all the lower ones opposite to their own inclination. I leave to your judgment which has the more likelihood in it. (p. 396; italics added)

The Fourth Day's dialogue, the final one, is the most ironic in a life filled with irony because it presents what Galileo believed to be the most conclusive argument supporting the heliocentric universe, that the twofold motions of the earth explain the ebb and flow of the tides. He was so convinced of its validity that he had intended to have a reference to the tides included in the title of the *Dialogue*, had Urban not prohibited it. Since the explanation turned out to be false, I will deal with it briefly.

Although the correlation between the periodicity of the tides and the cyclical motion of the moon had convinced most past astronomers, especially Kepler, that it was a mutual force between them that caused the tides, this explanation is rebuffed by Galileo. Suspicious of mysterious or occult forces, Galileo rejected these for a purely mechanical cause based on the two contrasting motions of the earth. Like the rushing back and forth of the water in the hold of ships due to their pitching in rough waters, he believed the tides were due to the shifting of the oceanic basins produced by the axial rotation and annual revolution of the earth when they are opposed or coincide.

Admitting to some uncertainty and incompleteness of the explanation due to the complexity of the process, he nonetheless has Sagredo reply that after

> having read and listened to the great follies which many people have put forth as causes for these events, I have arrived at two conclusions . . . that if the terrestrial globe were immovable, the ebb and flow of the oceans could not occur naturally; and that when we confer upon the globe the movements just assigned to it, the seas are necessarily subjected to an ebb and flow agreeing in all respects with what is to be observed in them. (p. 417)

In contrast to the eighteenth-century philosopher David Hume who will deny that any "necessary connections" can be discovered in nature, this physical argument is reinforced by Galileo's belief that even though we do not know the underlying causes in nature, the uniformity of nature enables us to infer a "fixed and constant connection." "Thus I say if it is true that one effect can have only one basic cause, and if between the cause and the effect there is a fixed and constant connection, then whenever a fixed and constant alteration is seen in the effect, there must be a fixed and constant variation in the cause" (p. 445). This case illustrates how difficult it can be to discern what the "fixed and constant connection" actually is.

Yet once more he summarizes the evidence he considers "very convincing" in support of the heliocentric system, this time expressed by Sagredo.

> In the conversations of these four days we have, then, strong evidences in favor of the Copernican system, among which three have been shown to be very convincing—those taken from the stoppings and retrograde motions of the planets, and their approaches toward and recessions from the earth; second, from the revolution of the sun upon itself, and from what is to be observed in the sunspots; and third, from the ebbing and flowing of the ocean tides. (p. 462)

Regardless of all the disclaimers, it is clear that he believed the evidence for the Copernican system was "very convincing," as opposed to the Aristotelian (or Ptolemaic) system, despite the pope's insistence that he treat both systems "impartially." Even having Simplicius, the rebuked Aristotelian, express the pope's disclaimer that "God in His infinite power and wisdom" could have used "some other means" than the two motions of the earth to cause the tides, could be interpreted as dissembling—although it also could be seen as the most appropriate. As Simplicius states:

> As to the discourses we have held, and especially this last one concerning the reasons for the ebbing and flowing of the ocean, I am really not entirely convinced; but from such feeble ideas of the matter as I have formed, I admit that your thoughts seem to me more ingenious than many others I have heard. I do not therefore consider them true and conclusive; indeed, keeping always before my mind's eye a most solid doctrine that I once heard from a most eminent and learned person [Pope Urban VIII], and before which one must fall silent, I know that if asked whether God in His infinite power and wisdom could have conferred upon the watery element its observed reciprocating motion using some other means than moving its containing vessels, both of you would reply that He could have, and that He would have known how to do this in many ways which are unthinkable to our minds. From this I forthwith conclude that, this being so, it would be excessive boldness for anyone to limit and restrict the Divine power and wisdom to some particular fancy of his own. (pp. 464)

The following reply by Salviati concluding the *Dialogue* is puzzling because though it agrees with Galileo's skepticism regarding the *current* discovery of any actual causes in nature, it would seem to undermine his confidence in the truth of any explanations, along with the *future* progress of science which he has ardently maintained. Was Galileo sincere in declaring this or was it one last effort to appease the pope?

> An admirable and angelic doctrine, and well in accord with another one, also Divine, which, while it grants to us the right to

> argue about the constitution of the universe (perhaps in order that the working of the human mind shall not be curtailed or made lazy) adds that *we cannot discover* the work of His hands. Let us, then, exercise these activities permitted to us and ordained by God, that we may recognize and thereby so much the more admire His greatness, however much less fit we may find ourselves to penetrate the profound depths of His infinite wisdom. (p. 464; italics added)

Although at the urging of Prince Cesi the book was supposed to be published in Rome under the auspices of the Lincean Academy, the death of the Prince intervened. But after two years of difficult negotiations it was finally agreed that the book would be printed in Florence under the supervision and censorship of the Florentine Inquisitor. Lavishly praised by his friends when it appeared in February 1632, Castelli writing "I still have it by me, having read it from cover to cover to my infinite amazement and delight" (Drake, p. 336), it aroused a storm of protest from the Jesuits who wanted the book prohibited and an even fiercer reaction from the pope himself.

Initially angered by the fact that only one of the three arguments he had requested was included and that one stated by Simplicius, the weakest of the three disputants, he reacted furiously when he learned that Galileo had been instructed in the edict of 1616 not to "hold, defend, or teach in any way orally or in writing" the motion of the earth and the stability of the sun, "on pain of punishment." Not having informed him of the edict in their meeting in 1624 when he had agreed to Galileo's writing the *Dialogue*, Urban concluded that Galileo had deliberately deceived him and betrayed his trust.

THE TRIAL, CONVICTION, AND ABJURATION

As a consequence the Florentine Inquisitor the following September sent the original manuscript to Rome for examination, prohibited all sales of the book, and ordered Galileo to appear before the Roman Inquisition in November. Being bed ridden at the time he did not arrive in Rome until February 13, 1633. Not withstanding the intense reaction of the Jesuits and the pope, Galileo apparently believed that Cardinal Bellarmine's "affidavit of assurance" (which his accusers were unaware of) would exonerate him and that he could "justify himself" before his accusers. Yet despite my tremendous admiration of Galileo and revulsion at the Church's censorship and the Inquisition, I find it difficult to understand how he could have justified himself. As I previously wrote:

> With a deep feeling of disappointment and sadness, I have concluded that he was guilty of bad faith or self-deception. Even if the unsigned edict of 1616 did not have the proper legal standing and even though Bellarmine's affidavit did not (as the edict did) contain the words "or teach in any way orally or in writing, the said propositions" regarding the motion of the earth and stability of the sun, Galileo did agree to abide by Bellarmine's affidavit stating that the Copernican opinion, which is contrary to Sacred Scripture, "may not be defended or held."[7]

This is the letter that he intended to use in his defense, but in all honesty could he have done so? Did he now really believe that he did not hold the Copernican system to be "more probable" than the Aristotelian contrary to what he had written to Kepler and stated in the *Dialogue*? Despite his frequent disclaimers, could he sincerely maintain that he had not intended to "defend" the two movements of the earth, especially to explain the tides? It is difficult to believe that he could have marshaled such acute critical arguments refuting the presuppositions of the Aristotelian system if he had not believed in the superiority, if not the certainty, of the heliocentric system, despite its being expressly prohibited by the pope!

On April 12 he was transferred from the Tuscan embassy to the Offices of the Inquisition, where he was received in a friendly manner and housed in a comfortable apartment with his attendant. The following day the interrogation began with questions about the edict and affidavit of 1616 during which he claimed "not to have contravened in any way the precept [of 1616], that is, not to have held or defended the said opinion of the motion of the earth and stability of the earth on any account."[8] Aware that this was not true, Vincenzo Maculano, the Commissary General of the Inquisition, proposed "the Holy Congregation grant me power to deal extra-judicially with Galileo to the end of convincing him of his error and bringing him to the point of confessing it when he understood" (p. 349).

Accordingly, Maculano met with Galileo privately and persuaded him that in his book he had contravened the edict and disobeyed the pope. Finally admitting this, he asked time to reexamine his book and compose his confession, which he presented on the last day of April. According to Drake,

> having reread it he realized that in many places a reader ignorant of his intention might think the arguments earned the day for the *position he meant to confute*, especially the arguments from the sunspots and from the tides. He could only excuse himself on grounds of vanity and ambition, every man liking to show himself cleverer than others in his own subtleties. He had not meant any disobedience but confessed vain ambition, ignorance, and inadvertence. (p, 350; italics added)

This confession, I believe, offers further evidence of his bad faith or self-deception in maintaining that his *real intention* in the *Dialogue* had been to "confute" the Copernican system and thus had not meant any "disobedience." One wonders if the effects of the proceedings, including the threat of torture and imprisonment, had so disoriented the ailing, aged scholar that he was convinced this had been his purpose? Or perhaps under the pressure of the trial he had undergone some kind of conversion in which he disowned all that he had previously fought for, such as the right of natural philosophy (science) to decide empirical questions, even if they disagreed with Holy Scripture. As recorded by Drake, this at least appears to be the case in a letter he wrote to Francesco Rinuccini shortly before he died denying all that he had previously ardently defended and upheld.

> "The falsity of the Copernican system must not on any account be doubted, especially by us Catholics, who have the irrefragable authority of Holy Scripture interpreted by the greatest masters in theology, whose agreement renders us certain of the stability of the earth and mobility of the sun around it." (p. 417)

The "irrefragable authority of Holy Scripture" was certainly denied in his Letter to Castelli. Despite his signed confession, on 16 June the pope "ordered Galileo's examination on intention, to be followed (if he sustained this) by imprisonment for an indefinite term at the pleasure of the Holy Office, confiscation of the *Dialogue*, and mandatory public reading of the sentence to professors of mathematics throughout Italy and elsewhere" (p. 351). On June 21 he was examined as to his intention during which he again claimed that after the edict of 1616 "he had adhered to the fixed earth and movable sun" and had "considered no argument as conclusive and the decision of 'sublime authority' as binding" (p. 351). Finally, asked on "pain of torture" if he spoke the truth, he replied: "I am here to obey, and have not held this opinion after the termination made, as I said" (p. 351). At a formal ceremony on June 22 before the cardinals of the Inquisition and witnesses, "the sentence of life imprisonment" was read to him "after which he had to abjure on his knees before them" (p. 351).

At the intervention of Cardinal Barberini, the pope's nephew, who had stood by Galileo throughout the trial defending him from more severe treatment and who was one of three cardinals that refused to sign the sentencing document, his place of imprisonment was directly transferred to the Tuscan embassy in Rome. Shortly thereafter the pope permitted him to accept an invitation to stay in the custody of Archbishop Ascanio Piccolomini of Siena. So ends the sorry ordeal of the trial that has been called "the disgrace of the century."

The trial ended, Galileo left for Siena in July where he was comforted by the warm hospitality of Cardinal Piccolomini, an ardent admirer of Galileo who had studied mathematics under his disciple Cavalieri. Initially severely distraught by what he believed to be the unfairly harsh sentencing of the pope considering his extra-judicial confession, Galileo gradually regained his physical and mental health under the affectionate care of his host, and even began writing his second great book, *Dialogues Concerning Two New Sciences*. Again due to the intercession of Cardinal Barberini and Ambassador Niccolini, "on 1 December the pope recommended to the Holy Office that Galileo be permitted to return to his villa at Arcetri, in the hills beyond Florence, provided he receive few visitors and refrain from teaching" (p. 356). So by mid-December he was in his own villa near to the convent of his beloved daughter, Sister Maria Celeste, but a cruel fate even took away this precious consolation when she died suddenly the following year.

DIALOGUES CONCERNING TWO NEW SCIENCES

Considering his mental breakdown at the time, it is a sign of Galileo's extraordinary resilience that he could recover sufficiently during his stay at Siena to begin work on the *Two New Sciences*. Despite the earlier *Dialogue*'s renown for its incisive criticism of the Aristotelian geocentric system in support of the heliocentric theory, perhaps because of the dire consequences he considered his second book "superior to everything else of mine hitherto published . . . contain[ing] results which I consider the most important of all my studies. . . ."[9] Even though he had previously described his experiments on motion in several of his earlier works and had given his most definitive statement of his scientific methodology and theory of knowledge in *Il Saggiatore*, it is true that his treatise on the *Two New Sciences* presents his most outstanding scientific discoveries.

Not knowing the internal structure of physical substances nor the actual causes of motion as he repeatedly maintained, he was restricted to demonstrating experimentally the mathematical ratios of the correct parameters, like space and time, of motions such as gravitational fall and projectiles which throughout the past had eluded most investigators. While the number of laws is not impressive, that he found the correct method for discovering and proving them is. As he states in a another famous passage:

> My purpose is to set forth a very new science dealing with a very ancient subject. There is, in nature, perhaps nothing older than motion, concerning which the books written by philosophers are neither few nor small; nevertheless I have discovered by experiment some properties of it which are worth knowing and which have not hitherto been either observed or demonstrated. Some superficial observations have been made, as, for instance, that the free motion of a heavy falling body is continuously accelerated; but to just what extent this acceleration occurs has not yet been announced; for so far as I know, no one has yet pointed out that the distances traversed, during equal intervals of time, by a body falling from rest, stand to one another in the same ratio as the odd numbers beginning with unity. (p. 147)

The odd number law had been discovered by Nicole Oreme in the fourteenth century, but this apparently was unknown to Galileo.

His greatest discovery being the law describing gravitational acceleration, he next refers to his second law, expressing his belief that this was just the beginning of a whole new world of exploration, indicative of a renewed optimism about the future progress of science despite some of his previously stated reservations, and probably reflecting his own recent discoveries.

> It has been observed that missiles and projectiles describe a curved path of some sort; however no one has pointed out the fact that this path is a parabola. But this and other facts, not few in number or less worth knowing, I have succeeded in proving; and what I consider more important, there have been opened up to this vast and most excellent science, of which my work is merely the beginning, ways and means by which other minds more acute than mine will explore its remote corners. (pp. 147–48)

This new science will describe motion not as an effect of a motive cause in the way of the Aristotelians, but as due to an inner state or *impetus* and momentum or *velocitas* that can be quantified as ratios between two magnitudes, space and time. Thus he abstracted motion from its physical manifestations and causes whose two essential parameters could be quantified and represented geometrically, allowing the science of motion to be treated in the manner of Euclid and Archimedes. Although a limited kinematic approach that would be completed by Newtonian dynamics, it was an essential preliminary stage.

As is the common practice today, Galileo proceeds from the simpler to the more complex cases, dividing the discussion "into three parts; the first part deals with motion which is steady or uniform; the second treats of motion as we find it accelerated in nature; the third with the so-called violent motions and with projectiles" (p. 148). Starting with uniform motion, he defines it as "one in which the distances traversed by the moving particle during any equal intervals of time, are themselves equal" (p. 148). This is followed by a number of axioms and theorems with geometrical illustrations to bring out the "properties" or implications of his definition.

Next he turns to naturally accelerated motion eschewing arbitrary definitions for "a definition best fitting natural phenomena . . . and to make this definition . . . exhibit the essential features of observed accelerated motions" (p. 154). That he has succeeded, he says, is confirmed by "the consideration that experimental results are seen to agree with and exactly correspond with those properties which have been, one after another, demonstrated by us . . . employ[ing] only those means which are most common, simple, and easy" (p. 154; brackets added). He thus proposes that the increase in speed be proportional to the increase in time so that a "motion is said to be uniformly accelerated when starting from rest, it acquires, during equal time-intervals, equal increments of speed" (p. 155).

Again, using various geometrical diagrams to illustrate and support his reasoning, he then presents his famous "odd number law" that he had discovered earlier.

> It is thus evident by simple computation that a moving body starting from rest and acquiring velocity at a rate proportional to the time, will, during equal intervals of time, traverse distances which are related to each other as the odd numbers beginning with unity, 1, 3, 5; or considering the total space traversed, that covered in double time will be quadruple that covered during unit time; in triple time, the space is nine times as great as in unit time. And in general the spaces traversed are in the duplicate ratio of the times, i.e., in the ratio of the squares of the times. (p. 170)

Admitting that he found the mathematical reasoning "rather obscure," Simplicius nonetheless states that he is "convinced that matters are as described," but "as to whether this acceleration is that which one meets in nature in the case of falling bodies, I am still doubtful," suggesting "that this would be the proper moment to introduce one of those experiments—and there are many of them, I understand—which illustrate in several ways the conclusions reached" (pp. 170–71). Incredibly, given the detailed and precise description of the experimental procedure the French savant Mersenne declared "I doubt whether Galileo actually performed the experiments . . . since he does not speak of them" and the philosopher Descartes (both contemporaries of Galileo) "denied all of Galileo's experiments," while Koyré stated that "in spite of Galileo's assertion, one is tempted to doubt this,"[10] illustrating the distrust of experimental results at the time. Yet this is his description:

> A piece of wooden molding or scantling, about 12 cubits long, half a cubit wide, and three finger-breadths thick, was taken; on its edge was cut a channel a little more than one finger in breadth; having made this groove very straight, smooth, and polished, and having lined it with parchment, also as smooth and polished as possible, we rolled along it a hard, smooth, and very round bronze ball. Having placed this board in a sloping position, by lifting one end some one or two cubits above the other, we rolled the ball . . . along the channel, noting, in a manner presently to be described, the time required to make the descent. We repeated this experiment more than once in order to measure the time with an accuracy such that the deviation between two observations never exceeded one-tenth of a pulse-beat. (p. 171)

He then meticulously describes discovering the correlations between the distances and times.

> Having performed this operation and having assured ourselves of its reliability, we now rolled the ball only one-quarter the length of the channel; and having measured the time of its descent, we found it precisely one-half of the former. Next we tried other distances, comparing the time for the whole length with that for the half, or with that for two-thirds, or three-fourths, or indeed for any fraction; in such experiments . . . we always found that the spaces traversed were to each other as the squares of the times, and this was true for all inclinations of the plane, i.e., of the channel, along which we rolled the ball. (pp. 171–72)

Apparently neither Mersenne nor Descartes had read Galileo's book or they could not have doubted that he performed the experiments, but there is no excuse for Koyré's reservation.

Thus his incline plane experiment demonstrated what could not be directly observed because of the rapidity of the velocity of a free falling body (less than three seconds from a ten story building), that it accelerated proportional to the squares of the times, not proportional to its weight or distance from the earth, as the Aristotelians claimed. But could these results be extrapolated to apply to free fall? That was the crucial question? Galileo's last statement addresses this problem as he asserts that what holds for "all [angles] of inclinations of the plane" must be true also when the angle of inclination is 90 degrees. He also answers the objection of how the measurements of the times could be so precise by describing measuring the differences and ratios of the water accumulated during the descent "with such accuracy that although the operation was repeated many, many times, there was no appreciable discrepancy in the results" (p. 172).

Having proved the law of free fall, he then sought to show that objects in a vacuum would fall with the same speed regardless of their weights, not instantaneously as the Aristotelians also claimed. Relying on earlier experiments in hydrostatics showing that the *variations in the rates* of descent of objects of different weights *decreased* as the densities of the media decreased, he concluded that in the extreme or limiting case of a vacuum, where there would be no resistance of the medium, the rate of fall would be the same despite their different weights.

> Because if we find as a fact that the variation of speed among bodies of different specific gravities is less and less according as the medium becomes more and more yielding, and if finally in a medium of extreme tenuity, though not a perfect vacuum, we find that, in spite of great diversity of specific gravity, the difference in speed is very small and almost inappreciable, then we are justified in believing it highly probable that in a vacuum all bodies would fall with the same speed. (p. 70)

Even earlier as a young man he had decided this would be true when observing the pendular movements of differently suspended lanterns in the Cathedral of Pisa, noting their synchronous periodic swings. He then devised an experiment to prove this by suspending a ball of cork and a much heavier ball of lead from strings of equal length, demonstrating that the lengths and times of their pendular arcs were identical if released from the same distended position. From this he deduced that "if these same bodies traverse equal arcs in equal times we may rest assured that their speeds are equal" (p. 82).

His final major discovery pertained to projectile motion which was a consequence of his investigation of the "perpetual" horizontal motion of entities propelled along a frictionless flat surface which anticipates inertial motion, but does not fully describe it.

> Imagine any particle projected along a horizontal plane without friction; then we know . . . that this particle will move along this same plane with a motion which is uniform and perpetual, provided the plane has no limits. But if the plane is limited and elevated, then the moving particle, which we imagine to be a heavy one, will on passing over the edge of the plane acquire, in addition to its previous uniform and perpetual motion, a downward propensity due to its own weight; so that the resulting motion which I call projection, is compounded of one which is uniform and horizontal and of another which is vertical and naturally accelerated. (p. 234)

Although illustrating the conservation of momentum if one eliminates the resistance of the air along with the friction of the table, it does not fully describe inertial motion because when unsupported the gravitational force draws the object down producing a curved trajectory, nor does it distinguish between mass and weight, the latter produced by gravity. For inertial motion to occur the object must have an inherent mass and velocity, be free of gravitational forces, and thus move perpetually in a straight line.

But though this does not define inertial motion, it did enable Galileo to claim that projectile motion was compounded of two motions, "of which one is horizontal and uniform and of another which is vertical but naturally accelerated" which produces "the path of a projectile, which is a parabola." (p. 248) I believe this was the first time in history that projectile motion was described as parabolic. Thus he was able to conclude that when "the motion of a body is the resultant of two uniform motions, one horizontal, the other perpendicular, the square of the resultant momentum is equal to the sum of the squares of the two component momenta" (pp. 246–47). Given this mathematical discovery he could make a number of additional calculations and predictions, including the angle of elevation of the cannon that would produce the longest trajectory.

Even though the *Two New Sciences* did not contain any discussion of the Copernican system nor material objectionable to the Catholic Church, when inquiries were made about its possibly being published in Venice or Florence the Inquisitor replied that "there was an express order prohibiting the printing or reprinting of any work of Galileo, either in Venice or any other place, *nullo excepto*" (p. xi). This harsh disposition of the Holy Office toward Galileo is evident also in the refusal of the Inquisitor to allow him to go to Florence to have a painful hernia treated, there being no doctors available in Arcetri, informing him that "any more petitions from him would result in imprisonment" (Drake, p. 360). Nonetheless, he was finally successful in having the completed edition of the *Dialogues Concerning Two New Sciences* published in 1638 by the founder of the famous publishing firm in Amsterdam, Louis Elzevir, which is still in existence.

Although blind and physically and mentally exhausted, he still continued his research in the four years preceding his death following the publication of the *Two New Sciences*. He died on January 9, 1642 in Arcetri less than two months before his seventy-eighth birthday. He was privately interred in the church of Santa Croce in Florence where the Grand Duke wished to erect "a sumptuous tomb across from that of Michelangelo," but was prevented by the Catholic Church which "forbade any honors to a man who had died under vehement suspicion of heresy" (Drake, p. 436). Yet despite the Church's intervention then, there now exists in the church of Santa

Croce a monument paying homage to Galileo fully as splendid as the one across from his honoring Michelangelo.

CONCLUSION

Although this discussion has been brief compared to that of Drake, I hope it has succeeded in conveying the significance of Galileo's groundbreaking achievements. Where in the history of science can one find another who nearly single-handedly demolished an entrenched cosmology (that of Aristotle), along with its authoritative methodology, and replaced both with an entirely new conceptual framework? As Maurice Clavelin aptly states in his authoritative book on Galileo:

> The reason . . . why no scientific problem was ever the same again as it had been before Galileo tackled it lay largely in his redefinition of scientific intelligibility and in the means by which he achieved it: only a new explanatory ideal and an unprecedented skill in combining reason with observation could have changed natural philosophy in so radical a way. No wonder then that, as we read his works, we are struck above all by the remarkable way in which he impressed the features of classical science upon a 2000-year-old picture of scientific rationality.[11]

All the later advances in Newtonian mechanics, as extraordinary as they were, were additions to or extensions of Galileo's experimental discoveries utilizing his explicit formulation of the proper methodology of science. Add to this his heroic effort to wrest the autonomy of science from the pernicious control of the Catholic Church, and one can begin to appreciate his achievements. From this time forward the influence of religion would vary inversely with the advance of science and technology.

NOTES

1. Cf. Stillman Drake, *Galileo at Work: His Scientific Biography* (Chicago: The University of Chicago Press, 1978), pp. 19–21. The immediately following citations in the text are from this excellent work.

2. Galileo Galilei, *Sidereus Nuncius (The Sidereal Messenger)*, trans. by Albert Van Helden (Chicago: The University of Chicago Press), 1989, p. 37. The immediately following references are to this work. Following the custom of the time, I will capitalize the names of the planets as long as the authors do.

3. Stillman Drake, op. cit., p, 199; brackets added. The immediately following quotations again are from this work.

4. Galileo Galilei, *The Controversy on the Comets of 1618*, trans. by Stillman Drake and C. D. O'Malley (Philadelphia: University of Pennsylvania Press, 1960), p. 53; brackets added. This volume also contains Galileo's *Il Saggiatore*, thus the immediately following page references are to this work until otherwise indicated.

5. Stillman Drake, *Galileo at Work,* op. cit., p. 291. The immediately following citations are to this work.

6. Galileo Galilei, *Dialogue Concerning the Two Chief World Systems—Ptolemaic and Copernican*, trans. by Stillman Drake (Berkeley: University of California Press, 1962), pp. 5–6. Unless otherwise indicated, the following page references in the text are to this work.

7. Richard H. Schlagel, *From Myth to Modern Mind: A Study of the Origins and Growth of Scientific Thought*, Vol. II, *Copernicus through Quantum Mechanics* (New York: Peter Lang Publishing Inc., 1956), p. 163.

8 Stillman Drake, *Galileo at Work,* op. cit., p 347. Unless otherwise indicated, the following references are to this work.

9. Galileo Galilei, *Dialogues Concerning Two New Sciences*, trans. from the Italian and the Latin by Henry Crew and Alfonso de Salvio (New York: McGraw-Hill Book Co., reissued in 1963), p. ix; brackets added. The following references are to this work unless otherwise indicated.

10. Alexander Koyré, *Galileo Studies*, trans. by John Mepham (New Jersey: Humanities Press, 1978). For the quotation and source for Mersenne, see p. 126, n. 177, for Descartes' quotation p. 107 and source p. 126, n. 176 and for the statement by Koyré page 107.

11. Maurice Clavelin, *The Natural Philosophy of Galileo*, trans. by A. J. Pomerans (Cambridge: MIT Press, 1974), p. 383.

Chapter 10

NEWTON'S CELESTIAL MECHANICS AND OPTICS

THE EARLY YEARS

Unlike the contributions of the previous investigators which, though extremely significant and essential, were preliminary or preparatory, Newton's achievements were cumulative and culminating. It was he, for example, who introduced the necessary forces and laws that created modern celestial mechanics finally replacing Aristotle's cosmology and who discovered the light spectrum with his prism experiments. In addition, he invented differential calculus (which he called "the method of fluxions") independently of Leibniz which proved so essential for calculating the ratio of the increment of change of a variable, especially applied to motions, using integrals and differentials. Because of his incomparable achievements in the three crucial areas of science—experimental, theoretical, and mathematical—I believe he deserves to be called "the greatest scientist that ever lived."[1] But Alexander Pope's "Epitaph to Newton" says it best: "Nature and Nature's Laws lay hid in Night: God said, Let Newton be! and all was light." No scientist, not even Einstein, was so universally acclaimed during his lifetime as Newton.

Because of restrictions of space and the great diversity of even his early inquiries, I shall be brief in describing them in order to focus on his two major works, the *Principia* and the *Opticks*.[2] In any case, unlike Galileo but like Copernicus, Newton was an extremely sensitive, suspicious, and guarded person who shunned controversy and notoriety, especially in his early years, so that his personal life does not have the dramatic persona of Galileo's, despite the international fame he acquired early in his youth.

In one of the greatest historical sequences, Newton entered the world the year Galileo departed it, in 1642. He was born on Christmas Day in the family manor house of Woolsthorpe, in Lincolnshire, the only child of his mother Hannah and father Isaac. His father having died three months before he was born, his mother remarried three years later leaving Isaac in the care of his grandmother. After her second husband died eight years later she returned to Woolsthorpe with her two children by her second marriage to rejoin her elder son. The death of his father whom he never knew and the early departure of his mother until he was ten years old must have had a distressing effect on Newton, especially as he did not seem fond of his grandmother.

When he turned twelve he entered the Free Grammar School of King Edward VI in Grantham, an excellent school with a three centuries old tradition where he received the usual religious education, along with a good classical background in Greek and Latin. According to Westfall, he was described as "a sober, silent, thinking lad," who "never was known scarce to play with the boys abroad" (p. 59). Recognizing his exceptional intellectual gifts, the headmaster of the school pointed them out to his maternal uncle who persuaded Newton's mother to let him take the necessary courses to prepare him for the university. As a result he was admitted to Trinity College, Cambridge in the summer of 1661, where his uncle had studied thirty years earlier, then reputed to be "the leading academic institution in England" (p. 189). As was required to receive a degree, he also matriculated at Cambridge University.

The curriculum at Trinity continued his previous religious studies along with Greek and Latin, plus courses in Aristotle's logic, ethics, and rhetoric. He also studied Aristotle's mechanics, physics, and cosmology, indicating the continued influence of Aristotle on natural philosophy, but gradually became attracted to the works of more recent scholars, such as Kepler, Galileo, Descartes, Gassendi, Boyle, Hobbes, and Henry More. Surprisingly, it was Descartes' works more than Galileo's that were being discussed at the time in Cambridge, so Newton began studying Descartes avidly, especially his writings on optics and theory of vortices. Becoming increasingly independent in his studies, by 1664 his new interests were listed under the heading of "Quaestiones," specific topics related to mechanics and optics that foreshadow his later publications. They especially contain his criticisms of Descartes' explanations of light and colors, of gravity, and of the tides, along with his theory of vortices which Newton rejected because they could not account for eclipses nor Kepler's three laws.

In contrast to his dissatisfaction with the views of Descartes, he was drawn to the atomic and mechanistic interpretation of nature of Gassendi, Boyle, and perhaps More. In addition, he began studying mathematics with the same dedication and brilliance he brought to his study of natural philosophy. Yet there was no assurance of his receiving a fellowship at Trinity necessary for attaining a permanent position at Cambridge, which seems to have been his goal even then. At the time academic advancement did not depend upon intellectual ability, but on social position and connections, patronage, and seniority. Lacking the latter, despite his growing reputation as a gifted student, Newton's chances looked bleak, leading to one of the "disorders" that he suffered throughout his life when undergoing stress, but then a seemingly benevolent fate intervened, though scholars are not sure in what form, bringing about his election to a fellowship in 1664.

After this ordeal, the next two years between 1664–1666 are referred to as his "*anni mirabilis*" because of the intense concentration that he was able to bring to his studies resulting in three papers applying his newly acquired mathematical skills to the problem of motion. As described by Westfall, the

> tract of October 1666 on resolving problems by motion was a virtuoso performance that would have left the mathematicians of Europe breathless with admiration, envy, and awe. As it happened, only one other mathematician in Europe, Isaac Barrow, even knew that Newton existed, and it is unlikely that in 1666 Barrow had any inkling of his accomplishment. The fact that he was unknown does not alter the other fact that the young man not yet twenty-four, without benefit of formal instruction, had become the leading mathematician of Europe. (pp. 137–38)

It was not just in mathematics that he had this burst of creativity, but while reading Descartes' theory of vortices in *Principles of Philosophy* he also became interested in the mechanics of planetary motion. He had learned of the principle of inertia from his studies of both Galileo and Descartes, but the principle that bodies remain at rest or if in motion continue to move indefinitely *in a straight line*, unless acted upon by another force, could not explain the elliptical orbits of the planets. By his own account it was during the plague years of 1665–1666, when Cambridge University was forced to close causing Newton to return to Woolsthorpe, that one day as he was sitting in the orchard and saw an apple fall it occurred to him that if the earth's gravity extended to the moon, then this could explain its deviation from a rectilinear path to an elliptical one, as had been surmised by Kepler.

With that insight he then began considering what the magnitude of the gravitational force reaching to the moon would have to be to cause this deviation. Knowing Kepler's third law that the size of a planet's orbit varies with the 3/2th power of its average distance from the sun, he inferred that the strength of the gravitational force must vary inversely with the square of the distance. As he recounts,

> I began to think of gravity extending to ye orb of the Moon & . . . from Kepler's rule of the periodical times of the Planets being in sesquialterate [3/2th] proportion of the distances from the center of their Orbs, I deduced that the forces wch keep the Planets in their Orbs must [be] reciprocally as the squares of their distances from the centers about wch they revolve: & thereby compared the force requisite to keep the Moon in her Orb with the force of gravity at the surface of the earth, & found them answer pretty nearly. All this was in the two plague years of 1665–1666. For in those days I was in the prime of my age for invention & minded Mathematics and Philosophy more than any time since. (p.143; brackets added)

Two additional theoretical developments at this time helped lay the foundation of the theory of dynamics presented later in the *Principia*. First, he corrected Descartes' laws describing the impact of bodies, instead claiming that they interact by mutually reciprocating forces. Second, extending the principle of inertia, he introduced "the principle of the conservation of angular momentum for the first time in the history of mechanics: 'Every body keeps the same reall quantity of circular motion and velocity so long as tis not opposed by other bodys" (p. 153). After describing these discoveries in a paper titled "The Laws of Motion," he began experiments on what he called the "celebrated Phenomena of Colours."

Colors, especially as seen in rainbows, had intrigued natural philosophers throughout history. Having studied Descartes' *Météores,* Boyle's *Experiments and Considerations Touching Colors*, and Hooke's *Micrographia*, Newton was well versed in the current theories of light. The consensus was that the light striking the retina was normally homogeneous producing the experience of ordinary "white" or "luminous" light, while the perception of colors was due to the modification of this homogeneous light by its reflection or refraction when stimulating the retina. Furthermore, it was generally believed that red and blue were the dominate colors caused by stronger or weaker impulses of ordinary light striking the retina, the other colors being mixtures of these, none of which were produced by separate waves or rays corresponding to each color, but by distortions of the homogeneous light hitting the retina.

Unconvinced by these explanations, Newton decided to perform his own prism experiments which even at this early age display his ingenuity and dexterity, along with his meticulous attention to detail. The results led him to the opposite of the accepted theory: rather than ordinary light being homogeneous, it was composed of a mixture of "rays" which the refractive powers of the prism "analyses" or "separates." It is these separate rays that produce the colors when refracted by the retina, though when undifferentiated they are experienced as transparent and luminous. One experiment in particular seems to have directed him to this conclusion.

> That ye rays wch make blew are refracted more y^{n} ye rays wch make red appears from this expirem^{nt}. If one hafe of ye thred *abc* be blew and ye other red & a shade . . . be put behind it y^{n} lookeing on ye thred through a prism one halfe of ye thred shall appeare higher y^{n} ye other & not both in one direct line, by reason of unequall refractions in ye two different colours. (p. 160)

The separation of the two colors by refraction into higher and lower levels was the key.

He then found that when a beam of light emerging from a pin hole was shone through a prism onto "a wall twenty-two feet aways," it produced "a spectrum five times as long as it was wide" (p. 164). He then showed that the original beam of light could be reconstituted by redirecting the dispersed rays through another prism, reinforcing the conclusion. So rather than ordinary light being basic and homogeneous and colors formed by its distortion, Newton's experiments indicated the refraction of ordinary light by a prism (or droplets of water) produces rays that cause the various colors of the spectrum (or of the rainbow) and when blended constitute ordinary luminous light, indicating the rays were fundamental. After all the millennia of investigations and conjectures, the youthful Newton had attained a partially correct explanation. As Westfall states: "No other investigation of the seventeenth century better reveals the powers of experimental inquiry animated by a powerful imagination and controlled by rigorous logic" (p. 164).

This is not to say that he realized this immediately. Retaining part of the accepted explanation, at first he thought red and blue were the primary colors, yellow, green, purple or violet being a blend of several of these: purple, for example, being a mixture of red and blue. However, he eventually realized his mistake, it being difficult to derive yellow from a combination of blue and red. He also tried to measure the dimensions of the different colors using thin transparent films, discovering the rings of color produced by circular lenses, now called "Newton rings." Having been attracted to the corpuscular-mechanistic theory of explanation, he rejected the wave theory of light held by Hooke and Huygens in favor of the corpuscular theory, persuaded by the fact that light casts sharp shadows, rather than the diffuse pattern of waves. Moreover, the interpretation of red and blue colors as caused by strong and weak rays suggested a mechanistic explanation in terms of the velocity of the corpuscles.

As was true two years earlier when he confronted the uncertainty of being awarded a scholarship to Trinity College, he now faced the comparable obstacle of being elected a fellow at Trinity if he were to gain a permanent position at Cambridge. This time, however, though his social background remained unchanged, he seemed unconcerned. As the choice depended upon the Master of the college and eight Senior Fellows, it would appear that he knew of a strong backing from one of the Senior Fellows, with Humphrey Babington the likely person, according to Westfall, though because of the secrecy of the proceedings it is not known with certainty. Regardless of who was responsible, he was elected a Minor Fellow on October 1, 1667 and then automatically a Major Fellow when he became Master of Arts nine months later.

Previously Isaac Barrow, who then held the Lucasian Professorship of Mathematics at Cambridge, had been asked by Newton's tutor to examine him on Euclid with regard to his election to the fellowship at Trinity. Though Newton received the fellowship, he believed Barrow had not been impressed because he had been studying Descartes' geometry rather than Euclid. But several years later when he showed Barrow his method for calculating infinite series, the latter was so taken that he sent a copy to John Collins identifying the author as 'Mr Newton, a fellow of our College & very young . . . but of an extraordinary genius & proficiency in these things" (p. 202). Collins "was a mathematical impresario who . . . functioned as a clearinghouse for information, attempting by his correspondence to keep the growing mathematical community of England and Europe abreast of the latest developments" (p. 202). Newton could not have been introduced to a more favorable person to appreciate his genius and advance his reputation. Impressed with the first mathematical paper sent by Barrows, Collins began sending Newton additional difficult mathematical problems the solution to which he would return within a month or so later. Even more pleased by his ingenuity, Collins send copies to the leading mathematicians in England, Scotland, and Europe. But when he generously offered to have several of Newton's articles published, "observing a wariness in him to impart," Collins finally decided to "desist, and doe not trouble him any more . . ." (p 226). Nonetheless, Newton's mathematical brilliance was becoming known which probably contributed to Barrow's decision to resign as Lucasian Professor, recommending Newton be appointed his successor.

So at the young age of twenty-seven Newton acceded to the prestigious and lucrative chair of Lucasian Professor. As Westfall states, the professorship

> ranked behind the masterships of the great colleges and the two chairs in divinity...as the ripest plum of patronage in an institution much concerned with patronage. On 29 October, 1669, this

> plum fell into the lap of an obscure young fellow of peculiar habits, apparently without connections, in Trinity College—to wit, Isaac Newton. (p. 206)

Yet the problem of his continuing at Trinity College and Cambridge University persisted.

When he had accepted the Lucasian Professorship he had sworn that he would "embrace the true religion of Christ with all my soul . . . and will take holy orders when the time prescribed by these statutes arrives, or I will resign from the college" (p. 179). With two exceptions "the sixty fellows of the college were required to take holy orders in the Anglican church within seven years of incepting M.A." (p. 179), and to remain celibate. The latter posed no problem for Newton because of his sexual orientation, but the oath he was required to take for ordination into the Anglican clergy did. As the time of decision approached, Newton devoted a number of years to studying early church history to decide whether in good conscience he could swear to uphold the Trinitarian doctrine, as required for ordination into the Anglican clergy, or be expelled from Cambridge.

After careful examination of the fourth century Council of Nicaea's decision to adopt the Athanasian Creed over the Arian doctrine, he concluded that "a massive fraud, which began in the fourth and fifth centuries, had perverted the legacy of the early church. Central to the fraud were the Scriptures, which Newton began to believe had been corrupted to support trinitarianism" (p. 313). So not believing the Athanasian Creed that God the Father, Christ the Son, and the Holy Ghost were consubstantial or of one substance, but like Arius holding that Christ and the Holy Spirit were created by God and therefore subordinate to Him, to avert being expelled from Trinity a special dispensation was necessary to avoid his having to swear the oath required for ordination into the Anglican clergy. Once again a favorable fate intervened, probably the influence of Isaac Barrow, so on "27 April, the dispensation became official. By its terms, the Lucasian Professor was exempted from taking holy orders unless 'he himself desires to . . .'" (p. 333). Worded "to give all just encouragement to learned men who are & shall be elected to ye said Professorship," it was not granted to Newton personally but to "the Lucasian professorship in perpetuity . . ." (p. 333).

Much more could be written about Newton's scholarly activities during these early years, but enough has been mentioned to indicate his brilliance and the growing recognition of his contributions. Yet one further example of his multifaceted intelligence should be mentioned, his inventiveness, as it brought him additional immediate recognition. As early as his school days at Grantham he had been fascinated by mechanical devices, so when he discovered during his optical experiments that chromatic aberration was inevitable in refracting telescopes, he decided to build a reflecting telescope to eliminate the distortion. Without the help of an artisan, Newton "cast and ground the mirror from an alloy of his own invention," along with "the tube and the mount" (p. 233). Only about six inches long, the telescope nonetheless magnified objects nearly forty times in diameter (Westfall has a drawing of the telescope on p. 235).

Learning of his invention, the Royal Society of London asked that it be sent to them so they could examine it. At the end of the year 1671 Barrow, who was a member, delivered it to the Society. Delighted when they tested it, Henry Oldenburg, who was the Secretary of the Society, wrote to Newton in January asking for the exact specifications which would be sent "in a solemne letter to Paris to M. Hugens, thereby to prevent the arrogation of such strangers, as may perhaps have seen it here, or even wth you at Cambridge . . ." (p. 236). Huygens, who was the outstanding scientist in the field of optics in Europe and who could help prevent anyone there from taking credit for its invention, when informed of the telescope in his reply admiringly referred to the "marvellous telescope of Mr. Newton."

In recognition of his skillful invention Newton was "proposed candidate" for the Royal Society and duly elected on January 11, 1672, the first of many honors to be awarded him. In his letter to Oldenburg thanking him for the nomination, Newton wrote that to show his gratitude he would submit to the Society an account of a discovery that was of more scholarly significance than the telescope, referring to his theory of colors. Sending the paper describing his experimental conclusions, he soon received a reply from Oldenburg "filled with lavish praise" and informing him that when it was read to the Royal Society it "mett both with a singular attention and an uncommon applause," such that the "Society had ordered that it be printed forthwith in the *Philosophical Transactions* if Newton would agree" (p. 239). For once Newton's reticence to recognition was overcome and he agreed to its publication on February 6, 1672, which, because of the criticism it drew from defenders of the traditional interpretation, including Hooke and Huygens, was regretted. But with that we shall now turn to the presentation of his major discoveries in his three important treatises.

EVENTS PRECEDING THE PUBLICATION OF THE *PRINCIPIA*

Before turning to Newton's creation of celestial mechanics, it might prove helpful to recall the previous advances in natural philosophy that led up to it. Copernicus began the process by adopting Aristarchus' heliocentric system, along with the two motions attributed to the earth, the diurnal rotation by Heraclides and the annual revolution by Philolaus, on the basis that it was simpler and more harmonious. Next were the contributions of Kepler based on Tycho Brahe's exact naked eye astronomical observations and the Copernican system that enabled him to discovery his three laws: (1) that the planetary orbits are elliptical, not circular; (2) that a radius vector joining the planet to the sun sweeps out equal areas in equal times so that the motions of the planets are not uniform, but vary from faster at the perihelion to slower at the aphelion; and (3) that the planetary periods vary in the ratio of their 3/2th distance from the sun: that is, as their *distances* are squared their *orbital periods* are cubed. In addition, he replaced Aristotle's "intelligences," the medieval "crystalline spheres," "angels" and "souls," plus his earlier "*anima motrix*" as the cause of the planetary motions with two physical forces, magnetism and gravity forming, as he stated, a mechanistic, clockwork celestial system.

Galileo had the honor of substantially demolishing Aristotle's 2000 year old system with its ontological distinction between the celestial and terrestrial worlds involving their different substances and contrasting circular and terrestrial motions. In addition, he replaced Aristotle's methodological reliance on ordinary observations, logical deductions, and four causes with the search for mathematical correlations in nature and the use of experimentation to discover new regularities and to test theories and laws, the methodology adopted by modern science. Using these revised methods he discovered and confirmed the gravitational law of free fall, that objects in a vacuum fall with the same velocity regardless of their weights, and that the pattern of a projectile is parabolic.

In 1679 Robert Hooke wrote to Newton asking to renew their correspondence and inviting his comments on Hooke's proposed celestial mechanics based on an earlier essay in 1674, *Attempt to Approve the Motion of the Earth.* Republished in 1679 as *Lectiones Cutlerianae,* it included a theory of planetary motions titled "System of the World" that is significant because it also is the title of Volume 2 of the *Principia*, and therefore would be the basis of Hooke's later charge of plagiarism. Hooke's "System" contains "three suppositions" basic to an explanation of planetary motions, but does not include the crucial formulation of the exact law of universal gravitation, which he should have derived, like Newton, from Kepler's law. In his letter he stated that

> I shall explain a System of the World differing in many particulars from any yet known, answering in all things to the common rules of mechanical motions. This depends upon three suppositions: first, that all celestial bodies whatever have an attractive or gravi-

> tating power towards their own centers, whereby they attract not only their own parts, and keep them from flying from them, as we may observe the earth to do, but they do also attract all the other celestial bodies that are within the sphere of their activity; and consequently that not only the sun and moon have an influence upon the . . . motion of the earth, and the earth upon them, but that Mercury, also Venus, Mars, Jupiter, and Saturn, by their attractive powers, have a considerable influence upon its motion as in the same manner the corresponding attractive power of the earth hath a considerable influence upon every one of their motions also. The second supposition is this: that all bodies whatsoever that are put into a direct and simple motion, will so continue to move forward in a straight line, till they are by some other effectual powers deflected and bent into a motion, describing a circle, ellipse, or some other more compounded curve line. The third supposition is: that these attractive powers are so much the more powerful in operating, by how much the nearer the body wrought upon is to their own centers. Now what these several degrees are I have not yet experimentally verified; but it is a motion, which if fully prosecuted as it ought to be will mightly assist the astronomer to reduce all celestial motions to a certain rule, which I doubt will never be done true without it.[3]

Although Hooke's three suppositions were not exactly original since they duplicated Kepler's first two laws, they did provide the general conceptual framework of the new celestial mechanics. Unfortunately, Hooke could not transform his "suppositions" into exact mathematical laws, which was left to Newton.

While identifying all the dynamic components essential to explaining planetary motions—that all the planets exert a gravitational attraction to their centers on all the other planets, that this attractive force is what deflects their inertial tangential motion into a curved path, and that the strength of this force is inversely proportional to the distances separating the planets—Hooke, unlike Kepler, was unable to calculate what the magnitude of such an attractive force varying with the distance would have to be to transform the inertial tangential motions into a continuous elliptical orbit.

Recall that thirteen years earlier Newton, using his recently discovered method of fluxions (his differential calculus), had "deduced" from Kepler's third law describing the periods of the planets as varying with the 3/2th power of their distance from the center of their orbs, "that the forces wch keep the Planets in their Orbs must [be] reciprocally as the squares of their distances from the centers about wch they revolve . . ." (Westfall, p. 143). But distracted by other mathematical pursuits, his prism experiments, his research into church history in connection with the problem of ordination, and a growing interest in alchemy, he put aside his investigations of planetary motions.

Then a visit to Newton in 1684 by Dr. Halley (of Halley's comet fame) reignited his interest in the problem. During the conversation Halley asked what he

> thought the Curve would be that would be described by the Planets supposing the force of attraction towards the Sun to be the reciprocal to the square of their distance from it. Sr Isaac replied immediately that it would be an Ellipsis; the Doctor struck with joy and amazement asked him how he knew it, why saith he, I have calculated it. . . . (p. 403)

Once his interest was aroused Newton was not satisfied with demonstrating how he derived the elliptical orbits from his inverse square law, but was determined to create the dynamic framework (that is, the forces involved) of celestial mechanics. Two months later he sent Halley a short nine page essay, *De Motu corporum in gyrum* (*On the Motion of Bodies in an Orbit*) showing not only that elliptical orbits can be derived from the inverse square law, but added the crucial statement "that an elliptical orbit entails an inverse-square *force* to one focus" (p. 404; italics added). While most natural philosophers at the time rejected the concept of forces, especially if they implied action at a distance, according to Westfall Newton's "alchemical inquiries led him to be receptive to the efficacy and necessity of forces, whether they acted at a distance or through . . . some medium." Furthermore, he expanded his investigations to include not just the orbits of the planets, but also the orbits of the satellites of Jupiter and Saturn (its "horns" interpreted later as satellites), the moon's orbit, the causes of the tides, and a correct description of the motion of the comets.

This led to three successive revisions of the *De Motu* that show his progress in the clarification of concepts, description of the laws of motion, and formulation of the mathematics that foreshadow the presentation in the *Principia.* For example, he clearly distinguished between the "inherent force of inertia" and the "impressed force" of attraction, initially using Galileo's parallelogram diagram to calculate the combined forces. Noting that acceleration could be considered either a change in velocity or direction, he concluded that uniform circular motion was dynamically similar to the uniform acceleration of gravitational fall. Thus while an inherent inertial force results in a continuous *rectilinear* velocity, a constantly impressed force produces a continuous *change* in velocity proportional to the two forces, a significant insight as Westfall states: "[i]n this proportionality lay the possibility of a quantitative science of dynamics that would cap and complete Galileo's kinematics" (p. 417).

Newton termed the gravitational attraction to the center a "centripetal force" in contrast to Huygens's "centrifugal force" directed outward by rotary motion. In deference to all the natural philosophers who were unfamiliar with his method of fluxions (or Leibniz's differential calculus) which he used in his calculations, Newton demonstrated his results geometrically. Letting the arc segments of curved orbits represent time, using his fluxions he was able to determine that the tangential deviation proportional to the squares of the times corresponded to Galileo's law of free fall ($a = \frac{1}{2} gt^2$), along with Huygens's versed sign ratio computing the deviation from the tangent proportional to the centrifugal force. One can imagine his satisfaction at finding these agreements.

Distinguishing mass from weight, he defined the former as proportional to the density or quantity of matter, the latter produced in a gravitational field. Dissatisfied with his earlier use of Galileo's parallelogram of forces to compute the combined strength of inertial and gravitational forces, he replaced this with what became his third Law of Motion in Volume I of the *Principia: "To every action there is always apposed an equal reaction: or, the mutual actions of two bodies upon each other are always equal, and directed to contrary parts."*

In addition to these innovations, his experiments similar to those of Galileo's on the oscillations of pendulums of different weights convinced him that the attractive force of a material body consists of the totality of its constituent particles, measured not from the circumference but from the center of the body, a tremendous simplification because it could be more easily and exactly determined. Thus for purposes of calculation the entire planetary system could be represented as a collection of mass-points whose individual motions, due to the mutual gravitational forces among all the bodies calculated from their centers, actually varied inversely as the square of the distance from the nearest or greatest massive body. For example, as we now know as a result of Einstein's equation $E = mc^2$, the sun, because of its enormous energy, comprises 98% of the mass of the solar system, thus simplifying the calculation of the orbits of the planets, while the orbits of the moon and the satellites of Jupiter can be computed mainly from the gravitational effects of the earth and of Jupiter.

These improved discoveries and definitions expanded the nine-page essay he had sent to Halley the year before into two books more than ten times larger with a new title, *De motu corporum* (*On the Motion of Bodies*) which would become the heading of Volume I of the *Principia*. Early in the year of 1686 the two books were nearing completion which virtually contained the contents of Books I and III of the *Principia*, but Book I was so enlarged that he separated the latter portion into Book II. When completed he sent them to the Royal

Society. According to his own account written in a memorandum about thirty years later: "The Book of Principles [the *Principia*] was writ in about 17 or 18 months, wereof about two months were taken up with journeys, & the MS was sent to y[e] R. S. in spring 1686; & the shortness of the time in which I wrote it, makes me not ashamed of having committed some faults" (p. 444; brackets added). Although the contents were based on his previous two books except for the axiomatic organization, that such a colossal work was written in such a short time is another manifestation of Newton's incredible concentration and genius.

Deciding not to submit the manuscript himself, Newton had Dr. Vincent present to the Royal Society on April 28,1686, a manuscript treatise titled

> *Philosophiae naturalis principia mathematica*, and dedicated to the Society by Mr. Isaac Newton, wherein he gives a mathematical demonstration of the Copernican hypothesis as proposed by Kepler, and makes out all the phaenomena of the celestial motions by the only supposition of a gravitation towards the center of the sun decreasing as the squares of the distances therefrom reciprocally [somewhat of an oversimplification]. (pp. 444–45; brackets added)

A letter of thanks was sent to Mr. Newton by the Society with the decision as to its printing referred to the council of the Society, while the treatise was given to Mr. Halley.

When three weeks passed with no action by the council Halley raised the question of its publication at a meeting of the council on May 19, whereupon the society voted that "Mr. Newton's *Philosophiae naturalis principia mathematica* be printed forthwith in quarto in fair letter; and that a letter be written to him to signify the Society's resolution, and to desire his opinion as to the print, volume, cuts, &" (p. 445). Halley also oversaw the printing of the book which was completed on July 5, 1687. Unlike Copernicus who never acknowledged Rheticus' role in convincing him to write the *De revolutionibus* and arranging its printing, Newton graciously praised Halley (as well as the Royal Society) for his dedicated assistance in the publication of the *Principia*.

> In the publication of this work the most acute and universally learned *Mr. Edmund Halley* not only assisted me in correcting the errors of the press and preparing the geometrical figures, but it was through his solicitations that it came to be published; for when he had obtained . . . my demonstrations of the figure of the celestial orbits, he continually pressed me to communicate the same to the *Royal Society*, who afterwards, by their kind encouragement and entreaties, engaged me to think of publishing them.[4]

Not satisfied with acting as midwife to Newton's *Principia*, Halley also wrote an "Ode to Newton" that concluded with the moving lines:

> *Then ye who now on heavenly nectar fare,*
> *Come celebrate with me in song the name*
> *Of Newton, to the Muses dear; for he*
> *Unlocked the hidden treasuries of Truth:*
> *So richly through his mind had Phoebus cast*
> *The radiance of his own divinity.*
> *Nearer the gods no mortal may approach.* (p. xv)

This beautiful tribute to Newton, along with *Cote's Preface to the Second Edition* and Newton's Prefaces to the three editions, have been included in the present two-volume work. This consists of Andrew Motte's translation of it into English in 1729 as *Sir Isaac Newton's Mathematical Principles of Natural Philosophy and his System of the World,* revised by Florian Cajori in 1934, based on the third edition "prepared with much care by *Henry Pemberton*, M.D., a man of the greatest skill in these matters," as Newton wrote in the Preface to the Third Edition.

VOLUME I, "THE MOTION OF BODIES," OF THE *PRINCIPIA*

Turning then to VOLUME I, BOOK I, "The Motion of Bodies," of the *PRINCIPIA*, in a much quoted passage in the Preface to the First Edition, Newton presents a concise statement of the purpose:

> . . . I offer this work as the mathematical principles of philosophy, for the whole burden of philosophy seems to consist in this—from the phenomena of motions to investigate the forces of nature, and then from these forces to demonstrate the other phenomena; and to this end the general propositions in the first and second Books are directed. In the third Book I give an example of this in the explication of the System of the World; for by the propositions mathematically demonstrated in the former Books, in the third I derive from the celestial phenomena the forces of gravity with which bodies tend to the sun and the several planets. Then from these forces, by other propositions which are also mathematical, I deduce the motions of the planets, the comets, the moon, and the sea. (pp. xvii–xviii)

Following this description of what he succeeded in accomplishing, he continues by indicating what still remains unknown which he prophesizes will be discovered in the future with the same method. As he predicted, scientists centuries later would discover, using his mechanical principles, the subatomic particles and the strong and weak nuclear forces that would throw greater light on the problem.

> I wish we could derive the rest of the phenomena of Nature by the same kind of reasoning from mechanical principles, for I am induced by many reasons to suspect that they all depend upon certain forces by which the particles of bodies, by some causes hitherto unknown, are either mutually impelled towards one another, and cohere in regular figures, or are repelled and recede from one

> another. These forces being unknown, philosophers have hitherto attempted the search of Nature in vain; but I hope the principles here laid down will afford some light either to this or some truer method of philosophy. (p. xviii)

Emulating the rigorous geometrical style of Euclid and Archimedes, he begins Book I with precise definitions of the concepts he will use, 'mass,' 'momentum' (without using the term), and 'centripetal force,' followed by a Scholium (remarks) justifying his conception of the ontological coordinates of the universe, absolute space and time. Having concluded years earlier that Descartes' conception of relativistic motion encouraged atheism because it denied the absolutes associated with God's creation, he wrote the Reverend Bentley that "[w]hen I wrote my treatise about our Systeme I had an eye upon such Principles as might work wth considering men for beliefe of a Deity" (Westfall, p. 441), namely, absolute time, space, place, and motion.

According to his definitions:

> Absolute, true, and mathematical time, of itself, and from its own nature, flows equably without relation to anything external, and by another name is called duration: relative, apparent, and common time, is some sensible and external . . . measure of duration . . . which is commonly used instead of true time: such as an hour, a day, a month, a year.
>
> Absolute space, in its own nature, without relation to anything external, remains always similar and immovable. Relative space is some moveable dimension or measure of the absolute spaces; which our senses determine by its position to bodies. . . .
>
> Place is a part of space which a body takes up, and is according to the space, either absolute or relative. . . . (p. 6)
>
> Absolute motion is the translation of a body from one absolute place into another; and relative motion the translation from one relative place into another. Thus in a ship under sail, the relative place of a body is that part of the ship which the body pos-

> sesses. . . . But real, absolute rest is the continuance of the body in that same part of that immovable space, in which the ship itself, its cavity, and all that it contains, is moved. (p. 7)

These absolutes, which were accepted until the Michelson-Morley experiments and Einstein's theories of relativity, seem reasonable enough because they are abstractions from ordinary experience. But as we have continually learned, such abstractions usually are not true or just approximations of the independent universe. Because there is no direct evidence of absolute space or absolute time, any evidence of their existence in contrast to just inferential conclusions, had to be indirect. I know of no one a the time who attempted to experimentally confirm the existence of absolute time, and the best Newton could think of as evidence of absolute space was the changing shape of the surface of the water in a bucket which changed from a stationary to a rotating position. At rest the surface of the water is flat, but when the bucket is swiftly rotated the circular motion conveys a centrifugal force to the water that causes it to ascend the sides of the vessel indicative, Newton thought, of an absolute circular motion. As he states:

> The effects which distinguish absolute from relative motion are, the forces of receding from the axis of circular motion. . . . If a vessel . . . held at rest together with the water . . . is whirled about the contrary way . . . the surface of the water will at first be plain . . . but after that, the vessel, by gradually communicating its motion to the water, will make it begin sensibly to revolve . . . and ascend to the sides of the vessel, forming itself into a concave figure. . . . This ascent of the water shows its endeavor to recede from the axis of its motion; and the true and absolute circular motion of the water . . . becomes known. . . . (p. 10)

Because it is difficult to understand how this effect could be produced by an empty space, however absolute, Bishop Berkeley and later Ernst Mach argued that it was not motion relative to absolute space that produced the centrifugal force, but motion with respect to the stellar masses of the universe. Then in the latter nineteenth century A. A. Michelson and E. W. Morely devised an ingenious experiment to confirm the existence of absolute space. At the time the wave theory of light had gained ascendance and since waves are the configuration of a medium in which they are propagated, it was supposed that an ether pervaded the universe as the medium for the light waves. In addition, it was conjectured that since the earth revolves in the ether creating an "ether drift," this would alter the velocity of light depending upon whether the waves were in the direction of the ether drift or traversed it. Using an interferometer in which two light beams were sent in equal but perpendicular directions, one with and the other across the ether, they expected that the effect of the ether on the two velocities would result in their being out of phase when they were reflected back and measured in the interferometer.

Much to their consternation they never found the returning waves to be out of phase, even though the experiment was conducted very carefully in different locations on the earth to eliminate any fortuitous influences. There were various possible explanations of the null results, none of which seemed plausible: that the earth was not moving, that there was no ether, or that light was not a wave. Then in one of his famous four articles published in the *Annalen der Physik* in 1905, Einstein produced a surprising explanation. He argued that the asymmetries of Maxwell's electrodynamics "together with the unsuccessful attempts to discover any motion of the earth relatively to the 'light medium' [the ether], suggest that the phenomena of electrodynamics as well as of mechanics possess no properties corresponding to the idea of absolute rest [or motion]."[5]

Thus from the null results of the Michelson-Morley experiments which had predicted a *change* in the velocity of light due to the ether resting in absolute space, Einstein drew the unusual conclusion in the Special Theory of Relativity that the velocity of light was *invariant* or *constant* and therefore space (and time) were not privileged or absolute frames of reference. Moreover, he determined in his General Theory that the magnitudes of space (units of length), time (intervals of duration), and mass (compactness) were not velocity-invariant, but changed radically as physical systems either approached the velocity of light or entered dense gravitational fields. Also, the existence of the ether was denied as a necessary medium of electromagnetic propagation. Thus it was concluded that Newton's frames of the universe were not privileged or absolute, but an approximation when the velocity of a system is insignificant relative to that of light.

The defense of absolute space, time, and motion in the Scholium is followed by the next section titled AXIOMS, OR LAWS OF MOTION. LAW I is the law of inertia (without using the term), the key concept in the new celestial mechanics: "*Every body continues in its state of rest, or of uniform motion in a right {straight} line, unless it is compelled to change that state by forces impressed on it*" (p. 13; brackets added). LAW II describes the change in inertial motion due to impressed forces: "*The change of motion is proportional to the motive force impressed; and is made in the direction of the right line in which that force is impressed*" (p. 13). LAW III expresses the reciprocal impact of interacting bodies: "*To every action there is always opposed an equal reaction: or, the mutual actions of two bodies upon each other are always equal, and directed to contrary parts*" (p. 13). *The Laws of Motion* is followed by numerous *Corollaries*, *Lemmas*, additional *Scholia*, and *Propositions* supported by innumerable diagrams to further elucidate his system, but which are too extensive and complex to describe, but represent further evidence of his meticulous procedure.

The next section, Book One: *The Motion of Bodies*, presents Newton's exact mathematical demonstrations of the motion of bodies. Based on his conception of attractive forces Newton, as Galileo, states he is not going to speculate about the nature of the forces producing the motion of bodies, whether they are spirits, ether, or the air, but identify them and their mathematical proportions, as he writes in the Scholium.

> I here use the word *attraction* in general for any endeavor whatever, made by bodies to approach . . . each other, whether that endeavor arises from the action of the bodies themselves, as tending to each other or agitating each other by spirits emitted; or whether it arises from the action of the ether or of the air, or of any medium whatever, whether corporeal or incorporeal, in any manner impelling bodies placed therein towards each other. In the same general sense I use the word *impulse*, not defining in this treatise the species or physical qualities of forces, but investigating the qualities and mathematical proportions of them. . . . (p. 192)

Though he normally resorts to geometrical diagrams to elucidate his computations, he occasionally uses his theory of fluxions when he deals with ratios approaching vanishing limits: for example, Aristotle's paradox as to how a vertical projectile at its peak can reverse its direction of motion while momentarily coming to rest or how there can be instantaneous velocities since it would imply motion in durationless time intervals. His method of fluxions and Leibniz's differential calculus were created precisely to deal with problems as to how the rate of a dependent variable changes as the independent variable approaches the vanishing magnitude or "limit" of zero. Thus the differential notation ds/dt represents the infinitesimal distance magnitude as the time approaches zero. As Newton states in his reply to the Aristotelians,

> it may be alleged that a body arriving at a certain place, and there stopping, has no ultimate velocity. . . . But the answer is easy; for by the ultimate velocity is meant that with which the body is moved, neither before it arrives at its last place and the motion ceases, nor after, but at the very instant it arrives . . . and with which the motion ceases. And in like manner, by the ultimate ratio of evanescent quantities is to be understood the ratio of the quantities not before they vanish nor afterwards, but with which they vanish. (pp. 38–39)

This is where the concept of limit applies. "For those ultimate ratios with which quantities vanish are not truly the ratios of ultimate quantities, but limits towards which the ratios of quantities decreasing without limit do always converge . . ." (p. 39). This is an excellent example of the way in which the mathematical formalism provides the scientist with the means of computing such magnitudes and therefore why it is so indispensable in dealing with physical problems, as Newton attests.

> In mathematics we are to investigate the quantities of forces with their proportions consequent upon any conditions supposed; then, when we enter upon physics, we compare those proportions with the phenomena of Nature, that we may know what conditions of those forces answer to the several kinds of attractive bodies. And this preparation being made, we argue more safely concerning the physical species, causes, and proportion of the forces. (p. 192)

This function of mathematics is illustrated in his description of how planetary bodies are continuously deflected from a rectilinear trajectory to an oval one, due to the gravitating force of the nearest massive body, deducing the exact magnitudes of the distances, velocities, and forces to produce Kepler's first two laws. Thus he explained how the gravitational attraction of the sun deflected the orbital motion of the planets to an ellipse, unknown to Kepler. He then demonstrates that if the attractive gravitational force decreases with the square of the distance according to Kepler's third law, then "*the periodic times in ellipses are as the 3/2th power . . . of their greater axis*" (p. 62).

In another Scholium toward the end of Book I, he declares that the "attractions bear a great resemblance to the reflections and refractions of light" according to the investigations of light by Willebrord Snell and Descartes, adding that "it is now certain from the phenomena of Jupiter's satellites, confirmed by the observations of different astronomers, that light is propagated in succession, and requires about seven or eight minutes to travel from the sun to the earth" (p. 229). While this only presents a fraction of the contents of Book I, it should show how Newton's description of "The Motion of Bodies" completed Kepler's and Galileo's seminal discoveries with his precise definitions and mathematical laws of motion based on attractive forces.

Book II includes nine sections discussing problems such as the effects of resisting media (tenacity, attrition, and density) on the motion of bodies, the compression and density of fluid substances, the motion and resistance of pendular bodies, the properties of particles in liquids affecting their fluidity, diffusion, and resistance, the manner in which "tremulous" or oscillating bodies propagate their motions in an elastic medium, along with an attempt to find the "velocity of waves" and the "distance of the pulses." There seems to have been no limits to his attempted explanations. Book II ends with another refutation of Descartes' theory of vortices which he found implied that "the periodic times of the parts of the vortex to be as the square of the distances from the center of motion" (p. 393), rather than cubed according to Kepler's third law. He concluded "that the hypothesis of vortices is utterly irreconcilable with astronomical phenomena, and rather serves to perplex than explain the heavenly motions" (p. 396).

Nonetheless, the theory of vortices consisting of a purely mechanical explanation devoid of forces was preferred at Cambridge and had prominent adherents among continental natural philosophers, including Huygens, Perrault, Bernoulli, and others. When Voltaire visited England to familiarize himself with Newton's system he wrote: "a Frenchman who arrives in London finds himself in a completely changed world. He left the world *full*, he finds it *empty* [absolute space]. In Paris the universe is composed of vortices and subtle matter, in London there is nothing of the kind."[6] This is a perceptive contrast at the time between the English and the French outlook.

In the present edition, Book III constitutes Volume 2 of the *Principia*, titled *The System of the World*. Although Newton originally intended it to be a popular nonmathematical presentation of his system of the world, he later recast it in a more rigorous mathematical form to preclude controversy from those who were unable to follow the mathematical demonstrations and therefore found it unconvincing. As he states in the introductory paragraph:

> In the preceding books I have laid down the principles of philosophy; principles not philosophical but mathematical: such, namely, as we may build our reasonings upon in philosophical inquiries. . . . It remains that, from the same principles, I now demonstrate the frame of the System of the World. Upon this subject I had, indeed, composed the third Book in a popular method, that it might be read by many; but afterwards, considering that such as had not sufficiently entered into the principles could not easily discern the strength of the consequences, nor lay aside the prejudices to which they had been many years accustomed, therefore, to prevent the disputes which might be raised upon such accounts, I chose to reduce the substance of this Book into the form of Propositions (in the mathematical way). . . . (p. 397)

He then presents his Rules of Reasoning in Philosophy that explicitly state the principles guiding scientific investigations which had been followed by his predecessors and that would serve as canons for future research. Rule I, which Galileo had previously stated, affirms Occam's razor that explanatory principles should not be multiplied beyond necessity and "that Nature does nothing in vain:" "*We are to admit no more causes of natural things than such as are both true and sufficient to explain their appearances.*" Rule II proclaims the uniformity of nature that under comparable conditions similar causes will produce similar effects: "*Therefore to the same natural effects we must, as far as possible, assign the same causes.*" Rule III is an explicit justification of the primary qualities used by Galileo, Gassendi, Boyle, and Locke to define the atomic-corpuscular theory as the infrastructure of the physical world: "*The qualities of bodies, which admit neither intensification nor remission of degrees, and which are found to belong to all bodies within the reach of our experiments, are to be esteemed the universal qualities of all bodies whatsoever.*" Rule IV avers the empirical foundations of science that is not to be replaced by any nonscientific principles: "*In experimental philosophy we are to look upon propositions inferred by general induction from phenomena as accurately or very nearly true, notwithstanding any contrary hypothesis that may be imagined, till such time as other phenomena occur, by which they either be made more accurate, or liable to exceptions*" (pp. 398–400). Having replaced Aristotle's *Organon* as the basis of scientific reasoning, these principles have guided scientific inquiry ever since and can be considered the bedrock of modern science. It is unfortunate that they have not been read by defenders of intelligent design or special creation.

Rule III addresses the veridicality of scientific discoveries and worldview versus ordinary experience. Like the invention of telescopes that disclosed new phenomena in the astronomical world, the construction of microscopes in the latter half of the seventeenth century revealing a new dimension of microscopic entities reinforced the distinction between objects as they appear to us and as they exist in themselves, as did Boyle's explanation of his gas laws by the kinetic motion of the insensible particles composing the gas. Newton is justifying believing that certain independently existing physical properties belong to insensible objects in contrast to their sensory qualities.

> We no other way know the extension of bodies than by our senses, nor do these reach it in all bodies; but because we perceive extension in all that are sensible, therefore we ascribe it universally to all others also. That abundance of bodies are hard, we learn by experience; and because the hardness of the whole arises from the hardness of the parts, we therefore justly infer the hardness of the undivided particles not only of the bodies we feel but of all others. That all bodies are impenetrable, we gather not from reason, but from sensation. . . . The extension, hardness, impenetrability, mobility, and inertia of the whole, result from the extension, hardness, impenetrability, mobility, and inertia of the parts; and hence we

> conclude the least particles of all bodies to be also extended, and hard and impenetrable, and movable, and endowed with their proper inertia. And this is the foundation of all philosophy. (p. 399)

Thus began the renewed tentative acceptance of an atomic or corpuscular infrastructure that would await the identification of atomic elements and molecular substances in the eighteenth century, the discovery of the electron, proton, and neutron, along with alpha, beta, and gamma rays in the latter nineteenth and early twentieth centuries, and especially Einstein's explanation of Brownian motion owing to the impact of the water molecules agitating the dissolved pollen grains (analogous to Boyle) before the existence of atoms and molecules would be finally accepted.

Rule IV, the final rule, presents Newton's famous distinction directed at Descartes between "feigned" or "framed" hypotheses and those deduced from observations: that is, hypotheses that are introduced unsupported by inductive laws or experimental evidence, such as Descartes' theory of vortices and the earlier concepts used by Kepler to explain the structure and motion of the celestial orbs.

An important principle, Newton was not entirely clear in his distinction between "feigned" hypotheses and those which were "deduced from phenomena." In the famous General Scholium added to Book III in 1713, when the second edition of the *Principia* was brought out, he does make more explicit the difference between deduced and framed hypotheses, stating

> . . . hitherto I have not been able to discover the cause of those properties of gravity from phenomena, and I frame no hypotheses; for whatever is not deduced from the phenomena is to be called an hypothesis; and hypotheses, whether metaphysical or physical, whether of occult qualities or mechanical, have no place in experimental philosophy. In this philosophy particular propositions are inferred from the phenomena, and afterwards rendered general by induction. (p. 547)

Further clarification occurs in a letter to Oldenburg quoted by Cajori, indicating that hypotheses can be used to explain the properties of objects *after* they have been discovered by experiment, but not to predetermine them *before* the experiment.

> For the best and safest method of philosophizing seems to be, first diligently to investigate the properties of things and establish them by experiment, and then to seek hypotheses to explain them. For hypotheses ought to be fitted merely to explain the properties of things and not attempt to predetermine them except in so far as they can be an aid to experiments. (p. 673)

Yet he did not follow his own prescription. Though critical of Descartes' use of vortices that was not based on experiments but on "clear and distinct ideas," Newton endorsed alchemy as a means "to profit & edification. . . ."

> For Alchemy . . . is not of that kind wth tendeth to vanity & deceit but rather to profit & to edification inducing first ye knowledge of God & secondly ye way to find out true medicines . . . so y^{t} ye scope is to glorify God in his wonderful works, to teach a man how to live well, & to be charitably affected helping o^{r} neighbors. (Westfall, 298)

Recall that it was his alchemical investigations that induced Newton to introduce such nonphysical effects as gravity and the attractive and repulsive forces of minute particles believing that these provided better explanations than the "hooked shapes" or direct contact of atoms accepted by Locke, Hooke, and Huygens.

He seems explicitly to violate his previous disavowal of "framing hypotheses" *("Hypotheses non Fingo")* when, in the next paragraph, he introduces "a certain most subtle spirit" that might explain the action of these various forces, even though a spirit could not be inferred from inductive or experimental evidence but is entirely conjectural.

> And now we might add something concerning a certain most subtle spirit which pervades and lies hid in all gross bodies; by the force and action of which spirit the particles of bodies attract one another at near distances, and cohere, if contiguous; and electric bodies operate to greater distances, as well repelling as attracting the neighboring corpuscles; and light is emitted, reflected, refracted, inflected, and heats bodies; and all sensation is excited, and the members of animal bodies move at the command of the will, namely, by the vibrations of this spirit, mutually propagated along the solid filaments of the nerves, from the outward organs of sense to the brain, and from the brain into the muscles. But these are things that cannot be explained in a few words, nor are we furnished with that sufficiency of experiments which is required to an accurate determination and demonstration of the laws by which this electric and elastic spirit operates. (p. 547)

Later in the *Opticks* he will discard the concept of a "subtle spirit" for the more empirical or scientific concept of a pervading "Aether." Yet even if the concept of a subtle spirit does conflict with his previous rejection of hypotheses that are not deduced from phenomena nor derived from experiments, this statement could be seen as somewhat prescient, for it anticipates the current search among physicists for a grand unified theory (GUTS). We can forgive him at this early stage for thinking that this unified force was a spiritual one, rather than an "electrodynamic field," a "space-time continuum," or even a concatenation of strings, theories proposed by contemporary physicists. Knowing nothing about the present explanation of nervous discharges by changes in electrical potential among the ions in the outer membrane of neurons, he foresaw the propagation of nerve impulses "along the solid filaments of the nerves, from the outward organs of sense to the brain. . . ." Thus it could be argued that his foresight was remarkable even if he condoned alchemy and the use of spirit in his ultimate explanation, in opposition to his own injunction against feigned hypotheses.

In the General Scholium at the end of The System of the World he presents a final account of how the orbits of the superior and inferior planets with their satellites follow from Kepler's three laws of motion and are supported by the astronomical observations of John Flamsteed, the Royal Astronomer. In addition, he demonstrates that Kepler's elliptical orbits require the sun to be in the center, finally and conclusively confirming the Copernican heliocentric system. However, he shows that the orbital revolutions of the planets have a center of gravity that is at rest apart from the sun. The motion of the comets and the cause of the tides are again explained. With this he terminates the *Principia,* generally acknowledged to be the greatest scientific treatise ever written, which created the theoretical foundation for understanding the modern world for several centuries. Its impact was so astonishing that even some scholars wondered whether it could have been created by an ordinary mortal. For example, after reading the *Principia* Marquis de l'Hôpital, a French mathematician of considerable repute,

> cried out with admiration Good god what a fund of knowledge there is in that book? he then asked . . . every particular about S^r I[saac] even to the colour of his hair said does he eat & drink & sleep. is he like other men? & was surprised when . . . told . . . he conversed cheerfully with his friends, assumed nothing & put himself upon a level with all mankind. (Westfall, 473; brackets added with punctuation as in the original)

HIS COURAGEOUS STAND, NERVOUS BREAKDOWN, AND FINAL YEARS

A further example of Newton's exemplary character and moral courage occurred at the time when he was finishing his masterpiece. King James II who came to the throne of the United Kingdom in 1685 was determined to make Catholicism the established religion. Deciding that this could be most readily accomplished by Catholics attaining positions of power at the universities, on February 8, 1687, he tried using the traditional "letters of mandate" to confer higher degrees on Catholics to qualify them for positions of authority without their having to take "the oath of supremacy, in effect an oath to uphold the established Anglican religion."[7] When the King attempted to admit a Benedictine Monk, Alban Francis, to the degree of Masters of Art at Cambridge and the vice chancellor John Peachell decided to resist, Newton drafted a supporting letter "urging 'an honest Courage' which would 'save ye University" (p. 475).

He insisted that the mandate should be opposed because if this appointment violating university oaths were permitted, it would be a precedent for the King to appoint others. A second letter again producing no affirmation, the King "furious at being thwarted summoned Peachell and representatives of the University to appear before the Court of Ecclesiastical Commission," following which "the senate elected Newton (and Humphrey Babington) among the eight it designated to perform that duty" (p. 477). When the King proposed a compromise that would award Father Francis the degree on condition that this not become a precedent, Newton adamantly objected persuading the delegation that agreeing would be a dishonorable capitulation.

Thus the delegation headed by Peachell met with the Ecclesiastical Commission four times, the third with disastrous results. Lord Jeffreys, who headed the Commission, so intimidated Peachell that he was unable to forcefully defend the university's position, while any effort by the other members of the delegation to come to his aid was rebuffed. Declaring that Peachell was "guilty of 'an act of great disobedience,'" the commissioners "deprived him of his office, suspended him from the mastership of Magdalene [College], and stripped him of the income of his position" (p. 478; brackets added). With Peachell removed the defense of the university's opposition fell to the other members of the delegation. Newton again played a prominent role in drafting five letters preceding the final statement of refusal, even appending to one draft the assertion that a "mixture of Papist & Protestants in y[e] same University can neither subsist happily nor long together" which, however, was not included (p. 479).

The delegation then met with Lord Jeffreys and the Commission without knowing whether they would be submitted to the same fate as poor Peachell, but this time it was the Commission who gave way with Jeffreys accepting the opposition but admonishing the delegates in the future to obey His Majesty's commands. But this warning proved pointless because in eighteen months James II was disposed by William of Orange and fled to France. The successful confrontation proved especially beneficial to Newton whose fortitude and wise advice to the delegates was so admired that he immediately became prominent at the university and was duly rewarded. When two delegates were required to represent the University at the convention to ratify the Glorious Revolution, Newton was selected as one of them, and by acts of Parliament he was regularly appointed one of the Commissioners to oversee the collection "in Cambridge of aids voted to the government," a lucrative position indicative of his enhanced prestige. When the convention was reconstituted as a Parliament Newton moved to London for a year to serve in the new Parliament. It was then that he met Huygens and the philosopher John Locke with whom he corresponded extensively and formed a close friendship until Locke's death in 1704.

Newton's resolute opposition to the admission of Father Francis's appointment to Parliament, and the international acclaim accorded the publication of the *Principia* brought him a period of "manic euphoria" in the years from 1687 to 1693. Then in the autumn of 1693 there occurred the "black year" in his life due to his breakup with Fatio de Duillier, a young Swiss mathematician whom he had met in 1687. As apparently happened in 1677–1678 when the 'very close' relationship he had formed in his early years at Trinity College with his chamber-fellow, John Wickins, came to an end, Newton suffered another nervous breakdown. When Duillier, who was twenty years his junior, wrote him that "I could wish Sir to live all my life, or the greatest part of it, with you, if it was possible" (p. 533) and did not receive a mutual reply from Newton, he formed another attachment which terminated their relationship. This plunged Newton into a severe mental breakdown, as deep a state of depression as his former height of euphoria. But with the support of friends he regained his mental stability after being depressed for for eighteen months.

Once he had recovered from this ordeal he was determined to live in London which required that he seek a more lucrative position than he had at the university. Following his Trinity College tenure of thirty-five years Newton turned from the academic and scholarly world to a more socially and politically active life in London. Then failing to achieve several prestigious positions he finally succeeded, through the help of a friend Charles Montague, in being appointed Warden of the Mint on March 19, 1696. After shepherding the mint through a series of crises with his inimitable intelligence, mathematical skills, and dedicated leadership, he was raised to Master of the Mint three and a half years later. The elevation to Master of the Mint (a position he occupied until his death) finally resolved his financial needs because he would receive in addition to his salary "a set profit on every pound of weight troy that was coined," a sum that would prove to be considerable (p. 604). No longer in need of further income, he decided to resign his fellowship at Trinity College and the Lucasian Chair at Cambridge a year later, in 1701.

His position as Master of the Mint brought with it membership in the House of Commons. Then at the request of his friend, Charles Montague, who had now become Lord Halifax, he stood for parliament from Cambridge in 1701 and succeeded in being elected. Defeated the next year he stood a third time in 1705 at the urging of Lord Halifax. It was in this political context, not because of his scholarly achievements, that Newton was knighted to promote his candidacy when Queen Anne (who had succeeded her husband William to the throne upon his death) during her customary royal visit to Cambridge University knighted him Sir Isaac. The political significance of this gesture proved futile, however, since he came in last among four candidates ending his brief venture into politics.

Despite his duties at the mint, when Newton became warden in 1696 he had not lost his interest in mathematics. Johann Bernoulli, a leading European mathematician, published a challenge problem in Wallis' *Acta eruditorum* in June 1696 with the proviso that the problem had to be solved within six months. But having received only one response by December from Leibniz, who wrote that he had solved the problem but suggested that the time limit be extended to the following Easter and that it be announced and then published, both in the *Journal des scavans* and the *Philosophical Transactions* of the Royal Society, to ensure that they were brought to the attention of both European and British mathematicians, Bernoulli agreed. He also sent copies of the problem to both Newton and Wallis personally.

When Newton received his copy and solved the problems he returned them anonymously to Bernoulli who easily identified their author due his method used in solving them. Only three mathematicians at the time had submitted solutions: Leibniz, the Marquis de l'Hôpital, and Newton. Along with this achievement he was offered membership in the Académie Royale des Sciences of Louis XIV that would have brought him a large pension, but perhaps declined because his new income was sufficient. He also was offered the Masterships of St. Catherine's Hospital and of Trinity College which he also rejected, the latter because it required that he take the Orders.

But Hooke's death in 1703 providing the opportunity of seeking the presidency of the Royal Society since previously Newton had avoided Hooke because of their quarrel over Hooke's charge of plagiarism. Now he accepted the challenge yet his being elected was not a foregone conclusion, as one might have expected, because he barely received the necessary votes. But once elected he served with his usual dedication and distinction so that the attendance, membership, quality of discussions, and financial support greatly improved during his presidency.

For example, having lost their previous meeting accommodations in Hooke's apartment at Gresham College, Newton, with the aid of the Secretary Hans Sloane, persuaded the members to purchase the former house of Dr. Edward Browne in Crane Court, off Fleet Street, for its permanent meeting place. The house was bought in 1710 and according to the design and supervision of Christopher Wren was extensively renovated. "In less than six years, the society had fully paid for its new home and stood free of debt," so that for "the first time in the fifty years since its establishment, the Royal Society had its own home" (p. 677).

PUBLICATION OF THE *OPTICKS*

Following his recovery from his nervous breakdown Newton worked on presenting in book form the results of the optical experiments he had performed thirty years earlier, completing the final draft in 1694. When his friend Gregory saw a copy he declared that it "would rival the *Principia*." The Royal Society again offered to publish the work, but to avoid the controversy it had provoked earlier, Newton waited until Hooke's death to allow its printing. So after being elected president of the Royal Society he offered the *Opticks* for publication with Halley again instructed to read it and give his decision, with slight doubt about the outcome. Though it contained little that was new about his optical experiments, the kinds of questions he raised were so innovative and wide ranging that they were the basis of research throughout the eighteenth century.

Unlike the *Principia*, where controversial issues were minimized to avoid criticism, the *Opticks* opens whole new areas of inquiry pertaining to the nature of light, heat, radiation, magnetism, electrical attraction and repulsion, gravitational attraction, chemical reactions such as fermentation and combustion, the microstructure of substances, and the transmission of nervous discharges. It is as if the entire range of modern scientific inquiry was opened up by his searching investigations. Like Rutherford at the end of the nineteenth century who, after discovering alpha and beta particles, used them to probe the interior of gold foil revealing the proton, Newton used light to investigate the inner structure of substances, along with attempting to measure the dimensions of the light corpuscles and the distribution of the interior particles. As in the *Principia*, he speculates whether an extremely tenuous, *vibrating substance* might pervade the universe as a unifying medium for all the particles and forces, rejecting his previous conception of a "subtle spirit" for a physical material called the "Aether."

Although the new corpuscular theory he was advocating was accepted also by other mechanists (though not by the Aristotelians), his ingenious application of the theory in interpreting diverse phenomena was unique, introducing that method of research to future scientists. Unlike the *Principia* whose formidable mathematical demonstrations made it inaccessible to anyone except mathematicians and physicists, the less technical and more narrative style of the *Opticks*, along with its having been written originally in English rather than Latin, contributed to its becoming the primary source of research problems throughout the eighteenth century.

The contrasting influence of the two works is vividly described by I. Bernard Cohen in his excellent work, *Franklin and Newton*:

> Not primarily in the *Principia*, then, but in the *Opticks* could the eighteenth-century experimentalists find Newton's methods for studying the properties or behavior of bodies that are due to their special composition. Hence, we need not be surprised to find that in the age of Newton—which the eighteenth century certainly was!—the experimental natural philosophers should be drawn to the *Opticks* rather than to the *Principia*. Furthermore, the *Opticks* was more than an account of mere optical phenomena, but contained an atomic theory of matter, ideas about electricity and magnetism, heat, fluidity, volatility, sensation, chemistry, and so on, and a theory (or hypothesis) of the actual cause of gravitation.[8]

The *Opticks*, like the *Principia*, is divided into three Books but includes 31 Queries or Questions, of which only representative samples can be presented here. Book I begins with Definitions followed by Axioms, Propositions, and Observations. He states that he does not intend to *explain* the properties of light by hypotheses in that Book, but to *describe and prove* them by reason and experiments. He reintroduces his controversial theory that ordinary light consists of "Primary, Homogeneal and Simple Rays," each defined by its specific degree of refrangibility as disclosed in his prism experiments, while sunlight in contrast consists of these primary rays and therefore is "Heterogeneal." As one example of his detailed description, Axiom VII describes how light reflected from the surface of an object passing through the pupil is refracted by the crystalline lens and the humors to converge again "in the bottom of the Eye, and there to paint the Picture of the Object upon that skin (called the *Tunica Retina*) with which the bottom of the eye is covered. . . . And these Pictures, propagated by Motion along the Fibres of the Optick Nerves into the Brain, are the cause of Vision."[9]

He distinguishes between the physical rays reflected by the object which have the "Power and Disposition" to propagate their motion to the Sensorium and the "Sensations of those Motions" which are experienced "under the Forms of Colours" (cf. pp. 124–25). It could have been his discussions with Newton that led John Locke, in his *An Essay Concerning Human Understanding,* to refer to the "powers" of the "primary qualities" of physical objects to affect the senses, which then transmit the stimuli or sensations to the brain where they are experienced by the mind as colors. In addition, Newton discusses near and farsightedness and chromatic aberration in refracting telescopes due to the different refrangibility of the rays of light, defending the superiority of the reflecting telescope he constructed in 1668. This is accompanied by numerous intricate diagrams and measurements supporting the descriptions of the experiments.

Book II describes his many experiments refracting and reflecting light by various transparent substances such as air, water, prisms, and convex lenses, finding that if the lenses are very thin they display a series of color rings, now called "Newton rings." Meticulously measuring the width of the films of air separating the thin plates of "plano-convex glass," as well as the diameters of and distances between the successive rings of colors, he found that the squares of the diameters of the lucid colors were in the arithmetical progression of the odd numbers, while those of the less lucid were in the arithmetical progression of the even numbers. (Cf. p. 200)

Based on these previous measurements he attempted to calculate the magnitudes and distances between the opaque particles within the air and water, along with the forces that determine the pressure, heat, and dispersion of gases, the first attempt ever to analyze the microstructure of phenomena by optical probing. As he states, "there are many Reflections made by the internal parts of Bodies, which . . . would not happen if the parts of those Bodies were continuous without any such Interstices between them . . ." (p. 249).

He composed a Table "wherein the thickness of Air, Water, and Glass, at which each Colour is most intense and specific, is expressed in parts of an Inch divided into ten hundred thousand equal parts" (pp. 232–33). From this table he inferred the diameters of the interior particles or corpuscles from the color of the light they reflected:

> . . . to determine the sizes of those parts [particles], you need only have recourse to the precedent Tables, in which the thickness of Water or Glass exhibiting any Colour is expressed. Thus if it be desired to know the diameter of a Corpuscle, which being of equal density with Glass shall reflect green of the third Order; the number 16.25 shows it to be 16.25/10000 parts of an Inch. (p. 255; brackets added)

The reader is not expected to follow the reasoning behind the calculations, but to be aware of the intricacy of the experiments and extraordinary precision of the measurements. This reflection of light from the internal particles of water, air, and glass, along with Boyle's explanation of his gas laws in terms of the movements of insensible particles in the gas, should be seen as among the first experimental evidence of the corpuscular or atomic theory antedating Einstein's explanation of Brownian motion.

Yet despite these ingenious probing experiments and precise calculations, Newton had to admit that, as to the nature of the atomic structure of substances, "what is really their inward frame is not yet known to us" (p. 269). He suggests that with the development of high powered microscopes one might be able to see the particles on which their reflected colors depends, which is achieved today with tunneling microscopes. Also, this reference to more acute visual observations with the aid of microscopes could have been the source of Locke's term "microscopical eyes."

Though he adopted the corpuscular theory of light, his discovery of the diffraction patterns of light, that "Light reflected by thin Plates of Air and Glass" and then directed through a prism "appear *waved* with many Successions of Light and Darkness made by alternate Fits of easy Reflection and easy Transmission" (p. 281; italics added), seemed to be evidence of waves. Thus Newton might have anticipated the present conception of light as either corpuscular or undulatory depending upon the experimental conditions. Though there is much more to Book II than described, what has been presented should indicate again Newton's extensive curiosity and insight as to how these phenomena might someday be explained.

Book III contains the famous 31 Queries added toward the end of Newton's life which reveal the depth and extent of his private reflections. As Cohen states in *Franklin and Newton*: "To the eighteenth-century reader, as to us, these queries reveal the mind of Newton in its innermost thoughts just as the reading of the book of Nature revealed to Newton the mind of the creating God" (p. 177). Usually beginning with "Are not . . . ," "Do not . . . ," or "Is not . . . ," they were posed less as questions than as the conclusions of his own initial investigations which served to stimulate additional research on these topics by succeeding generations of scientists.

Their scope is truly extraordinary culminating years of intense astronomical, optical, and alchemical research, the latter evident in his analyses of chemical reactions and agents. Today when there are daily reports of new discoveries in medicine, chemistry, molecular biology, neurophysiology, physics, electronics, and astronomy, one forgets how unusual was Newton's confidence that future experiments would produce solutions to his queries. Recall that just a few decades earlier Galileo initially was very skeptical of ever discovering the causes of most natural phenomena, but that Newton, after reviewing his discovery of the spectrum of light noting that the broadest bands were made by red light and the shortest by violet light, anticipated the division of the spectrum into different wave lengths. Observing the change in luminosity and color of black bodies as they are progressively heated, he attributed the change to the intensities of the internal vibrations of the heated particles anticipating "blackbody radiation."

Noting that though there are right and left optic nerves, he determined that a single visual image is acquired by the unification of the two nerves in what we call the optic chiasm, although he erroneously concluded that the right and left visual fields are conveyed to the right and left hemispheres of the brain, rather than the reverse as we now know. Having distinguishing transverse waves produced when stones are dropped in water from horizontal waves caused by percussions, he explained the polarization of Island Crystal as due to the "Positions of the Sides of the Rays to the Planes of perpendicular Refraction" (p. 359). That is, the planes in Island Crystal put the refracted rays into definite alignments perpendicular to each other so that if refracted through a second crystal the rays will either be transmitted if the crystal is similarly aligned or deflected if aligned perpendicularly.

Because of his acceptance of the corpuscular theory as opposed to Huygens' wave theory, Newton attributed the polarization of light rays to the arrangement of the particles in Island Crystal. Even though partially incorrect because today polarized light is explained by the wave properties of light, Newton's preference for explanations by particles and forces enabled him to anticipate many future causes of natural phenomena, as he indicates in Query 31:

> Have not the small Particles of Bodies certain Powers . . . or Forces, by which they act at a distance, not only upon the Rays of Light for reflecting, refracting, and inflecting them, but also upon one another for producing a great part of the Phaenomena of Nature? For it's well known, that Bodies act one upon another by the Attractions of Gravity, Magnetism, and Electricity; and these Instances shew the Tenor and Course of Nature, and make it not improbable but that there may be more attractive Powers than these. For Nature is very consonant and conformable to herself. (pp. 375–76)

In what could be his greatest prophetic insight, he declares that whatever these agents and powers are "it is the Business of experimental Philosophy to find them out" (p. 394). Has not the whole advance of science, from Newton's day to ours, confirmed this?

His scientific intuition finally leads to his rejection of his earlier conception of "a certain most subtle spirit" diffused throughout the universe to account for everything from gravitational attraction to the transmission of nerve impulses, for an "Aether," a "medium exceedingly more rare and subtle than the Air, and exceedingly more elastick and active" that "pervade[s] all Bodies" and extends "through all the Heavens" (p. 349). As was true of Kepler, this reflects his increased preference for physical explanations rather than those in terms of spirits or occult forces because of their greater explanatory power. His utilization of particles and forces did not violate his rejection of "feigned" or "formed" hypotheses, convinced as he was that they were "principles" whose truth would be manifest from phenomena and confirmed by experiments. As he states,

> to derive two or three general Principles of Motion from Phaenomena, and afterwards to tell us how the Properties and

> Actions of all corporeal Things follow from those manifest Principles, would be a very great step in Philosophy, though the Causes of those Principles were *not yet discover'd*: And therefore I scruple not to propose the Principles of Motion above-mention'd, they being of very general Extent, and leave their Causes to be found out. (pp. 401–402; italics added)

CONCLUSION

Following the publication of the *Opticks* in 1704 Newton was to live another twenty-three years, nearly a quarter of a century. He generally remained in good health and intellectually alert until about five years before his death when both his physical and mental condition began to deteriorate, though he continued as president of the Royal Society and Master of the Mint until his death. No longer pursuing his mathematical and physical researches, his lifelong interest in theology became his main area of study. As Westfall states: "Pious he undoubtedly was, but his piety had been stained indelibly by the touch of cold philosophy. It is impossible to wash the Arianism out of his religious views. Newton set out at an early age to purge Christianity of irrationality, mystery, and superstition, and he never turned from that Path" (p. 826).

Reflecting on his long life of unrivaled achievements, shortly before his death he modestly described these with the following well-known analogy, a further tribute to his exceptional character.

> I don't know what I may seem to the world, but, as to myself, I seem to have been only like a boy playing on the sea shore, and diverting myself in now and then finding a smoother pebble or a prettier shell than ordinary, whilst the great ocean of truth lay all undiscovered before me. (Westfall, p. 863)

He died after a short illness on March 20, 1727, at age eighty-five and is interred in a prominent location in the nave of Westminster

Abbey. His monument bears the fitting inscription: "Let Mortals rejoice That there has existed such and so great an Ornament to the Human Race."

NOTES

1. I made this claim in my *From Myth to Modern Mind*, Vol. II, *Copernicus through Quantum Mechanics* (New York: Peter Lang Publishing, Inc., 1996), p. 233. For my discussion of Newton, cf. chaps. 6, 7.

2. For those interested in a fuller account, see Richard S. Westfall's superb biography, *Never at Rest* (Cambridge: Cambridge University Press, 1980). All the citations in this section are to this work.

3. Quoted from Thomas S. Kuhn, *The Copernican Revolution* (Cambridge: Harvard University Press, 1957), p. 254.

4. Sir Isaac Newton, Vol. I of the *Principia*, Motte's trans. rev. by Florian Cajori (Berkeley: University of California Press, 1962), p. xviii. Until otherwise indicated, the references are now to this volume.

5. H. A. Lorentz, A. Einstein, H. Minkowski, and H. Weyl, *The Principle of Relativity*, trans. by W. Perrett and G. B. Jeffery (New York: Dover Publishing, 1923), p. 37; brackets added.

6. Voltaire, *Eléments de la philosophie de Newton*, 1783. Quoted by Alexander Koyré, *Newtonian Studies* (Cambridge: Harvard University Press, 1965), p. 14.

7. Richard S. Westfall, op. cit., p. 474; until otherwise indicated the following references again are to this work.

8. I. Bernard Cohen, *Franklin and Newton* (Cambridge: Harvard University Press, 1966), p. 120. Further references will be cited by the author's name and page number.

9. Sir Isaac Newton, *Opticks*, prepared by Duane H. D. Roller with An Analytical Table of Contents based on the 4th ed. London, 1780 (New York: Dover Publications, 1952), p. 15. Unless otherwise indicated, all further citations are to this work.

Chapter 11

NEWTON'S LEGACY IN THE EIGHTEENTH CENTURY

ELECTRICAL INVESTIGATIONS

An example of Newton's legacy is the research in electricity which was initiated during his last years of the presidency of the Royal Society. As nearly all scientific investigations of which we have records began with the Greeks, this was true of electricity, the word itself derived from the Greek "*electron*," which meant amber. In the *Timaeus* (80c) Plato refers to such "marvels" as "the attraction of amber and the Herculean stones," later called "loadstones" or "magnets." It may have been Jerome Cardan in his *De Subtilitate*, published in 1550, who distinguished between the attractive effects of rubbed amber and the magnetic effects of the loadstone.[1]

The modern use of 'electric' began with the reintroduction of the term by William Gilbert in his book, *De Magnete*, published in 1600.[2] Although, as the title indicates, Gilbert was primarily concerned with magnetism, he was interested enough in electrification to include a comprehensive review of electrical investigations. His studies led him to conclude that many other substances beside amber were capable when rubbed of attracting straw and chaff. He believed that rubbing generated heat that releases a distinctive "effluvium," which is the source of the attraction.

When Newton became president of the Royal Society he immediately invited Francis Hauksbee, a known experimentalist, to demonstrate some of his experiments before the Society, especially those involving electrification. One in particular is important because it may have been the first in which electrical repulsion was observed, complementing the well-known attraction. When he rubbed a glass tube or spun a globe between his hands he noted that the "effluvium" of the "charged" glass tube or globe when brought close to "leaf-brass" sometimes attracted and sometimes repelled the leaf, without however attributing any significance to the repulsion. He also was surprised in feeling a sensation like an "electric wind" when he held an electrified tube close to his face (cf. pp. 563–64).

In another experiment he encircled a spinning globe with a semicircular wire with strands of thread hanging from it which did not reach the globe. When uncharged the breeze produced by the motion of the globe pushed the threads in the opposite direction, but when charged by rubbing the attractive force overcame the breeze causing all the threads to align toward the globe. Even more unusual, when he pointed his finger toward the end of the threads facing the globe they were repelled, but when he pointed toward the top of the threads near the semicircular wire they were attracted. Again he nearly discovered opposite or polar electric charges, but since his effluvium theory only took into account the attractive effects of electrical charges he did not recognize the significance of the repulsion, further evidence of the crucial effect that preconceptions have on interpreting experimental results (cf. p. 567).

Stephen Gray (1666–1736) performed a series of electrification experiments between 1720 and 1732 whose main importance was to demonstrate that the "electric virtue" (meaning 'force') from a rubbed glass tube could be transmitted over large distances, over 650 feet, if connected to a suitable receiver by a "packthread" (cf. p. 579). Discovering what is now known as electrical conduction, he also found that whether the "electric fluid" could be transmitted depended upon the nature of the connecting material, anticipating the distinction between electrical conductors and nonconductors.

But the person who explicitly distinguished between "*electrical per se*" and "*non-electrical*" was Jean Desaguliers who stated that the former could be produced by rubbing, patting, hammering, or any other action performed on an electrified body, but not the latter which only could be electrified by coming in contact with such a body.[3] Then Stephen Gray made the important discovery that whatever the nature of the electricity itself, whether an effluvium, virtue, fluid, or charge, once transmitted it exists as a separate entity comparable to magnetism, gravity, or heat. He also noticed (as Benjamin Franklin later will) that when a metal rod with a pointed end is brought into contact with an electrified body the transmission of electricity is continuous and silent, in marked contrast to a rod with a blunt end that produces a bright flash accompanied by a snapping noise. From this he concluded (again preceding Franklin) that this electric fire could be identical to lightning.

Charles du Fay (1698–1739), or as he usually is called, Dufay, is considered the next major contributor to the science of electricity, illustrating how gradual and cumulative scientific progress normally is. Unlike Hauksbee and Gray, who because of their impoverished backgrounds had no university education, Dufay had an excellent academic background. So within eight months of learning of Gray's experiments he sent to the Royal Society the results of his own electrical investigations under seven sections, of which the last two are the most significant. The sixth section presents a clear summary of the preceding discoveries that will guide his inquiries.

> This principle is that an electrified body attracts all those that are not themselves electrified, and repels them as soon as they become electrified by . . . [conduction from] the electrified body. Thus gold leaf is first attracted by the tube. Upon acquiring an electricity . . . [by conduction from the tube], the gold leaf is of consequence immedi-

> ately repelled by the tube. Nor is it reattracted while it retains its electrical quality. But if . . . the gold leaf chance to light on some other body, it straightway loses its electricity and consequently is reatttracted by the tube, while, after having given it a new electricity, repels it a second time. This continues as long as the tube remains electrical. Upon applying this principle to the various experiments on electrification, one will be surprised at the number of obscure and puzzling facts it clears up. . . . (p. 585; quoted as written)

This summary makes explicit for the first time that "repulsion will occur when both bodies are electrified, provided that the one body has become electrified by conduction from the other" (p. 585). To explain this he suggests that if the transferred electrical effluvium separates from the charged body and is "self-repulsive," then this would explain why the second body after being charged would repel the first body if in contact with it. He then asks if this repulsion is restricted only to bodies that have received the same effluvium or would apply also to differently charged bodies, declaring that an "examination of this matter has led me to a discovery which I should never have foreseen, and of which I believe no one hitherto has had the least idea" (p. 586).

What he is referring to is the discovery of two kinds of electrics: one indicated by the gold leaf being *repelled* by a rubbed glass tube having the same charge as that which originally electrified it and the other by the gold leaf coming in contact with a piece of charged copal (a resinous substance) and being *attracted* by it. Entirely unexpected, he wrote that the result "disconcerted me prodigiously," until he realized that it indicated

> that there are two distinct electricities, very different from each other: one of these I call *vitreous electricity*: the other, *resinous electricity*. The first is that of [rubbed] glass, rock crystal, precious stones, hair of animals, wool, and many other bodies. The second is that of [rubbed] amber, copal, gum lac, silk, thread, paper, and a vast number of other substances. (pp. 586–87)

This discovery of two different electrics, obviously named from the two kinds of substances involved, *vitreous* from glass and *resinous* from amber and copal, was a major advance in the investigation of electricity, establishing that bodies containing the same kind of electricity, either vitreous or resinous, will repel each other while those having the opposite kind will attract. Dufay also found that neutral bodies have the potential of being electrified by either of the two kinds of electrification. It was then determined that the type of electrification depended not only on the kind of substance rubbed, but also on the material used in the rubbing: glass, for example, acquiring vitreous electrification when rubbed by wool, silk, or cat's fur, but becoming resinous electrified when rubbed by rabbit's fur.

Having found that electricity was transferred as a separate entity, the earlier "effluvium" designation was gradually replaced by an "electric fluid" conforming to the common use of fluid in the eighteenth century to describe such ordinary phenomena as heat (a caloric fluid) and combustion (the release of phlogiston). Thus electrical phenomena by the mid-eighteenth century was being explained by the two-fluid theory of electricity described by Duane Roller and Duane H. D. Roller:

> (a) there are two distinct electrical fluids, one of which may be called "vitreous" and the other, "resinous"; (b) any *unelectrified* object possesses equal quantities of these two fluids, which neutralize each other; (c) rubbing electrifies an object by removing from it one or the other kind of fluid . . . (d) the larger the quantity of a particular fluid removed, the greater is the strength of the electrification. Notice that by "equal quantities" of the two kinds of fluid is here meant simply the quantities present in an unelectrified object and, therefore, the quantities that will completely neutralize each other's effects. (p. 590)

Performing electrical experiments by both natural philosophers and amateurs had become quite the rage throughout Europe by the middle of the eighteenth century, often at considerable risk to the experimenter. One such experiment is dramatically described by Pieter van Musschenbrock, a celebrated Dutch professor and experimentalist at the University of Leiden, an extract of which was contained in a paper submitted to the French Academy by J. A. Nollet and published in the Academy's *Mémoires* in 1746.

As described in the abstract, Musschenbrock suspended a gun barrel by two silk threads at both ends of the barrel and then charged the left end by an electrified glass globe. At the right end he attached a brass wire that extended into a glass flask partly filled with water which he held in his right hand and then with his left hand touched the gun barrel with the following painful result: "Suddenly my right hand was struck so violently that all my body was affected as if it had been struck by lightning. . . . The arm and all the body are affected in a terrible way that I cannot describe: in a word, I thought that it was all up with me" (p. 594). As the experiment was performed at the University of Leiden, the flask held by Musschenbrock became known as the "Leyden jar" still used in laboratories today.

The striking achievements of Newton and the popular interest in electrical experiments attest to the growing confidence in the intelligibility of nature and of natural philosophers to uncover and explain its secrets. As Duane and Duane H. D. Roller state:

> The impressive successes of physical science—especially mechanics and astronomy, first in Italy and, after the middle of the 17th century, in England—had a strong popular appeal, leading to a lively public interest in natural phenomena. With this came an increasing awareness of the orderliness of these phenomena, a sense of their independence of capricious and magical influences. And, since the human mind was showing itself capable of coping with nature, there was a growing feeling of confidence in the exercise of personal judgment and understanding, as opposed to reliance on the dogmatic authority of others. (p. 591)

The following major contributor to the understanding of electricity was not from continental Europe or England, but from the early north American colonies, known as "America's first great man of science." Benjamin Franklin (1706–1790) was in his thirties when, after attending several lectures in Boston by Dr. Spencer from Scotland in 1743, he became intrigued by electrical phenomena. But it was not until Peter Collinson, an English manufacturer, naturalist, and Fellow of the Royal Society, sent to the Library Company in Philadelphia a résumé of current electrical experiments in Germany, alongwith a glass tube with instructions as to how it could be used to replicate these experiments, that Franklin became involved in electrical investigations.

He had helped organize in Philadelphia a club called the "Juno" which evolved into the American Philosophical Society which is still active, the first in the colonies established primarily for the discussion of scientific topics. In addition, he founded with other members of the Juno the Library Company, the earliest circulating or lending library in America which had received Collinson's account of electrical experiments and gift of instruments. As Franklin later wrote, this gave him "the opportunity of repeating what I had seen in Boston," while in his letter to Collinson in 1747 thanking him for his gifts to the Library Company he said "I never was before engaged in any study that so totally engrossed my attention and my time as this has lately done . . ." (pp. 596–97).

Having become a successful businessman (as well as a prominent publisher and journalist), Franklin had sufficient leisure so that he immediately began duplicating the experiments he had witnessed in Boston and learned from Collinson. Among the latter's further correspondence was an important publication by William Watson, *A Sequel to the Experiments and Observations tending to Illustrate the Nature and Properties of Electricity*, published in London in 1746. Obviously influenced by Newton, Watson's theory of electricity was that of an "electrified aether," a kind of elementary fire "formed by the creator" whose tenuity and elasticity enabled it to pervade all bodies, its densities accounting for their attraction and repulsion. After reading Watson's book Franklin wrote to Collinson that he and his Philadelphia associates had not only performed experiments similar to those described by Watson, but they had discovered "some particulars not hinted in that piece," even claiming that Watson was "deceived" in some interpretations of his experiments (pp. 597–98). This was no mean accomplishment considering that Watson was known as "the leading electrician in England."

Included in the letter to Collinson were four statements and explanations describing electrical experiments using three persons, two standing on wax to insulate them and the third standing on the conducting floor, to demonstrate the various ways the "electric fire" or "fluid" generated by a rubbed tube was acquired and transmitted among them. Unaware of Dufay's distinction between vitreous and resinous electricity, it was in this letter that Franklin introduced the new terminology of "positive and negative" to designate an excess "sum" or reduced "minus" electrical fire to describe electrical charges and transmissions (cf. pp. 598–99).

According to this terminology, all neutral or unelectrified objects contain an "equal share" or balanced amount of the electric fire or fluid. Electrification consists of the transfer of the electric fire to another body. When a vitreous object such as a glass tube is rubbed by wool or silk, it draws *from the material used in the rubbing* some electricity becoming electrically charged or, if the object is resinous such as copal, the rubbing material draws electricity *from the object rubbed* to itself becoming positively electrified. The object that gained electricity was called "positively charged" having acquired a "plus" quantity of electricity, while the object that lost electricity was called "negatively charged" having obtained a "minus" charge.

A neutral object brought to a positively charged body will draw off the excess charge or if the body is negatively charged transmit some of its electricity to it until equilibrium is reached. This exchange of electricity is called "electrical conduction." The transfer of electricity from oppositely charged bodies is what produces electrical attraction, while two positively charged objects repel because they cannot transfer their excess charges. Neutral bodies neither attract nor repel because there is no excess charge to transmit. A limitation of the theory is that Franklin did not explain the repulsive effects of two negatively charged bodies.

The advantage of this "positive" and "negative" terminology which was adopted is that it facilitated quantifying the charges. Furthermore, the theory assumes that in an insulated body the quantity of electricity remains constant so that electrification involves a transfer of electricity which is neither created nor destroyed, later called the "*principle of conservation of electric charge*" (p. 601). This principle enabled Franklin to explain the electrification of the Leyden jar and of electrification by influence.

A Leyden jar is a capacitor, a device used to collect an electric charge consisting of a glass jar (usually containing water initially thought to store the electricity) with a metal coating on the inside and outside separated by the nonconductive glass and closed by a cork containing an extended wire that conducts the electricity. According to the principle of the conservation of electric charge, if the inner metal coating gains an electric charge through the wire then the outer coating must have lost an equal quantity of electricity by conduction, while if the two are connected equilibrium is immediately reestablished. To test the theory of the two opposite charges Franklin devised a kind of electroscope, a small cork ball attached to a thread which, when brought near the positively charged wire was drawn to the wire until becoming positively charged and consequently repelled. Then when it was brought near the negatively charged coating it was attracted confirming the duality of charges. The transfer of electrical charge was called "electrification by influence."

Dissatisfied with this superficial description, Franklin attempted a deeper explanation in a paper titled *Opinions and Conjectures concerning the Properties and Effects of the Electric Matter, arising from Experiments and Observations made at Philadelphia, 1749*. Supposing the electric fire or fluid, since it can penetrate all substances including metals, consists of very subtle particles that *repel* each other, then the repulsion of two positively charged bodies could be attributed to their having an excess of these positively charged particles. Yet if in ordinary uncharged bodies they are neutralized by being *attracted* by a comparable amount of "common matter" composing these ordinary bodies, then this would explain their neutrality. Thus if a conductor loses electrical particles by transmission, its common matter will attract electrical particles to restore equilibrium.

Although this theory did account for most of the electrical phenomena, it did not explain, as indicated previously, why two negatively charged bodies repelled since there was no reason on this theory why two bodies having a lesser amount of electrical particles should repel each other. The difficulty was alleviated, however, when Franz Aepinus, a Berlin investigator, introduced a new principle of explanation.

> The revolutionary idea of Aepinus was that in solids, liquids, and gases the particles of what Franklin called "common matter" repel one another just like the particles of the electric fluid in Franklin's theory. Aepinus's revision introduced a complete duality. The particles of common matter and of electric matter each have the property of repelling particles of their own kind while each kind of particle has the additional property of attracting particles of the other kind. (Cohen, p. 540)

But while this emendation explained why negatively charged bodies repel owing to their having fewer than normal electric particles (accounting for their negative charge), thus leaving an excess of positively charged particles inherent in the object producing the repulsion, it violated Newton's law that all material particles exert a mutual gravitational attraction accounting for the solidity of matter. Though not complete, this discussion is sufficient to show how difficult it usually is at a certain period in history to arrive at a perfectly consistent or complete explanation of certain phenomena, especially when the solution depends upon unforeseen discoveries that introduce unexpected explanations, in this case the future discovery of the negative electron and the positive proton accounting for the duality of charges.

One well-known discovery by Franklin should be mentioned before leaving this topic, his demonstration that storm clouds are the source of lightning because they are electrically charged. Recall that Gray had suggested that lightning and electrical charges were of "the same nature" and that he had discovered that a metal rod with a pointed end attracted electricity in a smooth, silent manner, while a blunt tipped rod produced a bright flash with a loud snap. Aware of these facts, Franklin devised an experiment to test whether lightning actually was an electrical discharge from storm clouds. While he proposed an experiment from a high tower or church spire, the demonstration he actually performed with the help of his son was the famous kite experiment.

Using a kite with an attached wire at the top and tied to a kemp cord to which was fastened a metal key and silk ribbon at the lower end of the cord, at the outbreak of a thunderstorm he and his son raised the kite and then ran into a shed allowing the cord to become wet making it a more effective conductor, but being careful that the silk ribbon they held remained dry to serve as an insulation against possible electrification. (G. W. Richman, a physicist at the Imperial Academy of Saint Petersburg in Russia, was instantly killed in 1753 when he attempted a similar experiment.) As Franklin expected, when the storm broke an electrified cloud discharged electricity to the wire on the kite that was conducted down the moistened cord to the metal key and then into a Leyden jar where it was collected.

In another similar experiment, Franklin drew an electric current from a charged cloud down a wire at the side of his house to a metal frame containing two iron bells with metal clappers which, when the current reached the inside of the bells, produced a ringing sound. Owing to these experiments "scientists of the eighteenth century, such as Watson, although conversant with pre-Franklin speculations, referred to the 'verification of Mr. *Franklin's* hypothesis'" (Cohen, p.

488). He thus acquired the reputation as the preeminent "electrician" of his age, his fame spreading throughout Europe.

His book, *Experiments and Observations on Electricity* published in 1752, had numerous printings and was translated into French and German. The following year he was awarded the Royal Society's highest honor, the Copley Gold Metal. Joseph Priestly, in his prominent *History of Electricity* (1767), wrote that Franklin's proof that lightning is an electrical discharge is "a 'capital' discovery—'the greatest perhaps, since the time of Sir Isaac Newton'" (Cohen, p. 489). In 1773 he was made a "foreign associate" by the French Academy of Sciences, an honor so unusual that it was not awarded to another American scientist until the next century. Though his diplomatic missions to Paris and other public obligations forced him to abandon his scientific researches, he maintained a lively interest in scientific matters until the end of his life in 1790.

Given Newton's success in placing celestial mechanics on a secure mathematical foundation and Franklin's law that electrical forces are conserved, along with his description of electrical charges as plus and minus implying they can be quantified, led the next generation of electricians to seek a mathematical framework to describe electrical phenomena similar to Newton's. This belief was reinforced when Daniel Bernoulli, a Swiss physicist, in 1760 invented an electrometer for measuring directly the strength of the electrical force between two charged metal disks, finding that the force varied inversely with the square of the distance, similar to Newton's law of gravitation, a remarkable coincidence.

Then Priestly, who had met Franklin in London in 1765, using insulated pitch balls suspended in an electrified tin quart vessel set on a wooden stool replicated his experiments, confirmed that they were not attracted to the sides of the vessel indicating they had a net charge of zero. Although neither Franklin nor he could explain why this was true, Priestly drew an analogy between this and Newton's demonstration that the net gravitational force on an object placed anywhere within the hollowed earth would be zero, inferring that the same inverse square law should apply to the attractive force of electric bodies, as well as the attractive force of gravity (cf., p. 612).

Dissatisfied with Priestley's argument by analogy, like Bernoulli Charles C. Coulomb (1736–1806) devised an instrument to test whether the inverse square law applied to electricity. The electrical torsion balance he invented was so exact that he could measure "with the greatest exactitude the electrical force exerted by a body, however slightly the body is charged" (p. 617), demonstrating that the *repulsive* force between two identically electrified bodies is inversely proportional to the square of the distance. But for the law to apply generally he had to show that this was true also of *attractive* forces, so he developed an electric torsion pendulum to determine whether the force of *attraction* between two oppositely charged bodies also varied with the square of the distance, with the same result. As he wrote in a second *Mémoires de l' Académie Royale des Sciences* for the year 1785:

> We have thus come, by a method completely different from the first, to a similar result. We may therefore conclude that the mutual attraction of the electrical fluid called positive and the electrical fluid ordinarily called negative is inversely proportional to the square of the distance; just as we have found in our first memoir, that the mutual repulsion of electrical fluids of the same sort is inversely proportional to the square of the distance. (pp. 620–21)

But for the analogy to Newton's law that the gravitational force between two bodies is proportional to the *product of their masses* (density per volume), as well as inversely proportional to their distances, to be complete, it had yet to be shown that this held for the product of electrical forces. But because in Newtonian mechanics mass is an essential property of matter, it must have occurred to Coulomb that an electric fluid also must have an "electric mass" as an inherent property. Thus the possibility of quantifying Franklin's electrical charge of plus or minus electricity was made possible by Coulomb's concept of electrical mass or electrical charge. Even more impressive, "Coulomb's Law" expressing the magnitude of the force of electrical repulsion and attraction in the same mathematical form (q_1q_2/d^2) that applied to gravitational forces reinforced the belief in the consilient development of scientific inquiry, the quantifiability of natural processes, and the underlying unity of nature (cf. p. 621).

DISCOVERY OF THE WAVE NATURE OF LIGHT

In the previous chapter we learned that Newton had discovered that ordinary monochromatic light consists of a spectrum of individual rays ranging in color from red to violet and to account for the sharp edges of shadows he had adopted the corpuscular theory of light. Because of his tremendous prestige this theory was predominant in the early eighteenth century despite its having been challenged by the wave theory supported by Descartes, Hooke, and Huygens. It was especially Descartes' influential conception based on his theory of vortices, that light consists of an *instantaneously transmitted pressure* among the contiguous vortical particles, that Newton was particularly concerned to refute in his *Opticks*. His rebuttal was based on the conviction that this explanation would require an infinite force, that it could not explain how light heated objects, and since it presupposed that light would bend around objects it conflicted with the sharp shadows cast by light. As he states:

> If light consisted only in Pression [pressure] propagated without actual Motion, it would not be able to agitate and heat the Bodies which refract and reflect it. If it consisted in Motion propagated to all distances in an instant, it would require an infinite force every moment, in every shining Particle, to generate that Motion. And if it consisted in Pression or Motion propagated either in an instant or in time, it would bend into the Shadow. For Pression or

> Motion cannot be propagated in a Fluid in right Lines . . . but will bend and spread every way into the quiescent Medium which lies beyond the Obstacle.[4]

Convinced that these optical phenomena plus his discovery of colored rings due to light striking dense transparent substances producing the "Fits of easy Reflection and easy Transmission," along with the double refraction of "Island-Crystal," could be more readily explained by considering light rays to be composed of minuscule particles that exerted forces, he defended the corpuscular theory.

> Nothing more is requisite for putting the Rays of light into Fits of easy Reflection and easy Transmission, than that they be small Bodies which by their *attractive Powers*, or some other *Force*, stir up Vibrations in what they act upon . . . and thereby put them into those Fits. And lastly, the unusual Reflection of Island-Crystal looks very much as if it were perform'd by some kind of *attractive virtue* lodged in certain Sides both of the Rays, and of the Particles of the Crystal. (Query 29, pp. 372–73; italics added)

The first to seriously challenge Newton's particle theory was Thomas Young who, in a paper titled "Outlines of Experiments and Inquiries Respecting Sound and Light" published in 1800 in the *Philosophical Transactions of the Royal Society*, argued that light, having properties analogous to sound, could be best explained by the wave theory. In contrast to the corpuscular theory, which considered light to be particles ejected by luminous bodies that traversed space at tremendous but finite velocities, if light were propagated analogous to sound then it would consist of expanding horizontal waves having intervals of compression and rarefaction that were not particles.

Thus the properties of waves and particles are just the converse of each other: particles characterized by mass, size, shape, momentum, and linear motion have a discrete location in space and interact by deflection involving a loss of energy. In contrast, waves have the properties of length, frequency, amplitude, and intensities and are diffused in space and interact either to reinforce if in phase or destruct if out of phase. Also, particles have a velocity which is a product of their mass and acceleration and emit and are affected by forces, while the velocity of waves is proportional to their wave lengths and frequencies and are indifferent to forces.

Though Young's paper defends the wave interpretation and points out several difficulties of the corpuscular theory previously overlooked, it was a later paper titled "On the Theory of Light and Colours" published in 1802, in which Young describes his experiments producing diffraction patterns that was crucial for the wave theory. In the first of these experiments he demonstrated that when monochromatic light is reflected on a screen that has a circular opening, the *pattern* of light emerging and shown on a second screen behind the first varies with the size of the opening. When the opening is large compared to the wave length of the light an illuminated circle appears which then changes to light and dark bands when the aperture size is smaller than the wave length. Young describes these bands as "constructive" in phase and "destructive" out of phase interference effects typical of waves, a confirmation of his theory.

But as usually occurs in science, the defenders of the corpuscular theory proposed a mitigating explanation attributing the diffraction patterns to the effect of the edges of the apertures on the passing corpuscles somehow aligning them in the alternating bands. But when this *ad hoc* explanation was put to the test by Augustine Jean Fresnel, a brilliant French engineer, by varying the shapes and masses of the aperture but maintaining its size, he found there was no modification indicating the diffraction pattern depended solely on the size of the opening in relation to the length of the light wave. In the "Memoir on the Diffraction of Light" for which he was awarded a prize by the Paris Academy in 1819, Fresnel devised methods to measure the physical properties of waves, their lengths, frequencies, amplitudes, and intensities adding to Young's description.

Furthermore, one of the major objections to the wave theory (based on the analogy with water and sound waves) that light waves were propagated longitudinally and therefore could not account for the polarization of light emerging from Island-Spar or Island-Crystal was removed when Fresnel demonstrated that light waves were transmitted transversely, caused by the up and down vibrations perpendicular to the light waves. He even removed Newton's crucial objection to the wave theory that it was inconsistent with the sharp boundaries of shadows cast by large objects by showing that the destruction of light waves at the fringes of their propagation explains their ray-like appearance, but when the size of the obstruction is proportional to the wavelength then light waves bend around the object as do sound and water waves.

While Fresnel's experimental confirmation of the diffraction patterns and mathematical representation of the physical properties of light were turning the tide in favor of the wave theory, the fact that contrasting predictions could be derived from the corpuscular and wave theories that could be tested experimentally proved decisive. One of the predictions pertained to Newton's color rings with the wave theory implying that the intervals between the rings would be dark, while this would not be true of the corpuscular theory. Fresnel devised an experiment to test this and found the wave theory confirmed. The second test involved the change in the velocity of light in passing from a less dense medium like air to a denser one such as water. Newton had argued that when light passes into a denser medium its velocity would *increase* due to the greater attractive force exerted by the larger particles of the denser medium on the minute light particles, while the wave theory implied that the secondary waves caused by the diffraction through the denser medium would impede the light *reducing* its velocity. In ingenious experiments performed in 1850 but not published until 1862, often referred to as an *experimentum crucis*, Jean Léon Foucault demonstrated that the velocity of light in water is less than in air, providing crucial evidence for the wave theory of light.

Then in 1849 Hippolyte Louis Fizeau determined the velocity of light to be 3.15×10^{10} cm./sec.,[5] while Foucault by a different experimental method found its velocity to be 2.98×10^{10} cm./sec., nearly 300,000 kilometers per second adding further confirmation to the wave theory. Yet despite these supporting developments, in an article in 1827 John Herschel summarized the results as indicating that "neither the corpuscular nor the undulatory, nor any other system which had yet been devised, will furnish that complete and satisfactory explanation of *all* the phenomena of light . . ." (Achinstein, p. 22). Herschel's astute appraisal was confirmed in 1905 when Einstein introduced discrete quanta of light called photons to explain the photoelectric effect leading to the present dual conception of light as manifesting either particle or wave properties under different experimental conditions. Thus despite the properties being contradictory, either can be manifested depending on the experimental setup, a surprising result typical of many scientific discoveries in the twentieth century.

DISCOVERY OF ELECTROMAGNETISM

Further experiments during the nineteenth century disclosed the interdependence of electricity and magnetism, another notable example of the gradual unification of scientific concepts and theories. Hans Christian Oersted, Professor of Natural Philosophy at the University of Copenhagen, in lectures given in 1819–1820 remarked that he had noticed that in an electric storm the position of a magnetic needle changed which led him to investigate the effects of an electric current on the deflection of a magnetic needle, demonstrating that a changing electric current produces a magnetic field.

Owing to Oersted's experiments, Michael Faraday, though trained to be a bookbinder's journeyman, as a young man became intrigued by electrical phenomena, especially the lines of magnetic force generated by a magnet. In 1821 he wrote a *Historical Sketch of Electro-Magnetism* based on his replication of the experiments of the individuals he was reviewing which was published in the *Annals of Philosophy* in 1821. Although his main investigations focused on electromagnetic forces, in a discovery complementing Oersted's detection that a *changing electrical current* produces a magnetic field, he demonstrated that a *changing magnetic field* induces an electric current, the theoretical basis of dynamos. Equally important, this discovery also led to the conception of an electromagnetic *field* having an independent existence, rather than being a fluid or the configuration of an underlying reality, such as Newton's Aether.

Although preceded by such outstanding investigators as Carl Friedrich Gauss, George Friedrich Riemann, and William Thomson who became Baron Kelvin, it was James Clerk Maxwell who devised the equations describing the spatial structure of this subsisting electromagnetic field and how it changed with time. A graduate of Edinburgh University he, like Newton, became a fellow of Trinity College, Cambridge, in 1855. After reading Faraday's summary of the historical researches in electromagnetism, though equally qualified to pursue experiments, as a gifted mathematician he was more concerned to give mathematical expression to the properties involved in the experiments, producing his famous electromagnetic equations.

IDENTIFICATION OF ELECTROMAGNETISM AND LIGHT

Then in 1888 Heinrich Hertz, professor of physics at Karlsruhe University, during experiments to confirm Maxwell's equations discovered that the speed of light is the same as the propagation of electromagnetic waves, implying that light also is an electromagnetic phenomena. So important was this determination that Einstein later declared that the "discovery of an electromagnetic wave spreading with the speed of light is one of the greatest achievements in the history of science."[6] This identification of light with electromagnetism further obviated

Newton's conception of an underlying aether to explain the propagation of light, as well as the gravitational attraction. As described previously, the null results of the Michelson-Morely experiments provided a final disproof of the existence of the aether.

Furthermore, unlike particles that exert attractive or repulsive forces over a distance, because electromagnetic waves involve an interaction of contiguous fields the conceptual framework of electromagnetism represented a potentially radical departure from the corpuscular-mechansitic framework of Newton. Like the previous wave-particle duality, this dichotomy of atomic forces and electromagnetic fields reappeared in the twentieth century in the contrast between quantum mechanics involving interactions of subatomic particles and Einstein's general theory of relativity which attempted to reduce all physical processes to deformations of the space-time continuum or field.

Although this summary of the experimental discoveries and theoretical unification of light, electricity, and magnetism in the aftermath of Newton only covers the highlights, it was included because it illustrates clearly the manner in which modern scientific inquiry develops, especially the gradual elucidation and refinement of the conceptual framework, along with showing the increasing reliance on experimentation and mathematical correlations in arriving at the correct explanations. These conceptual developments can be seen more clearly in these earlier stages than later when the experiments become so technical and the mathematics so demanding that it is extremely difficult to understand the basic concepts without extensive training in the derivation of the mathematical equations used to describe the phenomena.

Eluding their comprehension and erroneously thought to be less significant than religious beliefs in comprehending and coming to grips with the world, this is one reason scientific discoveries and explanations do not have a greater influence on the thinking of the general public and why, despite all the complex effort and reasoning involved in scientific discoveries, religionists can accept such simplistic pseudo explanations as "special creation," "God willing," or "intelligent design," which may be more congenial but offer only the illusion of an explanation.

This is also the reason our educational system should put more emphasis on mathematics and science as demanded in the modern world, a change that is in effect in India and China. Superstition and shallow thinking will be the undoing of the United States's prominence in science and technology, along with dominance of the world's economy, if it is allowed to persist, as was clearly illustrated in the previous religious orientation of the Bush administration. While we should accept the beneficial ethical teachings and inspiring exhortations of religions, we need to discard their outmoded supernatural framework that originated at a time in history when practically nothing was known about the universe, if we are to face the world and the future realistically.

NOTES

1. This discussion is based on "The Development of the Concept of Electrical Charge: Electricity from the Greeks to Coulomb," by Duane Roller and Duane H. D. Roller, *Harvard Case Histories in Experimental Science*, Vol. 2, ed. by James B. Conant and Leonard Nash (Cambridge: Harvard University Press, 1948), pp. 543-623. Unless otherwise indicated, the quotations in this chapter are from this work.

2. Cf. William Gilbert, *De Magnete,* trans. by P. Fleury Mottelay (New York: Dover Publications 1958), cf., pp. xvi–xvii.

3. Cf. I. Bernard Cohen, *Franklin and Newton* (Cambridge: Harvard University Press, 1966), pp. 376–84.

4. Sir Isaac Newton, *Opticks*, based on the fourth edition, London, 1730 (New York: Dover Publications, Inc., 1952), Query 28, p. 362; brackets added. The following quotation also is to this work.

5. Cf. Sir Edmund Whittaker, *A History of the Theories of Aether & Electricity*, Vol. I, *The Classical Theories* (New York: Harper Torchbooks, 1960), p. 253. I am indebted to this book and Peter Achinstein's *Particles and Waves*

(New York: Oxford University Press, 1991) for the later material presented in this chapter.

6. Albert Einstein and Leopold Infeld, *The Evolution of Physics* (New York: Simon and Schuster, 1951), pp. 155–56.

Chapter 12

THE ORIGINS OF CHEMISTRY AND MODERN ATOMISM

THE ORIGINS AND DEVELOPMENT OF CHEMISTRY

Not surprisingly, the origins and development of chemistry replicate the history of astronomy. Just as the Copernican Revolution grew out of an awareness of the deficiencies of Aristotle's astronomy, along with the emendations by Ptolemy, reviving the heliocentric system of Aristarchus to replace geocentrism, the origins of modern chemistry lie in the inadequacies of Empedocles' theory of the four elements and Aristotle's explanation of change, reviving ancient atomism as the foundation of the new science of chemistry. What drove the replacements in both cases was a radical change in the conception of natural processes and what constitutes an explanation, initiating a different method of inquiry and way of understanding the universe.

Though the Greek philosophers had largely discarded mythological and narrative accounts in their attempts to understand the world, still, even Aristotle's empirical inquiries based on inductive generalizations and astute observations included qualitative distinctions, deductive logic as the formalism, and metaphysical explanations. Yet that began to change from the fifteenth through the seventeenth centuries with the discovery of astronomical laws, acceptance of the uniformity of nature precluding ad hoc and non-natural explanations, the introduction of forces such as gravity and electromagnetism to replace animistic or occult causes, the acceptance of mathematics as the formalism, and the adoption of Newton's atomic-mechanistic framework and celestial mechanics to replace Aristotle's organismic cosmology. All of these tendencies were augmented in the succeeding centuries as scientific inquiry began to be more expanded, refined, and institutionalized.

Although astrologers had created a complex systems for predicting events and casting horoscopes based on the movements of the moon, sun, and planets through the constellations of the zodiac, they still required a preexisting astronomical system to be effective. This was true also of alchemy with its early reactive agents of salt, sulfur, and mercury which preceded chemistry and whose objective was the transmutation of base metals into gold. But neither of these systems proved effective because their theoretical foundations were false and had to be replaced.

One of the most influential of the early alchemists had various names, but is recognized primarily by his Latin name Paracelsus. A Swiss physician (1493–1541), he opposed the teachings of Galen and Avicenna, attempted to reform medicine and was noted for his surgical lectures. Although no specific scientific contributions can be ascribed to him, he is especially known for hypothesizing three "principles," salt, sulphur, and mercury, to replace the Empedoclean four elements, which came to be synonymous with alchemy. According to J. R. Partington, "[s]alt was the principle of fixity and incombustibility, mercury of fusibility and volatility, sulphur of inflammability. The last two had long been recognized by the alchemists, but Paracelsus seems to have been the first to add salt, making up the *tria prima*, which he compared with body, soul, and spirit."[1]

A later follower was Johann Baptista van Helmont (1579–1644), a Belgian who received the degree of M.D. from the Jesuit School at Louvain in 1609. A pious mystic, he became very influential after his death due to his son collecting his father's writings which were published as *Ortus medicinae* or *The Fount of Medicine* in 1648. Although strongly influenced by Paracelsus, he rejected the latter's three principles believing that water and air were the fundamental elements. He defended the conservation of matter by showing that substances dissolved in liquids, such as metals in acids, can be recovered. He opposed Aristotle's view that a vacuum was impossible declaring that it is "something quite ordinary," as demonstrated in his experiments.

Apparently deriving the term "gas" from the Greek word "*chaos*," he introduced the concept to refer to various components of the air, identifying such gases as nitrogen, carbon dioxide, carbon monoxide, and sulfur dioxide as they are now called. Because of his use of the balance to weigh the ingredients in his experiments and his advocacy of the principle of the indestructibility of matter, both of which became essential components of the science of chemistry, he represents an important transition to the later science. While only a brief summary of van Helmont's investigations and doctrines, it should be sufficient to show that they were a precursor of chemistry, even though alchemy was still a pseudoscience lacking an established methodology and a viable conceptual framework.

The next in importance is Robert Boyle who was born in Lismore Castle, Ireland, in 1627 and died in London in 1691. After completing his education at Eton, in 1654 he moved to Oxford where he began experiments with a vacuum pump along with investigating combustion and calcinations in his lodgings adjacent to University College. He chose as his assistant Robert Hooke who also was to become one of the leading scientists of the period. Boyle was an original Fellow of the Royal Society founded in 1644–1645 and was elected president in 1680, but declining to serve he was replaced by Christopher Wren. As Partington states, he was the founder of modern chemistry.

> Boyle has been called the founder of modern chemistry for three reasons: (1) he realized that chemistry is worthy of study for its own sake and not merely as an aid to medicine or as alchemy—although he believed in the possibility of the latter; (2) he introduced a rigorous experimental method into chemistry; and (3) he gave a clear definition of an element and showed by experiment that the four elements of Aristotle and the three principles of the alchemists (mercury, sulphur and salt) did not deserve to be called elements or principles at all, since none of them could be extracted from bodies, e.g. metals. (p. 67)

A skilled experimenter who considerably improved the existing apparatus, he wrote numerous works but is especially known for his book titled *Sceptical Chymist* published in London in 1661. Most of the following quotations are from this book as quoted by Partington. He was influenced by van Helmont "whose works he studied with care and to whom he frequently refers as an authority" (p. 69), even though disagreeing with him. Of his major accomplishments, his definition of an element was an important theoretical clarification of that imprecisely used concept.

> I mean by Elements, as those Chymists that speak plainest do by their Principles, certain Primitive and Simple, or perfectly unmingled bodies; which not being made of any other bodies, or of one another, are the Ingredients of which all those call'd perfectly mixt Bodies are immediately compounded, and into which they are ultimately resolved. (p. 70)

While this passage seems clear enough, his further description is somewhat ambiguous as to whether the elements themselves are the minute particles composing the "mixt Bodies" or whether they are the particles of a more "Common substance" that actually composed the composite bodies. As he states in the *Sceptical Chymist*, quoted by Partington.

> "The greatest part of the affections of matter, and consequently of the Phaenomena of nature, seems to depend upon the motion and the contrivance of the small parts of Bodies," so that "there is no great need that Nature should alwaies have Elements before hand, whereof to make such Bodies as we call mixts," and "the difference of Bodies may depend meerly upon that of the schemes whereinto their Common matter is put . . . so that according as the small parts of matter recede from each other, or work upon each other . . . a Body of this or that denomination is produced." (p. 71)

This passage presents the outlines of an atomic theory that anticipates compound molecular structures since by "perfectly mixt bodies" he apparently means what we now refer to as molecular substances, such as water, air, and salt, whose properties are formed by the "affinity" of the minute, indivisible particles which are the elements. Yet it is ambiguous as to what he meant by "nature not having to await the elements beforehand" and by "Common matter." The clearest explication I have found is that of Charles Singer, based on Boyle's *Origin of Forms and Qualities*, published in 1666.

> He assumes the existence of a universal matter, common to all bodies, extended, divisible, and penetrable. This matter consists of innumerable particles, each solid, imperceptible and of its own determinate shape. "These particles are the true *prima naturalia*." There are also multitudes of corpuscles built up from several such particles and substantially indivisible or at least very rarely split up into their *prima naturalia*. Such secondary "clusters" have each their own particular shape. "Clusters" and "*prima naturalia*" may adhere together. They thus form characteristic and similar groups which are not without analogy to molecules and atoms in the more modern use of these terms. Nevertheless, the resemblance of Boyle's atomism to either modern or ancient atomism is far from close.[2]

Boyle's second major contribution was his discovery in 1662 of the first gas law named after him, that the volume of a gas varies inversely with the pressure. The apparatus he used consisted of a U shaped tube with one capped outlet shorter than a longer open one. By pouring mercury into the open tube he could increase the pressure on the air contained in the tube measuring the proportional changes in the height of the air relative to that of the mercury, discovering that doubling the volume of the mercury reduced the volume of the air by one half, tripling the mercury reduced the air by one third and so forth, showing that the one varies in inverse proportion to the other. The same was true when he reversed the process. Among his other significant contributions was "the suggestion of chemical 'indicators' for testing the acidity or alkalinity of liquids, and his isolation of elemental phosphorus" (Singer, pp. 272–73). Indeed, his inquiries were so extensive that they touched nearly every aspect of modern chemistry.

As mentioned previously, Robert Hooke assisted Boyle in his early vacuum-pump experiments becoming one of the most skillful and original experimenters of his time, especially in his investigations of light. Born on the Isle of Wight in 1635, he studied at Christ Church, Oxford, after serving as assistant to Boyle becoming curator of experiments to the Royal Society in 1662. He served as Secretary to the Royal Society for decades until his death in 1703, offering his chambers at Gresham College as its meeting place. It was his misfortune to be born at the time of the "incomparable Newton," for even though he made numerous important contributions, he could not match the latter's mathematical, theoretical, or experimental brilliance. He was embittered in his later life because of the controversy with Newton over the discovery of the inverse square law of gravity which he claimed Newton derived from looking at one of his manuscripts and therefore should have acknowledged him as the co-discover, at least. Newton, however, made the counter-claim that he had discovered the law after reading Kepler, which is more likely in my opinion. Nor is there evidence that Hooke ever stated the law.

With Christopher Wren, Hooke was asked to draw up a plan for London after the Great Fire in 1666 sketching a model of the tower that now stands at the original cite of the fire. He invented the balance spring of the watch and foresaw the use of the pendulum in clocks. His *Micrographia* published in 1665 was famous for its optical research defending the wave theory of light and for his intricate sketches of microscopic organisms, such as the common house fly, body louse, and thin slices of plant tissues in which he discerned "cells."

INQUIRIES INTO COMBUSTION AND DISCOVERY OF OXYGEN

The attempt to explain something as common and seemingly simple as combustion had eluded experimentalists until the late eighteenth century. Each of those individuals just discussed had the explanation of combustion as one of their main objectives, but it was George Stahl who provided the first systematic, though false, explanation in terms of phlogiston. The origin of the phlogiston theory (but not the name) was due to Johann Joachim Becker who, in his *Physicae subterraneae* published in 1669, described combustion as the burning off of the "fatty earth" (p. 86). Influenced by Becker's explanation Stahl, in his *Fundamenta Chymiae* published in 1723, renamed Becker's inflammable *terra pinguis* 'phlogiston,' claiming that it was not fire itself but "the matter and principle of fire" which is contained in and released from burning or calcinated (oxidized) bodies, including metals (cf., pp. 86–87).

According to Stahl's theory, substances like charcoal and phosphorus burn as vigorously as they do because they contain large quantities of the *inflammable* phlogiston that they release when ignited. Burning in a closed vessel, a candle will be extinguished when the air becomes saturated with phlogiston, thus air normally has a large amount of phlogiston. Ores are transformed into metals when burned with charcoal because they absorb the phlogiston released by the charcoal, a "metalizing principle," while metals conversely emit phlogiston when burnt producing calyx as a residue. As Partington describes phlogiston, it

> was material, sometimes the matter of fire, sometimes a dry earthy substance (soot), sometimes a fatty principle (in suphur, oils, fats and resins), and sometimes invisible particles emitted by a burning candle. It is contained in animal, vegetable and mineral bodies, and is the same in all. It can be transferred from one body to another. It is the cause of metallic properties, of colours . . . and of odours. . . . (p. 87)

While Stahl's theory inverted the actual processes of combustion and calcination, since giving up phlogiston was actually taking in oxygen and adding phlogiston was removing oxygen, for a time it was nonetheless the best explanation even though he made no effort to quantify the reactions, ignored what had been discovered of gases, and did not attempt an atomic explanation. Based on this theory despite its weaknesses, the Swedish chemist Carl Wilhelm Scheele made important discoveries that he presented in his book translated as *Chemical Treatise on Air and Fire*, published in 1777. Based on his experiments he distinguished between two constituents of air, one highly flammable and the other inflammable, naming the first "Fire Air" and the second "Foul Air."

He described Fire Air as colorless, odorless, "in which a taper burned with a dazzling brilliance," thus he was the first to detect what later was named "oxygen," while Foul Air was later identified as "nitrogen." But his adherence to the phlogiston theory was so strong that he did not appreciate the significance of his experiments, illustrating again how preconceptions can prevent a full understanding of a new discovery. As will be true also of Priestly, in an experiment in which he burned a candle in an enclosed flask which also contained some water, he noted that as the candle burned the water rose in the flask indicating that something was being removed from the air by the burning candle, the opposite of what was predicted by the phlogiston theory. Moreover, even though he previously had observed that a candle burned more brightly in Fire Air, he failed to make the connection between this and the consumption of something in the air by the burning candle, that reduced the volume of the air, because it contradicted the predictions of the phlogiston theory. Such are the obstructions due to the preconceptions of thought.

THE IMPACT OF THE INDUSTRIAL REVOLUTION

Before discussing the next three contributors, it would be of interest to point out that a major social and cultural development occurred in England that not only led to the next phase in the development of science, but also created the technological and economic conditions for the advances which eventually would distinguish privileged from underprivileged societies, despite the often deplorable conditions of the industrial workers employed. For it was the industrial revolution which transformed previous cottage industries and other small enterprises into large manufacturing companies, along with initiating major scientific discoveries and technological innovations.

Imitating human fingers, spinning machines were invented using series of spindles or rollers to spin thread strong enough to be woven into cloth greatly increasing the productivity, along with the horrendous working conditions. One type of machine was invented in the 1760s by James Hargreaves while another was patented and financed by Richard Arkwright. "Within 15 years, Samuel Crompton had combined the two inventions in a machine which is, in its essentials, the spinning machine still used today."[3]

In addition, other new enterprises such as ironworks, steel foundries, coal mining, and the conversion of coke into coal inaugurated the factory system that created the industrial revolution in Birmingham and Manchester in England. Among other famous contributors were James Watts who perfected the steam engine, John Wilkinson an ironmaster, James Kier founder of the chemical industry, and Josiah Wedgwood founder of the Wedgwood potteries. Not belonging to the aristocracy nor the landed gentry, and called "nonconformists" owing to their unconventional religious affiliations as Unitarians, Quakers, or Presbyterians, rather than Anglicans, they were excluded from attending the prestigious "public schools" of Eton and Harrow, along with the celebrated universities of Oxford and Cambridge.

Rather than acquiring the traditional classical education based on Greek and Latin, they became highly skilled as experimenters, engineers, inventors, or entrepreneurs. Often acquiring considerable fortunes, they used their wealth not only to create mills and modernize the provincial cities of Manchester and Birmingham, but also to build excellent new educational institutions such as "Dissenting Academies" and "Colleges." They also established libraries and societies for the discussion of "experimental philosophy, technology, and literature," such as the Manchester Literary and Philosophical Society and the Lunar Society in Birmingham (so named because it met on the night of the full moon so that distant members would have more light when traveling to the meetings). As described by Bronowski and Mazlish:

> These dissenting academies were the first institutions which gave an education in the knowledge of their times: in medicine, in logic, in modern languages, and in science. They began as schools to train nonconformist ministers, who could not be trained in the universities, but they soon became broader; and the greatest of them, the Warrington Academy, from its foundation in 1757 set itself to give a modern education to laymen as much as to those who were going to preach. (p. 324)

One of the most distinguished teachers at the academy was Joseph Priestly whose investigations of electricity were previously mentioned and who is considered the identifier of oxygen, despite its earlier detections by Stahl and Scheele. Although the son of a cloth dresser, he was able to attend the Dissenting Academy at Daventry, where he acquired a knowledge of Hebrew, Greek, and Latin that enabled him to engage the more traditionally educated scholars on equal terms. He became a tutor at Warrington Academy in Lancashire shortly after it was established, teaching English, history, and modern languages, but was attracted by his colleagues' lectures to the new sciences of electricity and chemistry. "As a result, Priestley became the greatest chemist of his day, and, on his discoveries rests the modern system of chemistry" (Bronowski and Mazlish, p. 324).

He was elected F.R.S. (Fellow of the Royal Society) in 1766 and a year later published his *History of Electricity* mentioned previously. As a sign of his learning and reputation, in 1773 he became "literary companion" to Lord Shelburne (who became Marquis of Lansdowne and in 1782 prime minister) in which capacity he served for seven years. During his travels with Lord Shelburne on the Continent Priestly met Lavoisier in Paris in 1774 who conveyed important information to him. In addition to his electrical experiments, Priestly is famous for his discovery of oxygen (which had not yet acquired the name), though he did not immediately recognize it as a new gas.

While as early as 1771 he had noted the unusual effects of a peculiar air on burning candles and the respiration of animals, it was not until 1775 that he explicitly attributed these effects to a particular gas in an article titled *Experiments and Observations on Different Kinds of Air*. In 1774, using a magnifying glass with "a lens of twelve inches diameter, and twenty inches focal distance" to heat various substances, he extracted a large quantity of air from *mercurius calcinatus per se*, but "what surprised me more than I can well express, was, that a candle burned in this air with a remarkably vigorous flame" that he was "utterly at a loss how to account for it" (Partington, pp. 117–18). A few months later he found that a mouse lived longer in this air than in normal air, and that when he breathed it himself he felt light headed.

In an experiment on March 1, 1775, during which he distinguished this unusual gas from nitrous oxide, he was convinced that he had discovered something original, saying "that he was unaware of the real nature of the new gas until this date . . ." (p. 119). Noting like Scheele that when substances are burned in air they *gain* weight while the volume of air *decreases* contrary to the predictions of the phlogiston theory, he too failed to recognize its significance. He states in *Experiments on Air* published in 1790:

> For seeing the metal to be actually revived, and that in considerable quantity, at the same time that the air was diminished, I could not doubt, but that the calx was actually imbibing something from the air; and from its affects in making the calx into metal, it could be no other than that to which chemists had unanimously given the name of *phlogiston*. (p. 137)

Adhering like Scheele to the phlogiston theory, he erroneously ascribed the striking combustionable properties to the air being depleted of phlogiston and therefore able to absorb an abundance of it, calling it *dephlogisticated air*. So like his predecessors Stahl and Steele, he failed to interpret correctly what he had discovered, but since he clearly recognized that it was a new "air" he is given credit for the discovery.

Thus it was left to Antoine Laurent Lavoisier, the son of a wealthy advocate who graduated from the Collège Mazarin, to provide the correct interpretation. Though rigorous in his quantitative experiments, he was more of a theoretician being especially astute in interpreting the experimental results of others. He insisted that chemical analysis depended on the conservation of matter and on carefully weighing the reagents in all experiments. As he stated in the *Traité Élémentaire de Chimie*, published in 1789: "The whole art of making experiments in chemistry is founded on this principle: we must always suppose an exact equality or equation between the principles of the body examined and those of the products of its analysis"

(p. 124). He described his procedure in a famous sealed note deposited at the French Academy on November 1, 1772.

> About eight days ago I discovered that sulfur in burning, far from losing weight, on the contrary, gains it; it is the same with phosphorus. . . . This discovery, which I have established by experiments, that I regard as decisive, has led me to think that what is observed in the combustion of sulfur and phosphorus may well take place in the case of all substances that gain in weight by combustion and calcination; and I am persuaded that the increase in weight of metallic calxes is due to the same cause.[4]

At that time, however, he was unaware of the correct explanation, attributing the gain in weight to "fixed air" (carbon dioxide).

He was misled by the fact that when an oxide of mercury or phosphorus is heated with burning charcoal they form mercury or phosphorus plus carbon dioxide. In modern notation not yet developed: $2HgO$ (oxide of mercury) + C (charcoal) $\rightarrow$ (when heated yields) $2Hg$ (Mercury) + CO_2 (fixed air or carbon dioxide). Thus he inferred that what increased the weight of sulphur and phosphorus when heated alone to produce the calyx must be fixed air. It was at the gathering mentioned previously, attended by Priestley, Lord Shelburne, and Lavoisier and their wives in Paris on October 1774, that Lavoisier learned of his error. As recounted by Priestley, he "told Lavoisier at dinner of his discovery of dephlogisticated air [oxygen], saying he 'had gotten it from *precip* [of *mercurius calcinatus*] *per se* and also *red lead*;' whereupon, all the company, and Mr. and Mrs. Lavoisier as much as any, expressed great surprise" (Partington, pp. 126–27; brackets added).

The reason for the surprise was the fact that the "dephlogisticated air" produced when Priestley heated mercury oxide supported combustion and respiration, just the opposite reaction of fixed air or carbon dioxide. As these results were discovered by varying the experiments, Lavoisier recognized the importance of choosing the correct experiments, as well as weighing the ingredients. In two papers read to the French Academy of Sciences describing his replications of Priestley's experiments, the second one presented on August 8, 1778, he reported also finding a different gas from "fixed air" incorrectly deciding it must be a form of common air, even though describing it as "purer than common air." But additional experiments finally convinced him that what was *added* in the conversion of metals to calxes when burnt in air and *released* when the burnt calyxes returned to metals was an entirely new gas, not fixed air or carbon dioxide.

Having declared his identification of a new gas in his *Mémoire* of 1778, he still hesitated rejecting the phlogiston theory and delayed announcing his discovery of oxygen. Yet he summarized his new theory of combustion in a *Mémoire* "On Combustion in General" in 1777, subsequently naming the new gas "oxygène" in his *Traité de Chimie* published in 1789. As recounted by Partington:

> In 1782 Lavoisier says Condorcet had proposed the name "vital air" for pure air, but in a memoir received in 1777, read in 1779, and published in 1781, titled "General considerations on the nature of acids and on the principles composing them," Lavoisier called the base of pure air the "acidifying principle" or "oxigine principle" (*principe oxigine*), which he later changed to "oxygène," derived "from the Greek acid. . . ." (pp. 131–32)

Though Priestley himself never renounced the phlogiston theory he generously conceded the significance of Lavoisier's discovery in the following statement:

> There have been few, if any, revolutions in science so great, so sudden, and so general, as . . . what is now usually termed *the new system of chemistry*, or that of the *Anti*phlogistons. . . . Though there are some who occasionally expressed doubts of the existence of such a principle as . . . *phlogiston*, nothing had been advanced that could have laid the foundation *of another system* before the labors of Mr. Lavoisier. . . . (Conant and Nash, pp. 69–70)

In addition to explaining combustion and giving a better account of respiration, an even greater consequence of Lavoisier's experimental inquiries was the discovery, owing to his meticulous weighing of the chemical elements involved in the experiments, that it was possible to determine their proportion by weight in the compounds, antedating John Dalton. Nitrogen having been identified along with oxygen, it was found that ordinary air is composed mainly of nitrogen and oxygen with some carbon dioxide and water vapor and that water consists of two parts hydrogen and one part oxygen.

Furthermore, the composition of a number of compound gases such as nitric oxide, carbon dioxide, and mercury oxide was determined, as well as acids like nitric acid and sulfuric acid, along with bases, alkalis, and other compounds. As previously mentioned, Lavoisier had defined an element as "the last point which analysis is capable of reaching" and he and three of his French contemporaries, Claude Louis Berthollet, Guyton de Morveau, and Antoine François Fourcroy, published a *Méthode de Nomenclature Chimique* in 1787 in which the first modern list of elements was proposed (cf., Partington, p. 136).

Then noticing that any sample of a compound substance, such as water, normally contains its components such as oxygen and hydrogen in fixed ratios by weight, Joseph Louis Proust in 1799 and 1806 stated his "law of constant proportions:" "We must recognize an invisible hand which holds the balance in the formation of compounds. A compound is a substance to which Nature assigns fixed ratios, it is . . . a being which Nature never creates other than balance in hand, *pondere et mensurâ*" (pp. 153–54).

While Proust's law allowed "that two elements could combine in more than one proportion," called the "Law of Equivalent Proportions," it did not specify the ratio of the proportions in such combinations. But in a series of papers between 1792 and 1802 Jeremias Benjamin Richter formulated a general statement of this law as paraphrased by Leonard K. Nash: "*If, for any two substances, there are certain weights that are equivalent in their capacity for reaction with some third substance, the ratio of such weights is the same regardless of what the third substance may be.*"[5] Despite this progress there still was no explanation of how or why these proportional relations occur because of the volumes or weights involved. Thus we have reached the next great advance in the physical sciences, the development of the modern atomic theory.

JOHN DALTON AND THE ATOMIC THEORY

The background of John Dalton resembles that of Joseph Priestley in that his father was a cottage weaver, he was a Quaker nonconformist, and he left the small rustic village of Eaglesfield to seek a better life as an assistant and then principal of a Boarding School in Kendal. There he pursued studies in Latin, Greek, French, mathematics, and natural philosophy. He later moved to Manchester in 1793 where he became tutor in mathematics and natural philosophy in the New College for six years and then earned his living as a private and public teacher of mathematics and chemistry in Manchester, occasionally giving lectures by invitation in London, Edinburgh, Glasgow, and Leeds.[6]

Again like Priestley, he owed his early scientific development and advances to several benefactors. During his early schooling at Eaglesfield, Elihu Robinson, a wealthy Quaker and a man of considerable learning in natural philosophy and meteorology who corresponded with Benjamin Franklin, seeing that Dalton had an aptitude in mathematics after winning a mathematical dispute, tutored him in mathematics in the evenings after his school studies. Then during his twelve years at Kendal he met a most unusual and talented young man, another Quaker by the name of John Gough. Though Gough was blind and suffered from epilepsy, due to the intellectual interests and financial resources of his family he had been able to acquire considerable knowledge of the classics, mathematics, physics, botany, and zoology. Although nine years older than Dalton and considerably more advanced in his studies, when he learned of Dalton's eagerness to learn he became his intellectual mentor.

Becoming great friends, Gough shared his family's excellent library and fine collection of scientific instruments with Dalton while the latter reciprocated by serving as Gough's reader and amanuensis. Owing to Gough's instruction Dalton became quite proficient in algebra, geometry, and fluxions (Newton's calculus), along with physics, chemistry (including some writings in French), and astronomy. It was also Gough's unselfish recommendation of Dalton to Dr. Barnes (because it meant that he would have to move away), the principal of New College in Manchester, that was responsible for Dalton's appointment as Tutor to New College in 1794. It was a most fortunate development because it permitted him to spend the remainder of his life very happily in Manchester. On his arrival he was immediately welcomed by the prominent "Mancunians" and elected to the prestigious Manchester Literary and Philosophical Society which he attended from 1794 until his death in 1844.

He played a very active role in the Society presenting 117 papers of which 52 were printed, as well as serving as an officer for forty-four years. Unfortunately, New College suffered a declining enrollment so after teaching there for six years Dalton resigned, but when the Manchester Literary and Philosophical Society moved to new quarters in 1800 he was offered rooms for his tutoring and experiments indicative of the high regard in which he was held. His first inquiries and publications were in meteorology investigating the nature of water vapor in the air, along with confirming Lavoisier's measure of the ratios of nitrogen and oxygen in air. Then in four essays he described his further investigations in meteorology, chemistry, and of gases. Elizabeth Patterson states that the "wealth of material in these four essays is extraordinary. Even today they are hailed as 'epoch making' and 'laying the foundations for modern physical meteorology.'"[7]

It was his investigation of the combination of gases, along with their different solubilities in water, that was particularly significant because it redirected his attention from Newton's theory of atomic forces to questions regarding the number and weight of the atoms composing the various gases and substances asking, for example, why all gases are not equally soluble in water? The first explicit statement of the problem and his tentative solution was in a paper read to the Manchester Literary and Philosophical Society on October 21, 1803, titled "On the Absorption of Gases by Water and Other Liquids." In the paper he states: "This question I have duly considered, and though I am not yet able to satisfy myself completely, I am nearly persuaded that the circumstance depends upon the weight and number of the ultimate particles of the several gases . . ." (Leonard K. Nash, p. 222; the following quotations also are from this source).

This realization was reinforced by "the dynamical theory of heat—which related heat to the motion of submicroscopic particles" and his meteorological considerations of the nature and constitution of the atmosphere, especially as to "how a *compound* atmosphere, or a mixture of two or more elastic fluids, should constitute apparently a homogeneous mass, or one in all mechanical relations agreeing with a simple atmosphere" (pp. 222–23). Hypothesizing that these phenomena could be explained by *the weights and numbers of the particles* composing the various substances appears to have been the crucial insight of his new theory of atomism!

As he states in the same article "On the Absorption of Gases by Water and Other Liquids":

> An enquiry into the relative weights of the ultimate particles of bodies is a subject, as far as I know, entirely new; I have lately been prosecuting this enquiry with remarkable success. The principle cannot be entered upon in this paper; but I shall just subjoin the results, as far as they appear to be ascertained by experiments. (p. 222)

"Subjoined" to the paper was a "Table of the relative weights of the ultimate particles of the gaseous and other bodies." According to Nash, this was "the first published tabulation of atomic weights, and the figures cited make it plain that Dalton had by this time formulated all the essential parts of his theory" (p. 222).

It would be difficult to overestimate the significance of this transition. Except for Newton's substitution of attractive and repulsive forces for Democritus' hooks and irregular shapes to explain the adhesion of the atoms (as important as this was), his atomic theory was little different from that of the ancient Greeks. Like Leucippus and Democritus he held that the atoms were indivisible and *homogeneous in nature*, the variability in their primary qualities accounting for the diversity of the four major substances, fire, air, earth, and water. Thus the theory did not include the possibility of discovering a limited number of irreducible *heterogeneous* elements or atoms, such as oxygen, nitrogen, carbon, sulfur, and so forth, whose primary attribute of weight could be used to distinguish among them and explain how they could combine with other elements to form compound substances with their various macroscopic properties and interactions.

It was this entirely new conceptual framework that Dalton envisaged. As he stated in the chapter "On Chemical Synthesis" in his great work, *A New System of Chemical Philosophy*, the first part of which was published in 1808, the components of every sample of a substance such as water or salt, contain "*ultimate particles*" (like hydrogen and oxygen) that "*are perfectly alike in weight, figure, etc.*," ensuring their uniformity (p. 229). He maintained that while previous investigators had been concerned to ascertain the relative proportions of the constituents of mixed gases or compound substances, no attempt had been made to determine "*the relative weights of the ultimate particles or atoms . . .*" (p. 229). Nor had this knowledge been used to explain why these particles combine in the ratios they do, or, as Proust said, "the means which nature uses to restrict compounds to the ratios in which we find them combined" (p. 240).

Thus his system involved the following principles derived from his writings quoted by Nash. In contrast to the ancient Greek philosopher Anaxagoras, but as most scientists of his time, he adopted the theory that "matter, though divisible in an *extreme degree*, is nevertheless not *infinitely* divisible," so that its composition by "ultimate particles . . . can scarcely be doubted" (p. 228). In agreement with the uniformity of nature, he declared "that *the ultimate particles of all homogeneous bodies are perfectly alike in weight, figure, &*," so "every particle of water is like every other particle of water; every particle of hydrogen like every other particle of hydrogen . . ." (p. 229).

Continuing, he affirmed the conservation of matter declaring that "No new creation or destruction of matter is within the reach of chemical agency." Therefore "[c]hemical analysis and synthesis go no farther than to the separation of particles one from another, and to their reunion. All the changes we can produce, consist in separating particles that are in a state of cohesion or combination, and joining those that were previously at a distance." Consequently, all chemical investigations have as their objective to determine "*the relative weights of the ultimate particles, both of simple and compound bodies*," along with "*the number of simple elementary particles which constitute one compound particle*," such as a grain of salt, drop of water, or unit of gas (pp. 228–29).

The crucial problem was to determine in what ratios the ultimate particles combine to form the compounds, for even though Dalton found that about "6 grams of oxygen united with 1 gram of hydrogen to form 7 grams of water" (p. 230), calculating the relative weights of the atoms forming the oxygen and hydrogen depended upon knowing in what ratio the particles of oxygen and hydrogen were joined in water, such as HO, HO_2, or H_20, to determine whether the relative weight of hydrogen to oxygen was 1 to 6, (HO), 1 to 3, (HO_2), or whatever combination. Now that we are accustomed to the solution, it is easy to overlook what a number of combinations were possible.

It is a tribute to Dalton's physical intuition that he was able to devise "the rule of greatest simplicity" that initially proved very useful in conceiving how these atomic combinations might occur: for example, 1 to 1, 1 to 2, 2 to 1, 1 to 3, and so forth. He also listed the "general rules" that will serve "as guides in all our investigations respecting chemical syntheses" (p. 230). Guided by the laws of Chemical Proportions, Constant Composition, Multiple Proportions, and Equivalences, his rules were ingenious in that they in turn completely accounted for the laws. As Greenaway indicates, all the rules of chemical proportions follow from his simplifying principles.

> The composition of any substance must be constant (Law of Constant Composition). If two elements A and B combine to form more than two compounds then the various weights of A which combine with a fixed weight of B bear a simple ratio to one another (Law of Multiple Proportions). If two elements A and B combine separately with a third element C, then the weights of A and B which combine with a fixed weight of C bear a simple ratio to each other (Law of Reciprocal Proportions or Law of Equivalents). (p. 133)

He also composed a Table consisting of twenty known elements with their individual symbols and weights relative to hydrogen. While neither his symbols for the elements nor his relative weights were adopted, having been replaced by Jöns J. Berzelius' Atomic Weight Tables published in the early nineteenth century using the first letter or letters of the names of the elements as their symbols, the general schema was due to Dalton.[8] As Berzelius wrote in 1811 of Dalton's atomic hypotheses when he first learned of it, "supposing Dalton's hypothesis be found correct, we should have to look upon it as the greatest advance that chemistry has ever yet made in its development into a science" (returning to Nash, p. 248). Then a year later, having read Dalton's *A New System of Chemical Philosophy*, he wrote to Dalton that the "Theory of multiple proportions is a mystery but for the Atomic Hypothesis, and as far as I have been able to judge, all the results so far obtained have contributed to justify this hypothesis" (p. 249).

A BRIEF ACCOUNT OF LATER DEVELOPMENTS

Although the basis of the modern atomic theory had been laid by Dalton, a crucial weakness of the theory was Dalton's inability to place "its molecular formulas and atomic weights on a more strictly rational foundation" (Nash, p. 250). This was the challenge facing the investigators during the nineteenth century. Beginning in 1809 Joseph Louis Gay-Lussac, in a classic paper titled "Memoir on the Combination of Gaseous Substances with Each Other," announced that in analyzing the composition of gases using "parts by volume," rather than Dalton's "parts by weight," he had derived exact values for the combining ratios (cf. p. 255). Converting the combined weights of the gases into volumes by means of their densities, he then was able to demonstrate the exact ratios by which hydrogen and oxygen combine to form water vapor (H_2O), nitrogen and oxygen to form nitrous oxide (N_2O), and nitric acid (HNO_3). As he stated:

> It is very important to observe that in considering *weights* there is no simple and finite [integral] relation between the elements of any one compound; it is only when there is a second compound between the same elements that the new proportion of the element that has been added is a multiple of the first quantity. Gases, on the contrary, in whatever proportions they may combine, always give rise to compounds whose elements by *volume* are multiples of each other." (p. 260; italics added)

Although Gay-Lussac asserted that his law of combining gases was "very favorable" toward Dalton's theory, Dalton disagreed with his inference that the reason equal volumes of nitrogen and oxygen combine to form nitrous gas (NO) is that they each contain equal numbers of atoms, which Gay-Lussac believed was probably true of equal volumes of all gases. In contrast, Dalton thought the experimental evidence indicated that the same volume of different gases (under the same conditions) did not contain the same number of particles because of their different sizes. This was supported by the well-known fact that in some instances when different gases combine, the resultant compound is less heavy than the sum of the individual gases, which would not be true if they contained the same number of particles. To reconcile the differences Amedeo Avogadro, in an "Essay [in translation] on a Manner of Determining the Relative Masses of the Elementary Molecules of Bodies, and the Proportions in which They Enter into These Compounds," published in 1811, proposed that the elements of gases are not composed of single atoms but of different combinations of atoms and therefore were polyatomic (cf., p. 278).

Accordingly, under the same conditions equal *volumes* of gases, even though varying in their weights or densities, could contain the same *number* of particles if some were polyatomic rather than monatomic, hence heavier than others. If the "'particles' present in the gaseous elements *do not consist of the individual atoms* of the elements *but of groups of atoms of the same elements joined in a single molecule* of that element" (p. 284), then this could solve the problem. Despite these advantages there were three weaknesses in Avogadro's polyatomic explanation. First, there was at that time no empirical evidence to support the hypothesis. Second, even if some of the particles in the elemental gases were polyatomic it was not known how many were joined and what held them together. Third, there was no way of knowing how many of the polyatomic particles combined into "integral molecules" to form such composite gases or substances as water vapor, ammonia, or nitric compounds. However, the fortunate convergence of advances in electricity and chemistry offered a solution when it was found that chemical compounds could be decomposed into their elements by electrolysis.

ELECTROLYTIC DECOMPOSITION OF COMPOUNDS

In 1800 Volta invented the pile (a series of plates of dissimilar metals alternately stacked) that produced an electric current which, if the force binding the atoms into polyatomic particles were electrical, could be used to decompose them by electrolysis. Inducing opposite electric charges to two terminals or electrodes attached to the electric pile, if strong enough to overcome the electrical forces binding the atoms, would attract the oppositely charged particles to the positive or negative pole. As Sir Humphry Davy wrote in 1807, if "chemical union be of the [electrical] nature which I have ventured to suppose . . . there is every . . . hope that the new [electrical] method of analysis may lead us to the discovery of the *true* elements of bodies" (p. 296). Within a year he had decomposed alkali metals by electrolysis confirming that the binding force among the particles causing the cohesion and stability of chemical compounds was due to an electric force.

Previously we saw that Berzelius had improved on Dalton's Table of Elements with his own system of notation and more accurate atomic weights. Using the method of electrolysis to arrive at these atomic weights and Gay-Lussac's data on combining weights, he found that oxygen was attracted to the positive electrode and hydrogen to the negative electrode, along with demonstrating that H_2O is the correct formula for water. Noting that other compounds also decomposed into oppositely charged particles, he suggested a dualistic classification of chemical elements into "electropositive and electronegative." He inferred that the stability of compounds could be explained by the attraction of their *oppositely* charged particles which, when conjoined, were neutrally charged, but when decomposed by electrolysis retained their original charges.

With this method Berzelius was able to arrive at the correct chemical formulas for substances such as water, ammonia, nitrous gas, along with other gases and because ascertaining the correct combining weights and structures of these substances was a precondition for ascertaining their correct atomic weights, after a decade of devoted research he was able to publish his Tables of Atomic Weight. Yet there still was disagreement as to whether gases under similar conditions contain equal numbers of particles and whether they were polyatomic atoms. But by the middle of the nineteenth century new evidence supported Avogadro's theory of polyatomic particles, so by 1858, fifty years after the publication of Dalton's *A New System of Chemistry*, Stanislao Cannizzaro was able to resolve these disagreements by a different procedure involving "the comparison of the densities of the gaseous *compounds* of the elements" (p. 316).

Since the number of polyatomic atoms could not be determined from the densities of the gases without knowing how many atoms compose the constituent elements, he proposed calculating what fraction of the weight of a gaseous compound was due to an element by weighing the densities per unit volume of a number of gaseous compounds containing the element and determining what proportion of the compound the element constituted. Starting with the least quantity of the element contained in a compound, he discovered that the larger proportions of the element are always *integral multiples* of that amount, a tremendous simplification. He then calculated the relative weights of the elements by comparing the ratios in which these minimal weights are found in various compounds. If the particle contained one atom this established its atomic weight.

Furthermore, if the unit volume or the density of the pure element were twice that of the atomic weight one could infer that it is binary consisting of two atoms. Then comparing the various weights with the standard of hydrogen as 1, he could determine the relative weights of the other elements: for example, that the atomic weight of oxygen is 16, carbon 12, sulfur 32, and so forth. His derivation of accurate atomic weights in 1858 had an immediate and profound effect on chemists. Confirming Dalton's atomic hypothesis with his Periodic Table presented in the *Principles of Chemistry* published in 1869, Dmitri Ivanovich Mendeleyev (1834–1907) described the

Periodic Law as the law "that the properties of the elements are in periodic dependence upon their atomic weights" (Partington, p. 349). This Law was reinforced when several of the missing elements predicted in his Periodic Table were discovered a few years later. Then Julius Lothar Meyer (1830–1895) published his Periodic Table a year later, along with his well-known "atomic volume curve" the following year. As Partington also states: "the periodic dependence of a quantitative property was clearly shown as a function of the atomic weight; as the atomic weight steadily increases, the property alternately rises and falls over definite periods of the elements" (p. 349).

In a Faraday Lecture delivered to the Chemical Society of the Royal Institution in 1889 Mendeleyev described the significance of the Periodic Law.

> Before the promulgation of this law the chemical elements were mere fragmentary, incidental facts in nature; there was no special reason to expect the discovery of new elements, and the new ones which were discovered from time to time appeared to be possessed of quite novel properties. The law of periodicity first enabled us to perceive undiscovered elements . . . which formerly was inaccessible to chemical vision; and long ere they were discovered, new elements appeared before our eyes possessed of a number of well-defined properties.[9]

This is one of the clearest, most dramatic early examples of how scientific discoveries and advances can lead progressively to the formulation of a scientific framework disclosing an astonishing order in nature, at least partially explaining the basis of the order and leading to unforeseen predictions. Thus the remarkable foresight of Leucippus, Democritus, Galileo, and Newton that the tremendous diversity of the macroscopic world would someday be explained by combinations and interactions of simpler microscopic particles or atoms, the fundamental assumption of natural philosophy since the overthrow of Aristotelianism, was at last vindicated.

Much more would be discovered and learned, especially about the subatomic or nuclear particles and forces that would add to the explanation, but the foundation was laid. The extraordinary confirmation of the atomic hypotheses actually occurred with the tremendously challenging and improbable development of the atomic bomb by the most gifted group of theoretical physicists and electrical engineers ever assembled on a single scientific project. Even they were uncertain of its possibility until the successful climatic detonation at the Trinity site in New Mexico on July 16, 1945.[10] It is just heart-rending that this kind of scientific triumph was motivated by such a catastrophic, horrific use, whatever one may believe as to the justification.

NOTES

1. J. R. Partington, *A Short History of Chemistry*, 3rd ed. rev. and enlarged (New York: Harper Torchbooks, 1960), p. 44. The following page references are to this work until otherwise indicated.

2. Charles Singer, *A Short History of Scientific Ideas to 1900* (New York: Oxford University Press, 1959), p. 273.

3. J. Bronowski and Bruce Mazlish, *The Western Intellectual Tradition* (New York: Harper & Brothers, 1960), p. 312. Further quotations will cite the authors followed by the page number.

4. James B. Conant, ed., "The Overthrow of the Phlogiston Theory: The Chemical Revolution of 1775–1789," in James B. Conant and Leonard K. Nash, eds., *Harvard Case Histories in Experimental Science* (Cambridge: Harvard University Press, 1948), Vol. I, pp. 72–73.

5. Leonard K. Nash, ed., "The Atomic-Molecular Theory," in James Conant and Leonard K. Nash, eds., *Harvard Case Histories in Experimental Science*, op. cit., p. 242. Until otherwise indicated the following citations in the text will be to this work.

6. This description is based on his own account of his life written when he was sixty-six years old as recounted by Frank Greenaway, *John Dalton and the Atom* (Ithaca: Cornell University Press, 1966), p. 57. Further citations to this work will be preceded by the name Greenaway.

7. Elizabeth C. Patterson, *John Dalton and the Atom* (New York: Anchor Books, 1970), p. 94.

8. For facsimiles of Dalton's and Berzelius's Tables see Richard H. Schlagel, *From Myth to Modern Mind*, Vol. II, *Copernicus through Quantum Mechanics* (New York: Peter Lang Publishing, Inc., 1996), pp. 383 and 394, respectively.

9. Enrico Cantore, *Atomic Order* (Cambridge: The MIT Press, 1969), p. 39, n. 54.

10. Cf. Richard Rhodes, *The Making of the Atomic Bomb* (New York: Simon & Schuster, 1986).

Chapter 13

DISCOVERIES IN GEOLOGY, COSMOLOGY, AND EVOLUTION—REFUTING GENESIS, NOAH'S FLOOD, AND INTELLIGENT DESIGN

GEOLOGICAL DATING OF THE EARTH'S HISTORY

The discoveries and theoretical developments described previously were not a threat to Christianity because the scientists involved were Christians who believed they were discovering the astronomical and physical laws imposed by God at creation, thereby refining the simpler Genesis account that was then attributed to Moses. At least from the beginning of the eighteenth century, however, geological and paleontological discoveries raised serious challenges to the allegedly Mosaic version of creation that all the investigators, including John Ray, William Whiston, James Hutton, Compte de

Buffon, Carolus Linnaeus, and George Couvier, were determined to reconcile with Genesis.[1]

It is difficult to realize today that as recently as the early nineteenth century the adherents of the three Abrahamic religions believed that the universe was created by God in six days about six thousand years ago; that all the myriad kinds of existing genera and species were specially created in their present forms; that Adam and Eve were the progenitors of the human race tainted by their original sin, which brought about God's corrective punishment by the biblical flood; and that the earth's topography was formed by the force of that tremendous Deluge covering it—a striking example of how fanciful narratives fill explanatory voids left by the lack of empirical explanations.

Yet despite the prevailing ingenious efforts to fit the new discoveries into the biblical framework, three converging scientific developments in geology, paleontology, and evolutionary biology (and later astronomy) would erode the credibility of the biblical narrative. Although Genesis states that the universe was created in six days it does not state how long ago that occurred, but in the early seventeenth century Archbishop James Ussher (1581–1656), using biblical genealogies, calculated the date of creation to be October 23, 4004 BCE, thus about six thousand years ago. But in the nineteenth century geologists who were examining rock strata and discovering fossils at various levels gradually realized the implausibility of Ussher's calculation. As Eugenie C. Scott states in her book contrasting the evidence and arguments for evolution and creationism:

> By the mid-nineteenth century, the success of science as a way of understanding the natural world was clear. It was *possible* to explain geological strata, for example, by reference to observable forces of deposition, erosion, volcanism, and other processes, rather than having to rely upon the direct hand of God to have formed the layers. By the late nineteenth century, science was well on its way to avoiding even the occasional reliance upon God as imme-

diate cause and to invoking only natural cause in explaining natural phenomena.[2]

This evidence of varied layers of geological strata in the earth's history presented a serious challenge to the biblical story. The discovery of gradations of petrified or fossilized plants and organisms from simpler to more complex forms relative to the depths of the strata, along with the disclosure of extinct specimens such as woolly mammoths found in North American and Siberia, indicated a development over a much longer extent of time. Furthermore, as the oldest strata contained simpler morphological forms with some evidence of transitions to more complex structures, this suggested a development rather than a preformationist creation. Gradually the preponderance of this evidence overcame the biblical belief that in the beginning God, in accordance with his wisdom and providence, had created a paradise for human beings which, because of its perfection, could not be altered until the punishing Deluge. Yet, despite the accumulated evidence of the modification and extinction of genus and species, there was no way of determining the exact dates of their occurrences nor how they were brought about.

Then in the latter part of the nineteenth century the discovery of radiation provided ways of geological dating, just as previously the development of electrolysis had offered a method for determining the exact structure of compound substances and the atomic weights of their elements. This included Wilhelm Röntgen's discovery of X-rays in 1895, Henri Becquerel's detection of radiation in 1896, J. J. Thomson's identification of the electron in 1897, and Marie and Pierre Curie's discovery of radioactive substances, their first named "polonium" (in honor of Marie's natal country) in 1898, then thorium and radium. But it was their discovery that polonium spontaneously discharges half its radioactivity in a specific time, called "half-life," that was the decisive factor.

These discoveries revealed that atoms, rather than being indivisible, had an interior composition that was the source of the radiation and that their specific spontaneous radioactive transitions into half-lives could be used to date the ages of the materials involved. Specifically, it was discovered that the radioactive emissions alter the nuclear structure of the atomic material transforming it into forms of the same element with different atomic numbers. Then in 1905 based on Becquerel's discovery, Ernest Rutherford (who later will discover alpha and beta emissions) and Bertram B. Boltwood applied Becquerel's principle of radioactive decay to measure the age of minerals using the disintegration of uranium into helium as their chronometer. In 1907 Boltwood dated a sample of uraninite (an oxide of uranium) based on uranium/lead ratios.

In 1913 Frederick Soddy introduced the term "isotope" to designate the decayed elements of the radioactive materials that were identical in their chemical properties and occupied a similar place in the periodic chart, but after having radiated neutrons differed in their atomic mass and radioactive properties. For these discoveries Röntgen received the first Nobel Prize in physics in 1901, Marie and Pierre Curie shared the Prize in physics in 1903 (Marie earned another in chemistry in 1911), Rutherford in chemistry in 1908, and Soddy in chemistry in 1923. As a result, radioactive elements are used as geological clocks because each decays at a nearly constant rate of its own. By measuring the ratio of the stable "daughter elements" (or isotopes) to the original radioactive "parent element" geologists can estimate the length of time during which the decay occurred.

In the 1940s Willard F. Libby and his team developed the method of "radiocarbon dating" that enabled geologists to assign approximate dates to archeological specimens up to 50,000 years or through the Pleistocene period for which he received the Nobel Prize in chemistry in 1960. Improved techniques produced the following half-life estimates of the stable daughter products from the radioactive parents:

Carbon 14 to nitrogen 14, half-life 5730 years.
Uranium 235 to Lead 207, half-life 704 million years.
Potassium 40 to Argon 40, half-life 1.25 billion years.

Uranium 238 to Lead 206, half-life 4.47 billion years.
Thorium 232 to Lead 280, half-life 14 billion years.
Rubidium 87 to Strontium 87, half-life 48.8 billion years.

These figures, derived from the Internet under the heading "Radiometric Dating," obviously refute Ussher's reckoning of the age of the universe and the biblical limit of creation to six days. Yet "Young Earth Creationists," rejecting the scientific discoveries of geologists, paleontologists, physicists, chemists, and astrophysicists, claim that the earth is between 6,000 and 10,000 years old, with some willing to extend the date to 15,000 years. According to Scott: "The term Young Earth Creationists is often associated with the followers of Henry Morris, founder and recently retired director of the Institute for Creation Research (ICR) and arguably the most influential creationist of the second half of the twentieth century" (p. 60). An unfortunate influence one might add.

Despite all the extensive research, consilient evidence, and rigorous tests supporting radiometric dating and Geological Time Tables, in a book Morris edited titled *Scientific Creationism* (Creation-Life publishers, 1974), he offers a number of arguments for rejecting radiometric dating concluding that it is all based on "invalid assumptions." As he states in his book: "the highly speculative nature of all methods of geochronometry becomes apparent when one realizes that *not one* of the above assumptions is valid! None are provable, or testable, or even reasonable" (Scott, p.152).

An adjoining excerpt in Scott's book from Robert C. Weins's (somewhat misleadingly named) *Radiometric Dating: A Christian Perspective* (ASA Resources, 2002), presents Weins's detailed *rebuttal* of all the "misconceptions" in Morris's arguments justifying his rejection of radioactive dating, concluding as follows:

> The fact that dating techniques most often agree with each other is why scientists tend to trust them in the first place. Nearly every college and university library in the country has periodicals such as

> *Science, Nature*, and specific geological journals that give the result of dating studies. . . . Over a thousand research papers are published a year on radiometric dating, essentially all in agreement. (Scott, p. 157)

Though titles such as "Institute for Creation Research" and "Scientific Creationism" imply that Morris and other creationists are engaged in research that can provide scientific evidence for rejecting radioactive dating, this is not so. Their reason for rejecting radioactive dating is not based on the discovery of alternative scientific evidence, but solely on the fact that radioactive dating provides evidence for the age of the earth that is contrary to biblical doctrine. Creationists base their arguments on presumed weaknesses or gaps in the scientific evidence without providing any alternative evidence or explanations. In actuality, because they believe the existence of the earth and what is on it depends upon God's creation, there can be no empirical evidence. What kind of justification, other than scripture, could they cite to support their criticism?

If the assumptions of radioactive dating are invalid, then on what can the true age of the earth be based? Here Morris reveals his position, claiming that "[s]ince there is no way in which the assumptions [of radioactive dating] can be tested, there is no *sure* way (*except by divine revelation*) of knowing the true age of any geologic formation" (p. 151; brackets and italics added). But where is the revelation that gives the true date for the creation of the earth and its inhabitants? Surely not in Genesis. Morris claims that the assumptions underlying radioactive dating are not "valid," "testable," "provable," or "reasonable," but what about the assumptions underlying the biblical assertion that God created the universe in six days? How testable or reasonable are they? It is not even possible to evaluate these creationist assumptions because they depend entirely on the validity of a revelation which, *if it could be authenticated*, is impervious to any critical evaluation. Hence Morris's attempt to present his own views as scientific is blatantly unfounded.

COSMOLOGICAL DEVELOPMENTS PROVING THE AGE OF THE UNIVERSE

Turning to the determination of the age of the universe, this awaited Edwin Hubble's telescopic discovery, in the 1920s, of the shift of light waves emitted by the galaxies to the red end of the spectrum that was interpreted as evidence of their recession. This in turn was explained as due to the expansion of space produced by the initial explosion of the Big Bang, referred to as the "inflationary universe,"[3] refuting the then prevalent steady-state theory vigorously defended by Fred Hoyle. This red shift made it possible for astrophysicists to calculate the time of the Big Bang to be 13.7 billion years ago. Supporting evidence for the theory occurred in 1964 when Arno Penzias and Robert Wilson, working at the satellite communications center at the Bell Laboratories, first detected the background radiation left over from the original explosion (that had been predicted by George Gamow) for which they received the Nobel Prize in physics in 1978.

Then in 2006 two Americans, John C. Mather, a senior astrophysicist at NASA's Goddard Space Flight Center, and George F. Smoot, an experimental astrophysicist at the Lawrence Berkeley National Laboratory, won the Nobel Prize for producing more tangible evidence that the universe began billions of years ago with the Big Bang. As reported in the *Washington Post*:

> Using pioneering data from a NASA satellite they helped design and create, the two produced measurements that confirmed an essential aspect of the big-bang scenario—that a cosmic bath of microwaves emanated from that original event and has been expanding and cooling ever since.
>
> The Nobel committee praised their findings as a scientific-turning point, one that changed cosmology from a theoretical science into one in which precise measurements and conclusions are possible. . . . "They have not proven the big-bang theory, but they have given it very strong support," said Per Carlson, chairman of

> the Nobel Committee for Physics. "It is one of the greatest discoveries of the century; I would call it the greatest. It increases our knowledge of our place in the universe."[4]

Yet attributing the origin of the known universe to an original explosion or Big Bang which, according to the unconfirmed cosmological theory of "parallel worlds" or "multiple universes,"[5] is just a creation from a much more extensive universe, also is rejected by creationists as inadequate because it does not assign its cause to God. But the same arguments apply here as in the case of the earth. Where in all of history can one find evidence of a God creating a universe?

REFUTATION OF THE MYTH OF NOAH'S ARK AND FLOOD

Another pillar of the Bible that had persisted until the early nineteenth century was the story of Noah's flood accounting for the earth's topography. As related in Genesis:[6] "The Lord saw that the wickedness of man was great" and "was sorry that he had made man on the earth," so the Lord said, "I will blot out . . . man and beast and creeping things and birds of the air, for I am sorry that I have made them" (yet since "God had created man in his own image" and being omniscient should have foreseen the consequences, he had only himself to blame). So in "the six hundredth year of Noah's life, in the second month, on the seventeenth day of the month, on that day all the rain fell upon the earth forty days and forty nights."

Having instructed Noah to build an ark of three hundred cubits in length, fifty cubits in breadth, and thirty cubits in height as a refuge for Noah, his wife, their three sons and three wives, along with "seven pairs of all clean animals, the male and his mate; and a pair of the animals that are not clean . . . and seven pairs of the birds of the air also," the menagerie entered the ark while God "blotted out every living thing that was upon the face of the ground" preserving "only Noah . . . and those with him in the ark." Then at "the end of a hundred and fifty days the waters had abated; and in the seventh month, on the seventeenth day of the month, the ark came to rest upon the mountains of Ar'arat."

After another ten months "the tops of the mountains were seen," so "at the end of forty days" (a glaring time discrepancy in the fantastic story), Noah sent forth a raven, then three doves to test whether the water had subsided and the earth had dried. When the third dove did not return, "in the six hundred and first year and two months" (presumably of his life) Noah "removed the covering of the ark; and looked, and beheld the face of the ground was dry," although devoid of all living creatures and presumably most vegetation after such a long period of submersion.

An incredible legend considering the implausible age of Noah, the impossibility of the ark holding all those pairs of animals and birds plus the quantity of food and water to sustain them for such a lengthy time, the tremendous quantity of rain that would be required to fall at the same period to cover the entire earth, the time discrepancies in the story, and the near impossibility of resuming life in such a devastated environment, it is hard to believe that any sensible person could have believed it and yet many still do. It is a striking example of how in the past men (for it was only men) created fantastic stories to explain phenomena for which they had no rational explanation.

But like other myths and legends it seems to have some foundation in fact. There is mention of a similar deluge in the ancient Mesopotamian Gilgamesh Epic written a millennium before Genesis which could have been its inspiration. Or, more likely, it was inspired by a great inundation of the Mediterranean Sea overflowing the Black Sea at about the time the Bible was written. In any case, despite its fanciful nature it was not only the biblical story that sustained belief in Noah's flood, but also the fact that in the early nineteenth century geologists still thought that the prominent horizontal striations on cliffs, the grooved rocks occasionally polished as smooth as marble, the ancient "moraines" consisting of mysterious deposits of rubble and rocks, and the huge "erratic boulders" not indigenous to the area implying they had been carried a great distance before being deposited by an enormous force, all seen on the mountains of Switzerland, France, and Scotland, were evidence of a tremendous "*diluvium*" that was Noah's flood. As was true of geological dating, however, scientists sought a more natural explanation.[7]

That glaciers were the actual cause had been proposed by Ignace Venetz and Jean de Charpentier, but had not been taken seriously. The turning point occurred during what is known as the "Neuchâtel Discourse" presented in 1837 at the annual meeting of the Swiss Society of Natural Sciences in Neuchâtel, Switzerland by a young Swiss naturalist named Louis Agassiz. Well known for his study of fossils, he stunned the distinguished scientific audience by endorsing glaciation as the cause of the geological formations. When he declared "I have no doubts that most of the phenomena attributed to great diluvial currents . . . have been produced by ice," pandemonium erupted with Agassiz ridiculed and vilified at the end of the meeting (p. 87).

Indicative again of how difficult it is to accept extremely novel theories, even though most of the scientists at the meeting had seen Swiss glaciers, the idea of such an enormous shelf of ice covering all of Europe as recently as 10,000 years ago was too great to be accepted. It was not until the American sea captain and polar explorer, Elisha Kent Kane, returning from two harrowing years trapped in the colossal ice glacier covering Greenland, brought back irrefutable descriptions and exact sketches of its gigantic size that the anti-glacier faction was largely won over. If scientists have such difficulty accepting new theories, one can appreciate how hard it must be for religionists to accept empirical evidence and natural causes as refutations of their cherished beliefs.

As an interesting sequence, Hershel Shanks, editor of the *Biblical Archaeological Review* (indicating that there are some religionists who accept empirical evidence), recounts discoveries leading to a natural explanation of the biblical flood.[8] Two Colombian geologists, William Ryan and Walter Pitman, in their book *Noah's Flood*, proposed that the present Black Sea was formed 7,800 years ago, at about the time the Old Testament was written, when melting glaciers caused the level of the Mediterranean Sea to rise overflowing the Bosporus Strait and inundating the Black Sea. As the rising salt water poured into the lesser fresh water of the Black Sea a catastrophic flood occurred covering "thousands of square miles of dry land . . . killing thousands of people and billions of land and sea creatures."

Because of the time of its occurrence and location in the Near East, it seems reasonable to infer that such a horrendous deluge could have been the inspiration for the biblical story of Noah's ark and flood to "explain" its origin. Recently a team of deep-sea explorers led by Robert D. Ballard discovered evidence, based on radioactive dating of mollusk shells, of a tremendous flooding in the Black Sea that occurred about 7,500 year ago, at the same date indicated by Ryan and Pitman. Using sonar images, they located the submerged pre-flood shoreline exactly as predicted by them, indicating that these ancient legends or myths can have some factual basis.

Yet even today the biblical flood narrative is defended by some "New World Creationists" causing a controversy between them and the Grand Canyon National Park Service.[9] The dispute involves a book, *Grand Canyon: A Different View*, compiled by Colorado River guide Tom Vail that "contains a collection of essays by two dozen creationists who maintain the canyon is over a few thousand years old" and was created by "the flood told of in Genesis." Sold in a souvenir shop on park grounds, the National Park Service wants it removed because its being sold under their auspices implies condoning its pseudoscientific explanation of the formation of the Grand Canyon.

Claiming that "all science is theory" and that the creationist account "is just as valid as current geological theories," Vail threatened legal action to prevent removal of his book, claiming that such action is based on "religious discrimination." However, he ignores the difference between the *prejudicial* meaning of discrimination as opposed to *factual* discrimination based on evidential assessment. Thus the obvious fallacy in his argument is the equivocation between dismissing a theory because of religious bias, which would be religious discrimination, and rejecting it based on sound evidence.

One could hardly find a better illustration of the irrational unwillingness of creationists to accept later scientific evidence over archaic religious narratives to explain natural phenomena, as well as their ignorance in claiming that all science is just theory, arguing that creationist accounts have the same credibility as scientific explanations. Is it so difficult to understand that while all scientific explanations begin with hypotheses or theories and usually retain that designation after they are confirmed, this does not mean they all remain just suppositional? Are heliocentricism, universal gravitation, atomic fission, bacteria, and the decoding of the genome still just theories? If they were only tentative theories, how could we have orbited satellites, landed men on the moon, constructed the atomic bomb, cured diseases, cloned animals, and modified genetic structures?

DARWIN'S THEORY OF EVOLUTION VERSUS INTELLIGENT DESIGN

The last and most influential pillar of the Bible is the account of Adam and Eve, along with the creation of all living creatures, "according to their kinds." We have just discussed the Lord's restoration of life on the earth following his devastating flood due to his disgust at the "wicked" and "corrupt" nature of humankind, even though, as the Bible states, "man was created in the image of God." Thus either the image of God was not a very satisfactory model or the creation itself was botched. Nor can the wickedness be attributed, as it usually is, to the misuse of free will by human beings if God were omniscient and omnipotent, since he should have foreseen and precluded his flawed creation. Moreover, the second outcome was not any more successful than the first given the continued wicked behavior of human beings (Hitler, Himmler, Stalin, and Mugabe come to mind), which is to be expected if they evolved from primitive forms of life rather than having been created by an all knowing, all powerful, all loving God.

Continuing the creation myth, we all know the story, how God formed man from the dust in the ground breathing life into his nostrils, planted a garden in Eden in the midst of which was the tree of the knowledge of good and evil. He then commanded man to till the garden but forbad his eating of the tree of knowledge lest he become Godlike. Realizing that he would be lonely he created woman from his rib to be a helpmate, but beguiled by a serpent she ate fruit of the tree of knowledge giving some to man who also ate of it causing both to become aware of the knowledge of good and evil. Learning they disobeyed him, God punished them by putting enmity between them, intensified the pain woman would experience at childbirth, declared that man, now called Adam, would rule over the woman named Eve, and be condemned to sweat in toil until his death.

Fearing that having disobeyed him once they might also eat of the tree of everlasting life thereby becoming more Godlike, the Lord God drove them from the garden placing a cherubim brandishing a flaming sword to guard the tree. Although there is no explicit mention of transmitting their original sin to their descendents, this is implied in the punishment of all humankind. Moreover, the tendency to do evil occurs immediately after their giving birth to Cain and Abel, the latter killing his brother out of anger because he was favored by God. And so the tale is carried forth to the time of Noah. A charming story serving the same purpose as other creation myths to dispel the mysteries of life, it is hardly credible and yet it was literally believed by nearly all Christians in the early nineteenth century and even by hundreds of thousands today.

There then occurred a scientific development in the middle of the nineteenth century that would have a shattering impact on the Genesis account of creation and of humans' understanding of their origin. This was Charles Darwin's theory of evolution presented in

1859 in *The Origin of Species*, followed by the *Descent of Man* in 1871, that proposed a naturalistic explanation of the origin of human beings to replace Genesis.

According to the creation account in Genesis, on the fifth day God "created the great sea monsters and every living creature . . . with which the waters swarm, according to their kinds, and every winged bird according to its kind." Following this he said (creation by verbal command): "Let the earth bring forth living creatures . . . cattle and creeping things and beasts of the earth according to their kinds." Thus was the source of the persistent belief that God originally created each species of living creature in its present form. It was this conception of a distinct creation of all the various kinds of species, including human beings, that Darwin rejected with his theory of evolution claiming that all existing creatures arose by "descent with modification" from earlier forms.

By the middle of the nineteenth century the evidence from geology and the recently discovered ancient fossils had convinced some naturalists, such as Buffon, Lamark, and Darwin's paternal grandfather, the well-known Dr. Erasmus Darwin, that the existing forms of life were modified descendents of earlier species. Charles Lyell in his influential *Principles of Geology*, published in 1830, had declared that in the past some species had become extinct and been replaced by others, while Count Keyserling in 1853 with foresight suggested that new forms of life might have arisen from the "germs of existing species" that had been "chemically altered" by surrounding molecules, both assertions anticipating two essential aspects of Darwin's theory of evolution.[10] Yet they differed from Darwin in their conceptions of how the descent had come about.

Still, it was Darwin (1809–1882) and Alfred Wallace (1823–1913) who introduced the theory of evolution in its more explicit form, though only Darwin will be discussed. Following his graduation in 1831 from Christ's College, Cambridge, where he had been preparing for the ministry, Darwin made the critical decision to change course by enlisting on the H.M.S. *Beagle* as the ship's naturalist for a five-year voyage of exploration. During his stay in South America for several years while the *Beagle* continued on its journey, he was "struck" by the great proliferation of species on atolls and islands, particularly the Galapagos Islands, too isolated for the diversity to have been produced by interbreeding with species from other locales. Following his return to England he struggled for twenty years to make sense of his observations, reluctantly concluding that his startling discovery of the local diversity of species was evidence of an indigenous evolution caused by successively congenital mutable traits being selected for their greater competitive adaptability by the natural habitat and propagated.

An essential part of the interpretation was suggested by Thomas Malthus's (1766–1834) thesis that because populations of living organisms increase in geometrical ratio, while their food supply only increases arithmetically, they are in a constant struggle for survival suggesting to Darwin that in this competition those creatures that had inherited greater adaptive or competitive traits would have a better chance of surviving. He called the process by which the favorable variations are chosen "Natural Selection," although he said that Herbert Spencer's phrase "Survival of the Fittest" was more accurate (p. 49).

As he summarized his position at the end of chapter IV in *The Origin of Species*:

> If under changing conditions of life organic beings present individual differences in almost every part of their structure, and this cannot be disputed; if there be, owing to their geometrical rate of increase, a severe struggle for life at some age, season, or year, and this certainly cannot be disputed; then, considering the infinite complexity of the relations of all organic beings to each other and to their conditions of life, causing an infinite diversity in structure, constitution, and habits, to be advantageous to them, it would be a most extraordinary fact if no variations had ever occurred useful to each being's own welfare, in the same manner as so many variations have occurred useful to man. (p. 85)

Having described these empirically supported suppositions, he describes the consequences and how they might be explained.

> But if variations useful to any organic being ever do occur, assuredly individuals thus characterized will have the best chance of being preserved in the struggle for life; and from the strong principle of inheritance, these will tend to produce offspring similarly characterized. This principle of preservation, or the survival of the fittest, I have called Natural Selection. It leads to the improvement of each creature in relation to its organic and inorganic conditions of life; and, consequently, in most cases, to what must be regarded as an advance in organization. Nevertheless, low and simple forms will long endure if well fitted for their simple conditions of life. (pp. 85–86)

Darwin thus concluded that evolution involved two essential factors: (1) that organisms vary owing to their different parental inheritance, and (2) that those variations more conducive to survival would be propagated more plentifully over the years and if followed by additional adaptive variations could lead to new species, or if maladaptive, to their extinction. Thus extinction was the ultimate punishment for a failure to produce the most adaptive features. Having rejected any *inherent* tendency to improvement in favor of random propitious variations being selected by the environment and transmitted by inheritance, it was not known if this was sufficient to account for the great diversity of variations and if not, how the additional variations were produced.

Jean-Baptiste de Lamarck (1744–1829), as other naturalists at the time, proposed that in addition to the novel variations acquired at birth, developmental improvements during a creature's lifetime brought about by the special use of ordinary muscles or organs that enhanced its survival also could be passed on increasing the number of possible acquired characteristics. Even Darwin at first held this position but later recognized that the evidence did not support it, while the theory eventually was disconfirmed by the distinction between genetic variations which are inheritable and somatic improvements of the Lamarckian type that are not inheritable, though recent research has indicated that there could be additional origins of variation.

Gregor Johann Mendel (1822–1884), an Austrian Catholic monk and botanist, is considered the father of genetics because of his systematic experiments involving the generation of garden peas, keeping exact records of a great number of offshoots over several generations. He is known for his Mendelian Laws of inheritance declaring that congenital traits can be inherited as separate units, anticipating the gene, and proposing the first laws governing the transmission of the inherited characteristics and their random recombination occasionally producing hybridization. Unfortunately, his classic article describing his experiments published in 1866, translated as *Experiments in Plant Hybridization,* was unknown to Darwin and the rest of the world until it was rediscovered at the beginning of the twentieth century and then served as the basis for genetic studies.

With the discovery of the chromosome and the gene, Mendel's earlier suggestions of genetic units and hereditary laws were confirmed as scientists gradually discovered the biomolecular structures and genetic processes involved. In 1953 James Watson and Francis Crick announced their groundbreaking discovery of the now familiar twisted ladder structure of deoxyribonucleic acid, the fabled DNA consisting of the genes on the forty-six chromosomes of which half are contributed each by the male and female parents. This helical structure is composed of two twisted vertical supports made of phosphorus and sugar, while the winding series of horizontal rungs consist of a pair of nitrogenous bases joined at the center by mutual attraction.[11]

The paired bases or nucleotides are Adenine joined to Thymine and Cytosine joined to Guanine whose initial letters, A, T, C, G, constitute the "alphabet" of the genome that carry "instructions" contained in three letter "words" or "codons," such as ATG, CAC, etc., that run along the two outer helical supports at the insertion of the rungs. At last count the human genome consisted of 3.1 billion nucleotide pairs and about 22,000 genes, the latter tiny segments or "sentences" of the "text" ranging from 1,000 to 100,000 letters. As described by David Brown in the recent article noted:

> A gene has a section of DNA that marks its starting point, a body consisting of coding sequences (called exons) interspersed with filler areas (called introns), and a termination point. The three-letter words in the coding sequences specify the order of the different amino acids that will make up a protein: the gene's product.
>
> When . . . activated, it is first transcribed into an intermediate molecule called mRNA. The introns are clipped out and the exons spliced together, and the whole thing is then translated into the protein.

Previously it was thought that about five percent of the exons or coding DNA directed the production of proteins while the remaining ninety-five percent, called "junk DNA," had no discernable function. Now it is believed that though this function remains largely unknown, some of it consists of directing the activity of the coding genes activating or deactivating them, regulating their production, and influencing the kinds of proteins they produce. This conception of the genome answers the questions that faced Darwin pertaining to the cause and transmission of inherited variations that were unknown at the time. The variations were discovered to be caused by mutations within the DNA when a three lettered codon, altered by the deletion or exchange of one of the lettered bases during the transcription process by cosmic or atomic radiation or by the effects of virus and adjacent chemicals, produce the mutated amino acids and proteins.

As a result, the genetic causes of such disorders as epilepsy, colorblindness, deafness, muscular dystrophy, cystic fibroses, and Down's Syndrome have been identified and in some cases corrected or eliminated. Thus the practice of medicine is being revolutionized by the discovery of the genetic causes of maladies and disorders enabling their prevention, rather than just treating the damaging consequences, while the identification of criminals has markedly improved with the accuracy of DNA tracings.

In 1989 one of the most audacious scientific programs ever conceived, comparable to the Manhattan Project and the Lunar Landing, was launched to decipher the genetic code of the DNA. About three years later the research was nearing completion with the sequencing of the human genome's three billion base pairs and what is believed to be about 22,000 genes. What took 3.5 billion years of evolution to create required just 3.5 years to decipher using the most powerful computers and high speed robots operating twenty-four hours seven days a week. In December 1998 it was announced that the entire genome of the lowly worm, *Caenorhabditis elegans*, had been decoded. Confirming our evolutionary lineage, it was found that 70 percent of the worm's genes were very similar to human genes; in fact, determining how these similar genes function in much simpler organisms has facilitated learning how they operate in humans.

Then in December 2002 researchers produced a nearly final draft of the genome of the humble laboratory mouse. Once again the similarities with the human genome were striking with 80 percent of the human genes having close facsimiles in the mouse, while the remaining 20 percent had some resemblance. So similar is the mouse's genome to the human's that geneticists compare it to the Rosetta Stone, saying that what they learned from it can be used to translate previously unintelligible DNA strands in humans. Finally, the more than 98 percent identity of the human genome with that of the pigmy chimpanzee and the bonobo ape leaves little doubt of their common ancestry.

In fact, the evidence for evolution is now so extensive and confirmed in terms of fossil records, embryological development, and discovered genetic similarities that the argument has shifted from whether evolution actually occurred to whether it presupposes an intelligent designer. As for its occurrence, not only is there evidence of evolutionary changes occurring today, but a recent fossil discovery of the nearly complete skeleton of an older species of a young girl who died 3.3 million years ago, along with an older female figure named "Ardi," conclusively shows their partial transition from ape to human based on their composite skeletal structure. As described in the *Washington Post*:

> Just recently an international group of paleontologists after fifteen years of studying and reconstructing the shattered skeleton of another female transitional figure named "Ardi," derived from *Ardipithecus*, who lived 4.4 million years ago in East Africa, announced that she was another primate who could be a common ancestor linking humans and primates. Living more than a thousand years before the famous Lucy, . . . this helps bridge the evolutionary gap. As the author of the article from which this was taken states: "The origin of the human species via evolution from earlier primates is beyond dispute. Field work over the past century has shown that the human line has originated in Africa, and the fossil findings have been bolstered by laboratory analysis of the genetic codes of humans, chimpanzees and other primates."[12]

Thus given the irrefutable evidence for evolution itself, the controversy now is whether it required an intelligent cause. At a White House ceremony on June 26, 2000, announcing the completion of a draft of the human genome, President Clinton stated: "Today we are learning the language in which God created life. . . ." Given the enormous intricacy of the genome involving such cryptographic terms as "letters," "words," "sentences," and "reading instructions" for decoding complex molecular structures, it is tempting to acquiesce in Clinton's assertion that the genome must have been written by an intelligent God. Similarly, in 1802 William Paley had argued, in his *Natural Theology*, that such complex organs as the human eye could not have arisen by fortuitous natural causes and thus, like a watch, must have been designed consciously. As he stated: "The marks of *design* are too strong to be gotten over. Design must have had a designer. The designer must have been a person. That person is God" (Appleman, p. 9).

Darwin, however, was convinced that the indigenous diversity of species he had observed on the Galapagos Islands, human's embryological kinship with other animals despite the dissimilarity of their later characteristics, and the selective breeding of domestic animals analogous to "descent by natural selection," were sufficient to remove Paley's skepticism. In addition, his readiness to extend the duration of life on the earth to thousands of years of prehistory, as exhibited in the fossil record, convinced him that over such a lengthy period of time slight advantageous changes in the construction of the eye could have converged to eventually produce such a marvelously complex organ, in all its various forms. Furthermore, he would have been delighted with the discovery of the transitional skeletons of the 3.3 and 4.4 million-year-old-apelike females confirming his belief that evolutionary theory applied to humans as well as lower organisms.

In terms of what we have since learned about the biomolecular mechanisms whose functions are activated and controlled by purely chemical-electrical interactions, accepting the eye's complexity as the product of evolution is much more plausible than in Paley's time. Darwin may have still believed, as did most everyone at the time, that at creation God imposed the general laws governing organic life, but held that *secondary* natural causes, such as congenital variations, environmental selection, and genetic inheritance, can account for the evolutionary process itself. Yet Thomas Huxley, the ardent admirer and defender of Darwin's theory, could see "no trace" of a providential purpose in nature, declaring "the whole world, living and not living, is the result of the mutual interaction, according to definite laws, of the forces possessed by the molecules of which the primitive nebulosity of the universe was composed."[13]

Since Huxley's day developments in chemistry, molecular biology, genetics, and even neurophysiology have provided evidence showing that nature itself is capable of producing complex structures that formerly would have seemed to require an intelligent designer. Perhaps it will help if one realizes that all the ingenious devices created by human beings, such as Paley's watch, nuclear reactors, global satellites, and computers were not made possible by a soul endowed by a divine mind, as previously believed, but by the creative use of our *physical* brains that function entirely by neurophysiological processes involving the same biomolecular principles as evolution. There is no discontinuity. Since the neuronal functions that underlie our intellectual and creative acts are now explainable in terms of neurophysiological processes, even if not yet complete, it should not be so difficult to accept that similar processes produced evolutionary developments as well.

Since evolution occurred over billions of years usually at an incremental pace except for the Cambrian Explosion, it required less intelligent or purposeful capabilities than exist in our brains. If such a physical organ can be the basis of all humankind's marvelous creativity in literature, art, music, sculpture, architecture, philosophy, mathematics, and science, it certainly should be possible for nature, which produced the brain based on similar biomolecular processes, to have the inherent capacity of causing its own evolutionary development without supernatural intervention.

The "anthropic principle" introduced by advocates of intelligent design to support their position is equally spurious. As stated by Michael Denton in an excerpted passage from his *Nature's Destiny: How the Laws of Biology Reveal Purpose in the Universe*: "the laws of physics are supremely fit for life and the cosmos gives every appearance of having been specifically and optimally tailored to that end . . ." (Scott, pp. 158–59). As I have persistently argued, however, such explanations, as the "anthropic principle," do not explain anything. Considering that Homo Sapiens have existed on this insignificant planet for only about 40,000 years while the planet itself is 4.7 billion years old, does this indicate a coordinated or intelligent design?

None of the states or occurrences of nature would have existed if the antecedent physical conditions had not been what they were, but this does not presuppose they were designed for that purpose, anymore than evolution was—just that they had to exist as they do for the world to be as it is. Discovering what these conditions were, as scientists are committed to doing, is the explanation. If we followed the anthropic argument to its logical conclusion that whatever shows intelligent design must have had an antecedent designer, then we would have to ask what designed the existence of God? Explanations have to stop with something being given and it is more reasonable to accept the universe as that something, since we know it exists, than a questionable amorphous spirit.

This raises the broader question as to whether anyone who did not have a predisposition to believe in the existence of God based on ancient scripture, if impartially considering the history of the universe, the nature of the evolutionary process, and the existential condition of human beings would come to the conclusion that it *had* been created by an intelligent being? Does the world as we know it suggest, as Leibniz maintained, that "it is the best of all possible worlds"? Why would an intelligent being create such an extensive cosmos of 100 billion extragalactic universes, each of which contains billions of solar systems, to house human beings in this tiny speck of a world? Why did life appear only about 1.5 billion years ago when the earth is 4.5 billion years old and the Big Bang occurred 13.7 billion years ago?

Why was the emergence of life preceded by a catastrophic devastation due to meteors and volcanic eruptions? Why did human history begin when Cro-Magnon man evolved from a common ancestor with the apes about 40,000 years ago, while contemporaneous Neanderthal man disappeared? Why have evolutionary mutations been so accidental, wasteful, and destructive, even though eventually progressive, if they were directed by an intelligent designer?

Why did the winnowing selective process have to be so viciously dependent on a savage tooth and claw predation? To produce human beings why would an intelligent God design the genome subject to accidental mutations creating grotesque embryonic forms, crippling birth defects, wrenching mental disorders, along with vicious killers and tyrannical rulers? Since most of these facts were known in

Leibniz's time one wonders how he could have been so indifferent to them. He should have been aware of the horrific suffering of humanity caused by infant mortality, the mother's birthing death, childhood diseases, cancer, epidemics (like cholera and influenza), along with natural disasters such as plagues, draughts, earthquakes, erupting volcanoes, and tsunamis. Why did an all-knowing, all-powerful, and all-loving God not foresee and avert these disasters?

Why did He allow the anguish owing to the unjust inequities of birth, such as inferior natural endowments, uncontrollable addictions, abusive family conditions, poverty, and inhospitable geographical habitats? What loving God would have inflicted the cruelly debilitating genetic disease, adrenoleukodystrophy (ALD), on innocent children? Why did He condemn human beings to suffer from the dreadful physical and mental deprivations of aging, such as Parkinson's and Alzheimer's disease, multiple scleroses, and dementia? Visiting a nursing home for the aged can be like entering a theater of the absurd or having a grotesque nightmare. If God were the cause of human existence, he also should be held accountable for all its tragedies and deprivations. The creationist's reply is that we cannot understand God's actions so must take them on faith, but how can not understanding God's actions be an explanation or a consolation?

NOTES

1. These developments are described by John C. Greene in his excellent account in *The Death of Adam: Evolution and Its Impact on Western Thought* (Ames, Iowa: The University Press of Iowa, 1959).

2. Eugenie C. Scott, *Evolution vs. Creationism* (Berkeley and Los Angeles: University of California Press, 2005), p. 50. Until otherwise indicated, the immediately following references in the text are to this work.

3. Cf. Alan H. Guth, *The Inflationary Universe* (Reading, Massachusetts: Helix Books, 1997). See chapter 12 for the discussion of the "inflationary universe" and for the discovery of the background radiation by Penzias and Wilson, see pp. 58–69.

4. Quoted from a report by Marc Kaufman, "2 in U.S. Win Nobel Prize for Research of Universe's Origin," *Washington Post*, October 4, 2006, A3.

5. Cf. Michio Kaku, *Parallel Worlds: A Journey through Creation, Higher Dimensions, and the Future of the Cosmos* (New York: Doubleday, 2005), chap. 4.

6. These quotations are from the Holy Bible: *Revised Standard Version* (New York: Thomas Nelson and Sons, 1953), sec. 5–8.

7. Cf. Edmund Blair Bolles, *The Ice Finders* (Washington, D. C.: Counterpoint, 1999).

8. This account of the recent exploration of the Black Sea is based on an article by Gay Gugliotta, "For Noah's Flood, a New Wave of Evidence," *Washington Post,* November 18, 1999, pp. A1, A28.

9. This account is based on a news item by Kimberly Edds, "At Grand Canyon Park, a Rift over Creationist Book," *Washington Post,* January 20, 2004, A–17.

10. Cf. Philip Appleman, *Darwin*, selected and ed., 2nd ed. (New York: W. W. Norton & Company, 1970), pp. 7, 25 respectively. Unless otherwise indicated, the following references to Darwin are to this work.

11. This discussion is based on Matt Ridley's excellent book, *Genome* (New York: Harper Collins, 1999). It is supplemented by a recent article in the *Washington Post* by David Brown, "How Science Is Rewriting the Book on Genes," November 12, 2007, A8. The quotation is from the latter article.

12. Joel Achenbach, "'Ardi' May Rewrite the Story of Humans," *Washingto Post*, October 2, 2009, A1, A6.

13. Thomas Huxley, "On the Reception of the 'Origin of Species,'" *Life and Letters*, I, pp. 554–55. Quoted from Greene, *The Death of Adam: Evolution and Its Impact on Western Science*, op. cit., p. 304.

Chapter 14

THE REMARKABLE ACHIEVEMENTS OF TWENTIETH-CENTURY PHYSICS

THE ORIGINS OF QUANTUM MECHANICS

The beginning of the twentieth century witnessed as remarkable a series of developments as ever occurred in the history of science. Newton had discovered with his prism experiments that light consists of a discrete spectrum of rays. In Query 8 of the *Opticks* he had asked: "Do not all fix'd Bodies, when heated beyond a certain degree, emit Light and shine; and is not this Emission perform'd by the vibrating motion of its parts?" This prescient query pursued in the latter nineteenth century eventuated in the investigation of "black body radiation," the change of color a black body, like a heated poker, undergoes due to the spectral emissions emitted by the energized oscillators that culminated in Max Planck's introduction of quantum mechanics in 1900.

The spectroscope having been invented to analyze the discrete spectral emissions of solar radiation and that given off by heated bodies, it was discovered that the colored lines could be used to identify elements, whether material or gaseous, just as atomic weights were used to organize the elements in the periodic table, only more reliably, but without knowing how the signature lines were produced. Though the complete history is much more complicated,[1] one of the main contributors was Gustav Kirchhoff who, when investigating black body radiation in October 1859 to ascertain the functional relation between emission and absorption spectra, succeeded in discovering the exact numerical correlation. The discovered law was stated by Abraham Pais as "the ratio of emissive to absorptive power of a body in thermal equilibrium with radiation is a universal function of frequency v and temperature T."[2] Pais adds that "[t]he function in question is proportional to the spectral density p(v, T) of blackbody radiation. Forty years later Planck would decode p(v, T), thereby founding the quantum theory" (pp. 167–68).

Motivated by Kirchhoff's discovery, Planck set out to find the exact mathematical correlation and physical cause between the energy distribution of the emission density and the entropy distribution of the thermal equilibrium, entropy defined as the system's state of disorder. To his astonishment and dismay, Planck discovered that the thermodynamic or radiational processes did not conform to a fundamental law of classical physics that all energy exchanges were continuous, but that they were discrete or discontinuous. Instead of occurring like the continuous stream of water that flows smoothly to increases or decreases in pressure up to a limit, they were intermittent and discrete as when water is sprayed.

Thus contrary to classical thermodynamics where the energy exchange could acquire any value, Planck found that the electronic oscillators believed at the time to cause the spectral emissions could not be activated by any energy in a continuum, but only in discrete quanta that were integral multiples of hv: $\varepsilon = nhv$, where n = 1, 2, 3 etc., ε standing for energy, v for *frequency,* and h for Planck's constant, which he determined to be 6.55×10^{-27} *erg. per. sec.*[3] Planck's constant hv became a most famous constant and one of the greatest innovations of modern physics, initiating what is called the old quantum mechanics for which he received the Nobel Prize in physics in 1920.

From this law he was able to derive a number of additional constants and empirical values that is the strongest indication of a theory's truth. As Emilio Segrè states: "from the first paper, Planck pointed out that from Stefan's law and from Wien's thermodynamic law it is possible to infer the two universal constants *h* and *k* [from Boltzmann's law], and from these the charge of the electron, Avogadro's number, and more" (p. 73; brackets added). It is these surprising, extremely precise convergent derivations obtained from independent phenomena that provide such convincing evidence of the validity of scientific theories that creationists ignore.

Yet despite these derivations Planck was never reconciled to his discovery because it was inconsistent with the deterministic laws of classical physics. Later in life he stated that "I tried immediately to weld the elementary quantum of action somehow in the framework of classical theory. But in the face of all such attempts this constant showed itself to be obdurate. . . ."[4] Einstein was another who was unable to accept an inherent indeterminacy in nature, declaring in his "Autobiographical Notes": "All my attempts . . . to adapt the theoretical foundation of physics to this [new type of] knowledge failed completely. It was as if the ground has been pushed out from under one, with no firm foundation to be seen anywhere, upon which one could have built."[5]

EINSTEIN'S SPECIAL AND GENERAL THEORIES OF RELATIVITY

There have been only two physicists whose works immediately captivated the world, Newton and Einstein (1879–1955). We have seen in what esteem Newton was held throughout his life while Pais recalls that during a symposium of physicists at Princeton to mark Einstein's seventieth birthday, an awed hush momentarily pervaded the hall when Einstein entered.[6] His name has become synonymous with scientific genius and his wrinkled visage with its shock of white hair recognizable throughout the world. He also epitomizes the intense devotion of solitary individuals to attain a more unified conception of the universe—an ideal that may never recur now that scientific inquiry has become such a technologically driven, government sponsored, team undertaking. Yet, according to Pais, even his discontent with the statistical indeterminism of and distancing from quantum mechanics did not prevent physicists "from recognizing Einstein as by far the most important scientific figure of this century" (SITL, p. 15).

In what has been called his "*l'anno mirabile,*" in 1905 when he was employed in the federal patent office in Berne, Switzerland, and just twenty-six years old, he wrote six papers of which five were sent to the *Annalen der Physik* between March and December and subsequently published. The first contained his explanation by light quanta of the photoelectric effect supporting Planck's discovery of quanta of energy; the second, his doctoral thesis, presented a new calculation of molecular dimensions; the third and sixth explained Brownian motion (the observed random motion of pollen grains suspended in water) as due to the impact of the water molecules; and the fourth and fifth dealt with the electrodynamics of moving bodies in the special theory of relativity from which he derived his famous equation $E = mc^2$, E standing for energy, m for mass, and c for light.

Although his fame is based mainly on his theories of relativity, his introduction of light quanta to explain the photoelectric effect, reintroducing Newton's corpuscular theory of light, raised the problem of the dual nature of light to account for both its quantum and wave properties. It is for this explanation that he received the Nobel Prize in physics in 1921, rather than for his theory of relativity, since the latter was still somewhat controversial at the time. According to the traditional wave theory of light, the force or velocity with which the electrons are ejected from the oscillators in an illuminated metallic surface is due to the *intensity* of the monochromatic light (more superimposed waves), yet experiments showed that the velocity of the ejected electrons did not depend on the *intensity* of the light but on the color or *frequency*, ultraviolet light with a frequency twice that of red light ejecting the electrons with twice the velocity. Rather than affecting the velocity, increasing the intensity of light increases the *number* of electrons ejected.

But if instead of waves, light consists of discrete quanta whose energy is proportional to the frequency, according to Planck's equation $\varepsilon = hv$, this would explain why the specific colors of monochromatic light having different frequencies eject the electrons with their specific velocities owing to the differences in energy. Moreover, if light consists of discrete quanta, then increasing its intensity would increase the *number* of quanta causing *more* electrons to be ejected consistent with the experimental results. But while this explains the photoelectric effect by introducing a new particle, later called the "photon," it conflicts with the wave theory since their properties are different: quanta defined by their energies and velocities and waves by their wavelengths and frequencies.

Yet because the diffraction patterns produced by light demand a wave explanation while the photoelectric effect requires a quantum explanation, as noted previously, it seemed that both configurations, wave and particle, were required to explain the different experimental results. Because of their renewed adherence to the wave theory scientists initially were reluctant to accept the existence of photons, but in 1923 when Arthur Compton demonstrated experimentally that the scattering of X-rays by electrons followed the same conservation laws of deflection and momentum as colliding particles, with energy hv according to Planck's formula, the reality of photons was conceded.

The third and sixth articles on Brownian motion based on his doctoral thesis proved to be very influential, cited more frequently than the others. This is because at the time it provided the strongest empirical evidence for the existence of microscopic particles. Because the motion of particles is affected by their size and density, along with the viscosity and temperature of the water, utilizing the kinetic theory of gases Einstein was able to deduce the motion of the molecules necessary to producing Brownian motion, as well as derive new values for Avogadro's number and Boltzmann's constant k.

His fourth and fifth articles analyzing "the electrodynamics of moving bodies" introduced his special theory of relativity. In contrast to his first article that contributed to the most revolutionary development in physics in the twentieth century, quantum mechanics, the articles on relativity, while radically revising Newton's spatial and temporal "frames of the universe" should, he said, be seen as "the natural completion of the work of Faraday, Maxwell, and Lorentz" (Pais, IB, p. 250). In fact, his general theory of 1915 incorporates a revision of the two classical theories that prevailed at the end of the nineteenth century, Newtonian mechanics and Maxwell's electromagnetism.

Because Newton defined space and time as "absolute frames of the universe" to account for absolute rest and motion, it was decided in the nineteenth century that the universe must be filled with a stationary ether to account for the propagation of light waves and for Maxwell's electromagnetic waves introduced in 1864. But not satisfied with merely assuming the existence of explanatory entities, scientists have an obligation to test their existence. Thus, as mentioned previously, Albert A. Michelson (1852–1931) and Edward W. Morley (1838–1923) devised experiments to confirm the existence of absolute motion in relation to the ether at rest in absolute space.

Because absolute motion was not detectable mechanically Michelson turned to optics to devise a test. Since the motion of the earth through the stationary ether would create an ether "wind" or "drift" opposite to its motion, it was assumed this would affect the velocity of light depending on its direction, just as an object's velocity is either slowed or augmented according to the direction of the wind striking it. To detect and measure the effect, Michelson constructed an interferometer consisting of a single light source divided into two beams which, after traveling the same distance within the interferometer though transverse to each other during part of their trajectory, were reflected by mirrors back to the source. There the interferometer would detect that they were out of phase owing to their different velocities, thereby recording the effect of the ether drift on their speeds.

Specifically, it was predicted that the beam directed against the ether would have its frequency decreased and thus its wavelength increased by a fraction of 1/25th, but to his astonishment, even though successive experiments were performed in various places to eliminate any compensating factors due to the earth's movement, the beams returned *in phase* indicating no change in their wavelengths or times of return and thus in their velocities. Michelson repeated the experiment with Morely in 1887 with identical results. It was decided that either there was no ether or that the velocity of light was invariant, both conclusions later confirmed. Michelson was awarded the Nobel Prize in physics in 1907, the first American to receive one, though it was not for his ether drift experiment, but for his invention of ingenious optical instruments and his other experiments.

As consequential as this result was, according to Einstein it did not play a "decisive role" in his formulation of the special theory of relativity that held that the velocity of light was invariant or constant. What particularly had influenced him was Hendrik Lorentz's 1895 article on "The Electrodynamics of Moving Bodies" (the title of Einstein's first article on relativity theory) in which he stated: "*According to our theory the motion of the earth will never have any first order influence whatever on experiments with terrestrial light sources*" (Pais, SITL, p. 117). As Einstein later recalled:

> I took into consideration . . . the truth of the Maxwell-Lorentz equations in electrodynamics . . . show[ing that] . . . the so-called invariance of the velocity of light . . . should hold also in the moving frame of reference. This invariance of the velocity of light was, however, in conflict with the [Galilean] rule of the addition of velocities we knew of well in mechanics. (Pais, SITL, p. 139; brackets added)

Thus his conclusion follows with stunning simplicity from the two assumptions in his original paper: the equivalence of inertial frames and the constancy of the speed of light. In fact, his assumptions are just the converse of Newton's: rather than the length of measuring rods and rate of clocks being unaffected by their velocities and thus the calculations of the velocity of light must vary from different reference systems as in Newtonian mechanics, he deduced that since the velocity of light as measured from different inertial frames was invariant, the instruments must vary to produce the same measured velocity.

In classical mechanics two observers moving relative to each other would not find the velocity or distance of a projectile to be the same, but to be greater or less depending on their own velocities and distances from the projectile. However, using Galileo's "addition of velocities" or "transformation principle," if one knew the direction of velocity and distance of the other observer one could correlate the difference in their measurements by taking into account the other's velocity and distance. But while this is true of mechanical velocities, it is not true of light or electromagnetic propagation which is invariant from all frames of reference.

Because the velocity of light as measured by two observers moving uniformly in the same direction, but one with greater velocity than the other, does not result in a lesser calculation by the faster moving observer as in classical mechanics, this could be explained by assuming that the measuring rods contract and the clocks slow down in that faster system. Since $v = d/t$, if the distance measurement is greater because of the contraction of the measuring rods and the time retarded or dilated due to the slowing of the clocks proportional to their velocities, then the fraction d/t will have a larger value and so will the velocity—in fact, it would have a constant value for all velocities. Thus Einstein rejected Newton's conclusion in the Scholium of the *Principia* that "[a]ll motions may be accelerated and retarded, but the flowing of absolute time [along with the distance dimensions in absolute space] is not liable to any change"[7] (brackets added). While the space and time measurements are *reduced* in the new equations, mass *increases* explaining why nothing can exceed the speed of light. When the velocities of the two systems are slight compared to the velocity of light, as they are on the earth, the new equations reduce to the classical Galilean transformation equations.

While the Lorenz transformation equations correlate measurements from a system assumed to be at rest to another in uniform motion, in 1908 Herman Minkowski devised a formula $(cT)^2 - (R^2)$ that gives the *same* value for the *duration* of two events as measured by systems in uniform relative motion. As described by G. J. Whitrow:

> If, according to a particular observer, the difference in time between any two events is T, this associated spatial interval is cT. Then, if R is the space-distance between these two events, Minkowski showed that the difference of the squares of cT and R has the same value for all observers in *uniform motion*. The square root of this quantity is called the *space-time interval* between the two events. Hence, although time and three-dimensional space depend on the observer, this new concept of space-time is the same for all observers.[8] (Italics added)

It was this formula that gave rise to the conception of the universe as a four-dimensional continuum of events, three of space and one of time.

Since in the special theory the two observing systems are kinematically equivalent because they are inertial or moving uniformly, allowing an equivalence or reciprocity in the measurements, the differences are regarded as merely apparent. Similar to the apparent reciprocal reductions in dimensions in objects when seen from a distance owing to the change in perspective, the alterations in the special theory have been attributed to the "perspective of velocity." What has been described applies only to electromagnetic phenomena and also entails that the velocity of light is independent of the velocity of its source. The wavelength and frequency are affected by the velocity, but in such a way that the product of the two, the velocity, remains constant.

Not content with such a restricted formulation, during the following decade Einstein extended the theory to apply to nonuniform accelerated motions and dense gravitational fields that generate the forces which actually cause the contractions, retardations, and mass increases. Although initially sounding like science fiction, Einstein's prediction that the trajectory of light would be curved inward as it passed through the sun's gravitational force or field was confirmed by A. E. Eddington's famous eclipse expeditions in 1919 (which had been delayed by the First World War). Since then his predictions have been reconfirmed by extremely sensitive clocks placed in jet aircraft circling the earth and by the extended decay lifetimes of radioactive particles from outer space due to their great velocities. Thus these effects are now included in relativistic quantum field theory owing to the tremendous velocities of subatomic particles.

The vision of a unified conception of nature consisting of universal laws was driven by his "Cosmic Religion," a belief in a universe governed by quasidivine laws but not a personal God who would intervene in human affairs. To attain this unification he devised ingenious thought experiments utilizing his "principle of equivalence" to eliminate apparent dichotomies in nature comparing an object's suspension in the air, when released in an elevator in free fall that canceled the effects of gravity, with an object's suspension in gravity-free outer space.

Conversely, he compared the effects of being in a stationary elevator in the gravitational field of the earth with that of being in an elevator in gravity free outer space but accelerated upward at the same rate as the gravitational fall on the earth, showing their equivalence. Because the same effects could be achieved merely by shifting the frame of reference, he concluded that gravity and acceleration are neither independent nor irreducible, just as his formula had shown that mass and energy are equivalent. The realization that the "gravitational field has only a relative existence," the key to the general theory, he declared was "the happiest thought of my life. . . ." (Pais, SITL, p.178).

This striving for a reductive unification of all the basic concepts in physics to continuous fields devoid of any singularities or mechanistic interpretation was his lifelong goal. Just as Maxwell's equations showed that the radiation of electromagnetism did not require an underlying ether but constituted a spatial field and that Newton's gravitational force also might be equivalent to a gravitational field, Einstein sought to reduce matter and energy to a condensation of a gravitational field which in turn could be reduced to a Riemannian curvature of space-time. As Pais states:

> The scientific task which Einstein set himself in his later years is based on three desiderata, all of them vitally important to him: to unify gravitation and electromagnetism, to derive quantum physics from an underlying causal theory, and to describe particles as singularity-free solutions of continuous fields. . . . As Einstein saw it, Maxwell's introduction of the field concept was a revolutionary advance which, however, did not go far enough. It was his belief that, also, in the description of the sources of the electromagnetic field, and other fields, all reference to the Newtonian mechanical world picture should be eradicated. (Pais, SITL, p. 289)

Although such theoretical unifications have occurred, it has been in particle physics, rather than Einstein's field interpretation, that has provided the most dramatic example exemplified in the Standard Model. As Michio Kaku has written,

> by the mid 1970s, it was possible to splice three of the four forces together (excluding gravity) to get what is called the Standard Model, a theory of quarks, electrons, and neutrinos, which interact by exchanging gluons, W^- and Z^- bosons, and photons. It is the culmination of decades of painfully slow research in particle physics. At present, the Standard Model fits all the experimental data concerning particle physics, without exception.[9]

Nonetheless, Einstein's conception of a finite-infinite spherical universe that is finite in configuration but boundless in extent, because the bending of space prevents going beyond it, has been the leading theoretical framework of cosmological research.

Still, his unrelenting conviction of the incompleteness of quantum mechanics and that a more fundamental theory would be found restoring strict causality and the objective reality of conjugate properties such as position and momentum and energy and time, eliminating the probabilities and paradoxes of quantum mechanics, has not been vindicated. Furthermore, the amazing progress in particle physics to which we now turn, which has not reduced to fields but explains interactions in terms of the exchange of virtual particles, like photons, gluons, vector bosons, and quarks, has not confirmed his expectation that fields would turn out to be the basic physical reality.

RUTHERFORD'S DISCOVERY OF THE PROTON

It would be difficult to find a better example of the intellectual creativity of human beings than the collective endeavor of nuclear physicists to determine the interior structure of the atom. As previously reviewed, remarkable discoveries in physics in the latter nineteenth and early twentieth centuries, including radiation, radioactivity, X-rays, electrons, alpha, beta, and gamma rays, ions, and isotopes, attributable to an internal structure of the atom, created considerable competition in atomic modeling. Among the entries were J. J. Thomson's "plum pudding" image of the atom as a positively charged sphere of fixed radius on which were embedded electrons like plums in a pudding, the negatively charged electrons balancing and neutralizing the diffuse positive charge of the central mushy sphere. Another contender was the solar model of Jean Perrin, consisting of a positively charged nucleus with negatively charged electrons as orbiting planets, that proved more successful.

But it was Ernest Rutherford who realized that the a particles that he had discovered and identified as doubly ionized helium about four times as heavy as hydrogen with two units of positive charge were ideal for atomic probing. Hoping to disclose new evidence of subatomic structures he instructed two able assistants, Hans Geiger and Ernest Marsden, to direct particles at thin gold foil and measure the percentage of deflections striking the 360° scintillating screen. Expecting that the majority of the massive and fast moving particles would pass directly through the gold foil with a few deflected at slight angles, they were amazed to find that some were rebounding straight backwards into the eyepiece used to observe the scintillations, as if rebounding from some massive internal particle. As Rutherford described his reaction when informed of this: "It was quite the most incredible event that has ever happened to me in my life. It was almost as incredible as if you fired a 15 inch shell at a piece of tissue paper and it came back and hit you" (Pais, IB, p. 189).

From various calculations Rutherford was able to devise a formula enabling him to infer that the atom consisted of what he called a constricted "nucleus," determining its defining properties of atomic mass and positive charge, the latter standing for the atomic number of the element. This was independently confirmed by J. J. Thomson and H. G. Moseley based on X-ray emissions of the atom. Rutherford also surmised that the nucleus was surrounded by negatively charged orbiting electrons which were relatively massless but occupied most of the volume of the atom. In addition, he concluded that the size of their orbits was determined by the electrostatic attraction between the positively charged nucleus and the negatively charged electrons.

Once convinced that his calculations were supported by the experimental evidence and having moved from McGill University in Canada to Victoria University in Manchester, England, he presented the results of his investigations first to the Manchester Literary and Philosophical Society in March 1911—the very Society to which Dalton had introduced his atomic theory. Two months later he submitted a more detailed account to the *Philosophical Magazine* that was followed by the publication in 1913 of *Radioactive Substances and their Radiations* (Cambridge University Press). Then in February 1914 he sent the impressively titled article, "The Structure of the Atom," again to the *Philosophical Magazine.* In the latter article he referred to the "recent atomic theory" of Niels Bohr, then a postgraduate student who had joined him at Manchester to learn about his theory of the atom.

Although still lacking a complete conception of the internal structure of the atom and explanation of spectral emissions, Rutherford's experimental discoveries were crucial, undoubtedly providing the basis for Bohr's own investigations. Moreover, his probing of the atom produced independent confirmation of a number of constants, along with compelling evidence of the reality of atoms. As Segrè states:

> By counting atoms, Rutherford and Geiger [who developed the Geiger Counter for measuring radiation] had a means of determining Avogadro's number, the charge of the electron, and other universal constants that could also be found by entirely different experiments, for instance, by studying the blackbody radiation. The numbers derived from both methods corresponded very well, and these experiments convinced even the most skeptical physicists of the real existence of atoms, overthrowing the most obstinate, conservative rearguard. (p. 102; brackets added)

Rutherford's preliminary conception of the atomic nucleus, based on the hydrogen nucleus, consisted of a heavy positively charged particle that was named "proton," after the Greek word "*protos*" meaning first, because it was the initial atomic particle to be discovered, surrounded by orbiting electrons. This model enabled physicists to formulate a notational representation of the nuclear composition and charges. Having designated the *charge* as plus or minus e and the *number* of units as Z, then $+Ze$ represented the total charge of the nucleus while $-Ze$ stood for the total charge of the surrounding electrons with a neutral atom having an equal number of positive and negative charges. Ionization involved a loss or gain of electrons while radioactive transmutation resulted in a change in nuclear number due to the emission of alpha, beta, or gamma rays.

In fact, all radioactive emissions occur in the nucleus, as when uranium in emitting an alpha particle transmutes into thorium: ${}_{92}U^{238} \rightarrow {}_{90}Th^{234} + {}_{2}He^{4}$.[10]

Because the chemical properties of the elements depend on the number of electrons, -*Ze* indicates the place of the element in Mendeleyev's periodic table while the ordinal numbers of each element are the same as the atomic number Z. Atoms with identical nuclear numbers and the same chemical properties but different atomic weights are isotopes, the three isotopes of hydrogen symbolized as ${}_{1}H^{1}$, ${}_{1}H^{2}$, ${}_{1}H^{3}$. There was one obvious discrepancy in this conception because the mass of the combined protons did not equal the atomic weight of the atom. This was corrected when Rutherford's colleague at the Cavendish Laboratory, James Chadwick, in 1932, discovered a second nuclear particle with the missing mass and a neutral charge that he named "neutron," earning him the Nobel Prize in physics in 1935. When combined with the proton it constituted the mass of the nucleus, but having a neutral charge it did not add to the atomic number.

This method of probing nature to reveal its secrets and discover new particles has been one of the great success stories of physics. In addition to his previous contributions, Rutherford also produced nuclear *disintegration* by radiating particles into the air disintegrating nitrogen nuclei into oxygen and hydrogen: ${}_{7}N^{14} + {}_{2}He^{4} = {}_{8}O^{17} + {}_{1}H^{1}$ (cf. Segrè, p. 111). The atomic numbers and weights on each side of the equation remain equal, thus consistent with the laws of the conservation of mass and energy reaffirming the consilient physical and mathematical relations underlying natural processes.

LISA MEITNER'S EXPLANATION OF URANIUM FISSION

The discovery of protons and neutrons not only accounted for nuclear numbers and masses, their existence eventually provided an explanation of Einstein's powerful formula, $E = mc^2$, presented in one of his earlier papers in 1905. Though the formula was well-known, the physical explanation of how matter and energy could be equated remained a mystery: how could a pebble, a piece of metal, or even a tiny amount of an element be a materialization of such a colossal quantity of stored energy equivalent to the product of their mass times the velocity of light, 300,000 kilometers (186, 281 miles) per second, *squared*? It was not until 1938 that the mystery was dispelled.

To relate as briefly as possible a complicated and disillusioning but fascinating story, a gifted young woman named Lise Meitner—that Einstein later would refer to as "our Madame Curie"—who at age twenty-seven in 1906 was only the second woman to receive a PhD in physics at the University of Vienna, arrived in Berlin the following year to study under Max Planck at the University of Berlin, where Hertz, Einstein, James Frank, and Planck, among others, were teaching. Though no women were admitted to the university at the time, in a physics seminar she met a young chemist, Otto Hahn, with whom she formed an extremely close (but nonromantic) lifetime friendship. He had received his PhD at the University of Marlburg in 1903 and later did research in Rutherford's laboratory at McGill University discovering radioactive substances.

Upon arriving in Berlin in 1906, Hahn was offered a paid position as Scientific Associate in the Chemistry Institute of Emile Fisher, an organic chemist. The following year when Meitner arrived she found that because of gender prejudices Fisher excluded women from his laboratories on the spurious grounds that their hair might catch on fire. However, he did provide her with a converted carpenter's workshop in the basement with a separate entrance where she could pursue her research. Despite the poor conditions, she and Hahn began their collaborative work initially investigating Beta emitters.

When the Kaiser Wilhelm Institute was opened in 1912 on the outskirts of Berlin, Fisher acceded to an important position that enabled him to offer Hahn a professorship with an increased salary.

In contrast, although Meitner's scientific reputation was equal to Hahn's, again because of female discrimination Fisher excluded her from his laboratories, although she was permitted to work with Hahn as an unpaid guest in his laboratory. Then in the same year Planck made her his Scientific Associate at the University (the first woman in such an appointment) with a stipend, but whose only function was to grade the papers of his students, another indication of how little female scientists were respected at the time.

But the following year she too was appointed a Scientific Associate at the Kaiser Wilhelm Institute with her own laboratory—at a salary considerably less than Hahn's. Their adjoining labs were referred to as the Hahn-Meitner laboratory, though it was her research that was attracting the most attention. Then with the outbreak of World War I their collaboration was interrupted with Hahn leaving to serve as an army officer in a unit involved in gas warfare. When he witnessed the terrible agony of the gassed soldiers he realized the horror of war and yet felt compelled to support the German war effort. Alone, she continued her research at the Institute for another year before volunteering in the summer of 1915 as an X-ray technician in an Austrian hospital near the Russian border where she was shocked by the gruesome wounds of the soldiers.

In 1916 she returned to the Institute to head the uranium project. Hahn rejoined her the following year and together in 1918 they announced their discovery of uranium 91, the missing element between thorium 90 and uranium 92. The detection of a new element always bringing international acclaim, Meitner was appointed head of the theoretical physics division, called the Meitner division, which became the most prominent in the Institute. She also was given a teaching position at the university while Hahn continued to make important discoveries in chemistry in an adjoining lab.

Then in 1934 Enrico Fermi in Rome found that slowed neutrons could be used to probe the interior of the nucleus of uranium leading to the discovery of new radioactive substances. Learning of Fermi's research, Meitner began similar experiments and realizing that she required the aid of chemists to identify the new elements being discovered, she invited Hahn to her laboratory, along with chemist Fritz Strassmann. Their collaboration was progressing favorably when Hitler became Chancellor in 1933.

Though Meitner was Jewish (but had converted to Protestantism in 1908), she did not feel threatened at the time because she was an Austrian citizen. However, when Hitler annexed Austria in 1938 she became de facto a German citizen and thus subject to Jewish persecution. A law was passed on July 4 excluding all Jews from public office resulting in Meitner's being expelling from her teaching position at the university and preventing her from attending conferences or presenting papers there, though she did not lose her position at the Institute because it was privately endowed. However, the hostility toward Jews among some of her colleagues, one declaring "the Jewess endangers our Institute,"[11] resulted in her being told she no longer could attend the Institute. At age fifty-nine Meitner was faced with having to leave her life's work behind and face an unknown future.

Although both Hahn and Strassmann were anti-Nazis and deplored Hitler's oppression of Jews because so many of their colleagues were Jewish, and despite Hahn's very close relationship with Meitner, he was instrumental in her being dismissed from the Institute. He later admitted that he "left her in the lurch," undoubtedly because being a friend of a Jew in Nazi Germany was a threat to his position and career, as well as to that of the Institute. Although many in the Institute detested Hitler, they chose to collaborate with the Nazi regime to safeguard the Institute and their positions. After the war Meitner would harshly criticize Hahn and the directors of the Institute for not offering at least passive resistance to the Nazi policies.

Because Hitler began sending Jews to concentration camps Meitner's situation was perilous. She was being offered positions by Bohr and others throughout the world, but having lost her Austrian citizenship she had to apply for a German passport in order to leave the country which was denied on the grounds that it was undesirable that prominent Jewish scientists be permitted to leave Germany where they could divulge the ongoing war research and denounce the Nazi regime, as Einstein was doing. The directive came from the highest authority, Heinrich Himmler himself. Meitner regretted then that she had not left Germany in 1933 when she had the chance, despite the research and reputation she had achieved during her five years at the Keiser Wilhelm Institute.

Mercifully, when Bohr learned of her desperate plight he contacted a friend of hers in Holland, Dirk Coster, who came to her rescue bringing an illegal passport to Berlin that enabled them furtively under fearful conditions to cross the German boarder fleeing to neutral Sweden. Once in Stockholm one would have expected that with her reputation as one of the world's leading nuclear physicists she would have been welcomed to join the new Nobel Institute for Experimental Research in Physics directed by the Nobel Prize physicist Manne Siegbahn. But even there gender discrimination prevailed so that while she was given a small room in the Institute, she was not invited to join in the research.

Despite her betrayal by Hahn she corresponded actively with him from Stockholm during the fall of 1938 helping to guide his and Strassmann's continuing investigations (using her abandoned apparatus) directing streams of slowed neutrons into uranium nuclei. They learned of a report from Irene Curie (daughter of Marie and Pierre Curie) that her injected neutrons had produced a new radiational substance that had the chemical properties of barium, but in such minute quantities that it only could be detected by a Geiger counter. However, when Hahn and Strassmann replicated the experiments they concluded that the residue was not barium, but "radium isotopes," on the assumption that nuclear reactions involve only small changes, and since radium is chemically similar to barium and closer in the periodic table, it was more likely the product.

In November, four months after leaving Berlin, Meitner traveled to Bohr's Institute in Copenhagen where Hahn also was visiting. Discussing Irene Curie's findings and his and Strassmann's experiments with Bohr, Meitner, and her nephew Otto Robert Frisch, who was then an assistant in the Institute, Hahn found that each of them objected to his interpretation but were unable to provide their own explanations. Meitner later wrote suggesting that he should reexamine the radioactive isotopes to determine whether they were in fact radium. On the twenty-first of December Hahn replied indicating that after additional experiments using fractional crystallization, in which he and Strassmann were unable to separate the radioactive isotopes from the barium, they were forced to conclude that despite their earlier assumptions, the radioactive isotopes must be barium, not radium. Somehow the injected uranium had changed into barium. Baffled by the result they turned to Meitner, a nuclear physicist, to provide an explanation!

As it happened, her nephew Frisch was visiting her over the Christmas holidays, so naturally she discussed the problem with him. According to Bodanis's account, at breakfast the next morning she suggested the experimental results could be explained if "the uranium atom had somehow split apart. A barium nucleus is about half the size of a uranium nucleus. What if the barium they were detecting was simply one of the big halves that had resulted?"[11] Then taking a walk in the snow-covered forest near the village where they were staying, they continued their discussion. It was known that the uranium nucleus contained over 200 combined protons and neutrons bound by the strong force. How could a few penetrating neutrons "break through every one of those bonds" to produce the two units of barium (p. 108)?

Here is where Einstein's equation $E = mc^2$ comes in. Meitner first heard of Einstein's equation during a lecture he gave in Salzburg in 1909 during which he described the equivalence, and thus the possible conversion of mass into energy. At the time the nucleus was considered to be a compact mass, so how could it be energy? Since then, however, Bohr had proposed that the nucleus was more like a droplet of water whose fluid shape was due to the surface tension (analogous to the strong force), but which was ready to burst in two if the interior particles were sufficiently agitated—illustrating the importance of analogical reasoning in scientific explanations. While physicists previously had thought that adding neutrons would merely create new uranium elements with larger nuclear numbers, the analogy with water droplets suggested that when invaded by the slowed neutrons they created a reaction that divided the nuclei of uranium into separate, lighter units. Frisch had learned of the water droplet model at Bohr's Institute, but the problem was explaining how the nuclear division occurred and where the enormous energy came from to produce it.

Sitting on a log in the forest with the paper and pencil they had the foresight to bring with them, they attempted the calculations. As Frisch recounts, fortunately his aunt

> remembered how to compute the masses of nuclei . . . and in that way she worked out that the two nuclei formed by the division of a uranium nucleus would be lighter than the original uranium nucleus by about one-fifth the mass of a proton. Now whenever mass disappears, energy is created, according to Einstein's formula $E = mc^2$. . . . (p. 110)

Knowing that the protons of the uranium nucleus contained about 200 MeV (million electron volts) of energy, they knew that an equivalent amount of energy would be needed to weaken the strong force binding each of the uranium protons enabling them to divide into the two units of barium. The loss of the weight or mass of the barium appeared to be the only source of this tremendous amount of energy. As Frisch later recounted, when his aunt did the calculations she found that "[o]ne-fifth of a proton mass was just equivalent to 200 MeV. So here was the source for the energy; it all fitted!" (p. 111). The laws of the conservation of mass and energy implied that the loss of the proton mass must be equivalent to the emergent energy. Yet since the mass of one-fifth of a proton is an infinitesimally small speck of matter it seems an improbable source of the huge amount of required energy until one realizes that its conversion consists of multiplying the mass by c^2, an incredibly large number.

On his return to Copenhagen Frisch searched for a term to describe the separating process they had just explained and when a colleague suggested the word 'fission' used to describe cell division, he incorporated it into the paper he and Meitner published in the journal *Nature* the following February. In their article they acknowledged that it was Hahn and Strassmann's chemical experiments that led to the discovery of atomic fission, but *they* provided the theoretical explanation and calculated the quantities involved. In his chemical experiments. Hahn had never encountered such a huge amount of energy. Meitner had immediately informed him of their explanation and calculations expecting that he would include their contribution in his publication because in his December letter he had stated that "[i]f there is anything you could propose that you could publish, then it would still in a way be work by the three of us [she, Hahn, and Strassmann]!" (p. 106, brackets added), implying she would be given credit for her theoretical explanation.

However, when Hahn and Strassmann published their paper in January in *Naturwissenschaften* describing their chemical experiments that revealed the division of the uranium into barium there was no mention of Meitner or Frisch. In fact, Hahn did everything possible to retain exclusive credit for the discovery of nuclear fission, declaring that "fission is for him a heaven sent gift." He tried to justify excluding Meitner and Frisch's contribution by declaring that he and Strassmann had made the *experimental discovery* of fission without the aid of theoretical physics, while they had merely provided the *explanantion*, as if the two functions were not equally significant. Lacking the explanation of Meitner and Frisch, neither he nor Strassmann would have understood the nature of their discovery or the quantities involved, only the chemical division of the uranium nuclei into barium which had been anticipated already by Irene Curie.

Sadly, the omission was perpetuated by the Nobel Prize Committee that announced in 1945 that in 1944 they had awarded the

Nobel Prize in chemistry to Otto Hahn alone for the discovery of nuclear fission. Although the Royal Swedish Prize is considered the highest recognition scientists can receive for their discoveries, its reputation has been greatly tarnished by the fact that the prize was awarded to Hahn without any mention of Lise Meitner and Robert Frisch (or Strassmann), even though the scientific community was well aware that the discovery of nuclear fission was an interdisciplinary undertaking involving both physics and chemistry. To correct the omission, physicists such as Niels Bohr subsequently recommended Meitner and Frisch for the Nobel Prize in physics but were rebuffed by the discreditable Siegbahn, then Director of the Physics Institute in Stockholm and member of the Nobel Committee who previously had excluded Meitner from doing research in his Institute.

Frisch in England in 1939 was among the first to announce the possibility of atomic fission being used to create an atomic bomb, about the same time that Einstein informed President Roosevelt of the danger of its being created by the Germans. Meitner was invited to participate in the Manhattan Project but refused to aid in the creation of such a devastating weapon, even though she must have known that Werner Heisenberg was in charge of the atomic bomb project in Germany and what a horrific consequence it would have been if Hitler had been the first to possess it and thus able to dominate the world. As it was, Heisenberg was making it a close race until February 1944 when the Norwegian resistance sabotaged his crucial source of heavy water from the factory at Vemork, preventing his possible construction of the bomb before the United States.

Although the American military command, including General MacArthur, Admiral Leahy, and General Curtis LeMay, opposed dropping the bomb on Japan because they believed an invasion of Japan was unnecessary, President Truman's civilian advisors convinced him that such a devastating blow was essential to force the Japanese high command to surrender, thereby saving hundreds of thousands of lives which would have been lost if an invasion had proved necessary (cf. pp. 153–62).

Notwithstanding Hahn's earlier refusal to defend her position at the Institute and selfishness in keeping the discovery of atomic fission for himself, Meitner maintained their close friendship for the remainder of their lives. In 1946 Hahn was made President of the Kaiser Wilhelm Institute partially based on his stolen reputation. Although largely living in obscurity in Stockholm in her later life, in 1982 Meitner did receive recognition when the Society for heavy ion research in Darmstadt, Germany discovered the element 109 and named it "Meitnerium" in her honor. Both she and Hahn died at age eighty-nine. She is buried in an English country cemetery while he is interred in a graveyard in Göttingen.

On her tombstone is carved "A physicist who never lost her humanity," while on his is inscribed the fission reaction formula. There is little doubt which inscription is more apt.

BOHR'S INITIAL THEORY OF ELECTRONIC ORBITS

While Rutherford and Chadwick had discovered the nature of the nucleus and incorporated electronic orbits into their atomic theory, they had little to say about the precise arrangement of the electrons beyond noting that the orbits' sizes depended upon the electrostatic attraction between them and the positively charged nucleus. This latter half of the representation of the atom was left to the famous Danish physicist, Niels Bohr (1885–1962), one of the most prominent and esteemed theoretical physicists of the twentieth century. This was due primarily to his conception of electronic orbits, but also to his character and because of the acclaimed influence his Institute of Theoretical Physics in Copenhagen had in training physics students. The number of outstanding young physicists who passed through his Institute is legion, including Frisch, Kramer, Heisenberg, Schrödinger, Dirac, and Pauli.

During his graduate studies in physics in Copenhagen Bohr

became intrigued by the problem of electronic orbits and having studied J. J. Thomson's atomic model of the atom and received a stipend from the Carlsberg Foundation, he decided to go to Cambridge University to do postgraduate studies under the supervision of Thomson. However, during their first meeting when the twenty-six year old Bohr told Thomson that when writing his dissertation he had found something "wrong" with Thomson's atomic model, the relationship between the two, though civil, did not flourish. But meeting Rutherford in Manchester shortly after and then hearing him lecture on his theory of the atom at Cambridge, Bohr decided to spend part of his year abroad working in Rutherford's laboratory in Manchester. This was most auspicious because they formed a lasting friendship as long as Rutherford lived, based on mutual admiration and respect, that developed into a father-son relationship, especially as Bohr's father had died earlier. The following August when Bohr married, he took his bride to visit Rutherford and his wife in Manchester before continuing on their honeymoon to Scotland.

Arriving in Manchester in March 1912 after leaving Cambridge, Bohr took a course in radioactivity and then pursued his interest in determining the orbital structure of electrons. Of particular significance was his meeting Charles Galton Darwin, the grandson of the famous evolutionist, who described his experiments bombarding clouds of electrons, rather than the nucleus, with particles. His discussion with Darwin led Bohr to believe that electrons were "atomic vibrators" that radiated in agreement with Planck's quantum of action. As he wrote in one of his three famous articles published in 1913: "[a]ccording to Planck's theory of radiation the smallest quantity of energy which can be radiated out from an atomic vibrator is equal to v [times] k" (Pais, NBT, p. 128), with v representing the number of vibrations per second and k standing for what is now Planck's constant h. As Pais adds: "*Thus did the quantum theory enter the interior of the atom for first time in Bohr's writings*" (ibid.).

But even earlier, in his criticisms of Thomson's plum pudding model of the atom in his dissertation, he had concluded that a more adequate conception of electronic orbits could not be based entirely on the principles of classical electrodynamics, that implied rotating electrons would radiate energy smoothly, eventually spiraling into the nucleus. In a paper in 1906 Einstein had indicated that oscillators, still the model of the electron, "emitted and absorbed energy by *jumps*," which are integer multiples of Planck's constant hv that could have been the origin of the conception of "quantum jumps." Then in a paper published in 1912 John Nicholson wrote: "If therefore the constant h of Planck has . . . an atomic significance, it may mean that the angular momentum of a particle can only rise or fall by discrete amounts when electrons leave or return . . . quantiz[ing] angular momentum" (Pais, NBT, p. 145; brackets added).

Thus others were beginning to recognize the necessity of incorporating quantum effects in constructing the orbital structure of atoms. But it was not until a student friend, Hans Marius Hansen, advised Bohr, while he was speculating about the exact nature of electronic orbits to consider Balmer's formula predicting a definite regularity of discrete frequencies among the spectral lines of hydrogen, that Bohr noticed the resemblance between Balmer's formula $v = R\left(\frac{1}{n_1^2} - \frac{1}{n_2^2}\right)$ and Planck's formula $hv = E_1 - E_2$, after which he declared "the whole thing was immediately clear to me" (Segrè, p. 122). Recognizing in Balmer's formula that the spectral emissions of hydrogen only occurred in integral numbers of discrete frequencies, he suddenly realized that perhaps the angular momentum of orbiting electrons also can change only by discrete frequencies.

This enabled Bohr to introduce the novel idea that, contrary to classical mechanics that permitted a continuous series of orbital periods, the electronic orbits were restricted to discrete orbits or "stationary states" correlated with the electron's angular momentum and frequency or energy. Furthermore, he now realized that not all atomic instability originates in transmutations in the nucleus, but that some emissions are due to transitions or "jumps" from one energy level of an electronic orbit to another accompanied by the radiation or absorption of a quantum of energy according to Planck's equation $E_1 - E_2 = h\nu$. Thus the discreteness of the electron spectra indicated in Balmer's formula is a consequence of the discreteness of orbital states, the transitions between them producing the emissions.

Accordingly, an electron jumping from a higher, less stable energy state to a lower, more stable state produces the emission of a photon whose energy equals the difference between the two energy states; conversely, the energy necessary to propel an electron from a lower to a higher energy state will be in discrete integral quanta according to Planck's formula. All spectroscopic analysis to this day depends upon this explanation. To avoid the classical prediction that an electron in the lowest energy state will radiate its energy until its angular momentum is overcome by the electrostatic force, thereby drawing it into the nucleus, Bohr postulated that the lowest or "ground state" was "stable." Thus he resolved the paradox by *fait accompli*. As Pais states: "Bohr circumvented this disaster by introducing one of the most audacious postulates ever seen in physics. He simply declared that the ground state *is* stable, thereby contravening all knowledge about radiation available up to then" (NBT, p. 147).

This explanation was also "audacious" because it combined both classical and quantum theory in an extremely unorthodox manner: the *mechanical* frequencies of the stationary orbits were described with classical mechanics while the *optical* frequencies of the photons emitted during the transitions from one orbit to another were described in quantum terms. Thus his solution incorporated two contradictory tenets illustrative of his later thinking in quantum mechanics which would include the "correspondence principle" and the "principle of complementarity." The first principle applies in the limiting case where the magnitudes of the frequencies merge with classical mechanics and thus tend to "correspond." The second principle applies when two descriptive states of physics, that would be contradictory if attributed to the same experiments, are considered complementary when applied under different experimental conditions. The principle of complementarity especially will be used by

Bohr throughout his life to reconcile two phenomena that appear at first contradictory.

Bohr's three publications on "the constitution of atoms" in 1913 when he was twenty-eight years old revolutionized physics. Not only could he derive a number of formulas, constants, and magnitudes from his system, he also corrected some errors. Furthermore, they offered new explanations of spectral emissions. For example, dividing the electron orbits into inner and outer shells he attributed the origin of visible spectra to the radiation of the outer electrons, the origin of X-rays to the ejection of an electron from an inner shell, and the radiation of photons to the replacing of the ejected electron by one from an outer orbit. Thus X-ray spectra show emissions not revealed by visible spectra. He correctly attributed beta rays, later identified as electrons, to radioactive processes in the nucleus, while assigning most of the chemical properties of the atoms to the number of electrons in the outermost orbit.

Though these three papers had a tremendous impact on twentieth century physics, his theoretical innovations were so radical that not everyone accepted them. Indicative of how unsettling Bohr's theoretical system was, Otto Stern and Max von Laue declared that "if by chance it should prove correct, they would quit physics" (Segrè, p. 129). This did not reflect any personal animosity toward Bohr because he was universally admired and liked, but their initial frustration at the blatant inconsistencies and paradoxes in his explanations—though they later became converts. Einstein, who remained the severest critic of quantum mechanics but was a great admirer and close friend of Bohr, provided the most eloquent tribute to Bohr's originality in his "Autobiographical Notes":

> That this insecure and contradictory foundation was sufficient to enable a man of Bohr's unique instinct and tact to discover the major laws of the spectral lines and the electron shells of the atoms together with their significance for chemistry appeared to me like a miracle—and appears to me as a miracle even today. This is the highest form of musicality in the sphere of thought. (pp. 45–46)

Despite all the misgivings and reservations, expressed by Bohr himself, his theory agreed with the experimental data which is the highest confirmation a scientist could expect. As Pais assesses his achievement:

> Atoms had been postulated in ancient times. As the year 1913 began, almost unanimous consensus had been reached, after much struggle, that atoms are real. Even before that year it had become evident that atoms have substructure, but no one yet knew by what rules their parts moved. During that year, Bohr, fully conscious that these motions could not possibly be described in terms of classical physics, but that it nevertheless was essential to establish a link between classical and quantum physics, gave the first firm and lasting direction toward an understanding of atomic structure and atomic dynamics. In that sense he may be considered the father of the atom. (NBT, p. 152)

We have come a long way from the atomic conception of Leucippus and Democritus.

NOTES

1. Cf. Richard H. Schlagel, *From Myth to Modern Mind*, Vol. II, *Copernicus through Quantum Mechanics* (New York: Peter Lang Publishers Inc., 1979), pp. 424–28, for a more detailed account.

2. Abraham Pais, *Inward Bound: Of Matter and Forces in the Physical World* (New York: Oxford University Press, 1986), p. 167. In future citations I shall distinguish this work from his other books by abbreviating the title as IB followed by the page reference.

3. Cf. Emilio Segrè, *From X-Rays to Quarks* (San Francisco: W. H. Freeman, 1980), pp. 72–73. Future references to this work will cite Segrè and the page number.

4. Abraham Pais, *Niels Bohr's Times: In Physics, Philosophy, and Polity* (Oxford: Clarendon Press, 1991), p. 86. Future references to this work will cite Pais followed by the abbreviated title NBT and page number.

5. Albert Einstein, "Autobiographical Notes," in *Albert Einstein: Philosopher-Scientist*, ed. by Paul Schilpp (Evanston, Illinois: Library of Living Philosophers, 1949), p. 45; brackets added. Future citations will be preceded by Schilpp, then the page number.

6. Cf. Abraham Pais, '*Subtle is the Lord . . .*' (Oxford: Oxford University Press, 1982), pp. 7–8. In future quotations I shall cite Pais's name followed by the abbreviated title SITL and page reference.

7. The Lorentz transformation equations are described in Albert Einstein, *Relativity: The Special and General Theory* (New York: Crown Publishers, 1961), pp. 83–87.

8. G. J. Whitrow, *The Structure and Evolution of the Universe* (New York: Harper Torchbooks, 1959), p. 85. Any future quotation will cite his name and page number.

9. Michio Kaku, *Parallel Worlds: A Journey Through Creation, Higher Dimensions, and the Future of the Cosmos* (New York: Doubleday, 2005), p. 82. Future quotations will be followed by citing his name and page number.

10. Cf. Yuval Ne'eman and Yoram Kirsh, *The Particle Hunters* (Cambridge: Cambridge University Press, 1986), p. 14. Future citations will be indicated by the authors last names and page number.

11. David Bodanis, *$E = mc^2$: A Biography of the World's Most Famous Equation* (New York: Walker & Company, 2000), p. 107. This account is based partly on this book, as are the immediately following references. I also found that a PBS DVD on Lise Meitner and other scientists produced by Rosemarie Reed Productions, Ltd., issued in April 11, 2006, was very helpful in this discussion.

Chapter 15

APPROACHING THE PRESENT

THE NEW QUANTUM MECHANICS

While the initial contributions of Planck, Einstein, and Bohr had laid the foundation of what came to called the "old quantum theory," many unanswered questions remained. Though Bohr had achieved a partial explanation of the electron orbits of hydrogen supported by his prediction of certain constants, he had done so using two incompatible principles, one from classical physics and one from quantum theory, was unable to explain why the electrons in the hydrogen atom were restricted to particular orbits with an angular momentum equal to an integral of Planck's constant h, and could not predict the spectra of atoms with an electronic structure more complex than hydrogen.

The solution to these problems awaited the construction of the "new quantum mechanics" in the mid-twenties by Louis de Broglie, Erwin Schrödinger, Max Born, Pascual Jordan, Werner Heisenberg, Paul Dirac, and Wolfgang Pauli.

In addition to the previous unanswered questions there was the problem of the wave-particle duality owing to Planck's discrete quanta of energy supported by Einstein's explanation of the photoelectric effect with photons, while the diffraction of light presupposed the existence of waves defined by their lengths and frequencies. Intrigued by the dualism, Louis de Broglie (1892–1987) wondered whether the inverse of Einstein's discovery was true: that since waves display particle properties under certain experimental conditions, perhaps particles manifest wave properties under contrary experimental conditions. Applying Einstein's relativistic effects to subatomic particles because of their extreme velocities, in 1923–1924 he demonstrated experimentally that particles do exhibit wave properties, along with deriving formulas for their wavelengths and frequencies. As described by Peter Gibbins:

> Every particle, every electron, proton and so on was to have an associated wave, of frequency v_0 given by $m_0c_2 = hv_0$ when at rest, where m_0 is the rest mass of the particle. When moving a particle with momentum p has a de Broglie wavelength λ such that $p = h/\lambda$. Here, λ is clearly the wavelength of a monochromatic, a pure sine wave.[1]

Despite some theoretical difficulties, his localization of particles via a superposition of waves or wave packets enabled him to explain why Bohr's orbiting electrons were limited to specific stationary orbits. Moreover, Erwin Schrödinger built his quantum dynamics on de Broglie's discovery while Clinton Davisson and Lester Germer confirmed in 1927 that electrons display wave properties when radiated through a diffraction grating. De Broglie won the Nobel Prize in 1929 and Davisson and George P. Thomson the prize in 1937 for their independent confirmation of his theory.

HEISENBERG'S MATRIX MECHANICS

Even those unfamiliar with physics may have heard of Werner Heisenberg's (1901–1976) uncertainty or, more accurately, indeterminacy principles. An exceptionally gifted person who was an excellent pianist whose father was a professor of Greek, he read the Greek classics, including philosophy, in the original Greek and was attracted to physics by reading Plato's attempt in the *Timaeus* to reduce the four elements, fire, earth, air, and water, along with the entire universe, to the five Pythagorean geometric solids described previously that had such a strong influence on Kepler. He was, he declared, "enthralled by the idea that the smallest particles of matter must reduce to some mathematical form."[2] Earning his doctorate at the University of Munich at the young age of twenty-two, he studied the latest atomic theories, especially the Bohr-Sommerfeld model of the atom, but decided that a better theory could be devised by discarding the visual image of orbiting electrons, retaining only the mathematical formalism.

Dismissing the "fictitious orbits" representing quantum states, he correlated the electron states with frequencies in a series respectively of horizontal and vertical lists presented in a paper published in the *Zeitschrift für Physik* in 1925. In this purely mathematical array later identified by Max Born as matrix mechanics, Heisenberg considered the interior of the atom as a mathematical configuration somewhat analogous to Plato's five geometrical figures depicting the four elements and the universe. As he vividly states: "I had the feeling that, through the surface of atomic phenomena, I was looking at a strangely beautiful interior, and felt almost giddy at the thought that I now had to probe this wealth of mathematical structures nature had so generously spread out before me" (Heisenberg, PAB, p. 61).

One of the strange aspects of this matrix mechanics, identified by Born, was that multiplying matrices violated one of the fundamental laws of mathematics, the commutative law, that when multiplying two numbers the order of numbers multiplied makes no difference: $4 \times 6 = 6 \times 4$. However, multiplying matrices is not commutative: 4×6 does not equal 6×4 with the different result being the same or constant for all multiplications. Together with a student especially gifted in mathematics, Pascual Jordan, Born wrote a paper clarifying the basic principles of matrix mechanics which they also published in *Zeitschrift für Physik* in 1925.[3] At the time Born was intent on showing that quantum theory presupposed classical mechanics, thus by using a mathematical operator, the Hamiltonian, which stands for the quantified energy of the system, they showed that spectral matrices were related to classical mechanics in the same way Bohr's explanation of hydrogen spectra was related to classical mechanics, a development that must have pleased Einstein who also sought to base quantum mechanics on classical theory.

After reading their paper, Heisenberg suggested that the three of them collaborate to write a third paper again published in the *Zeitschrift für Physik* in February 1926. Together they described the new matrix mechanics as "*quantenmechanik*." But the full significance of the noncommutative parameters such as position q and momentum p (as well energy e and time t) was not apparent until Born and Heisenberg later showed that they represented a *limitation on the exact measurement* of such conjugate parameters, depicting the reciprocal uncertainty or indeterminacy of their measurements.

DIRAC'S TRANSFORMATION THEORY

Another scientist who was influenced by Heisenberg's original paper was Paul Dirac (1902–1984), also a mathematical prodigy who was studying mathematics at St. John's College in Cambridge. After reading Heisenberg's matrix mechanics at age twenty-four he composed a more general and rigorous version of quantum mechanics that he called "transformation theory." As described by Crease and Mann:

> A lengthy letter in Dirac's minute, fussy handwriting arrived in Göttingen on November 20, 1925. Written in English, it explained to the astonished Heisenberg how in a few concise steps his version of quantum mechanics could be reformulated in classical terms using a mathematical device [the Poisson bracket, a

> probability density function] he had never heard of. Moreover, working alone, Dirac had come up with a version of quantum mechanics much more general and complete than produced by Heisenberg, Born, and Jordan in their just-completed joint paper. (p. 79; brackets added)

Though taken aback by this *tour de force*, Heisenberg graciously wrote Dirac complimenting him on "your extraordinarily beautiful paper on quantum mechanics." Unexpectedly, the two formalisms were found to be mathematically equivalent.

SCHRÖDINGER'S WAVE MECHANICS

During the same period when Born, Jordan, and Heisenberg were formulating their matrix mechanics and Dirac was devising his transformational theory, a fifth entrant was creating still another version of the new mechanics, called "wave mechanics." As compared to the previous *wunderkinder* who were in their mid-twenties, at age thirty-eight Erwin Schrödinger (1887–1961) was relatively old and well-known. He shared Einstein's belief that the waves or fields of Maxwell should replace physical particles as the basic physical reality and that an explanation of emission spectra consistent with classical mechanics could be formulated that would supersede electron orbits and the detested quantum jumps. Furthermore, he would replace the obscure matrix mechanics with the familiar wave theory used to explain electromagnetic phenomena, but extended now to subatomic structures and interactions as well.

Avoiding relativistic effects as a simplification, he devised a formula using wave functions to displace Bohr's angular momentum explanation of electron orbits that produced Rorschach-like images of the electronic stationary states of the hydrogen atom. Initially, these figurative solutions to his wave equation were interpreted as "electron clouds," diffuse condensed electric charges similar to Einstein's

"*Gespensterfeld*" or "ghost field" guiding the path of electrons. His new wave mechanics was presented in four papers in *Annalen der Physik* from January to April in 1926, titled "Quantization as an Eingenvalue [single value] Problem." Because it was an extension of classical electrodynamics using the more familiar formalism of wave mechanics and provided a physical representation with waves or fields, it was immediately acclaimed by physicists who preferred the more traditional approach.

MAX BORN'S MEASUREMENT PROBLEM AND PROBABILITY INTERPRETATION

After complimenting Schrödinger for "the crisp, clear conceptual formulations" of his theory in terms of classical mechanics, Born reversed himself presenting a radical probability interpretation of Schrödinger's wave mechanics in two papers in the *Zeitschrift für Physik* in the summer of 1926. Titled (in translation) "Quantum Mechanics of Collision Phenomena," the first paper addresses what came to be called the "measurement problem," attributing the uncertainty in the measurements of the position and momentum of a particle to the momentum exchange or scattering caused by the impact of the ray used in measuring the particle. According to Pais, "*It is the first paper to contain the quantum mechanical probability concept*" (NBT, p. 286). It was this uncertainty in the collision measurement that precluded quantum mechanics from having the strict causality and mathematical precision that prevailed in classical physics.

In the second paper Born interpreted Schrödinger's wave function $|\psi|^2$ as the probability distribution for locating the particle where the density is greatest, rather than as a description of an actual wave packet of a certain intensity, adding to the probability of *measurements* the probability of *states*. Though accepting Schrödinger's formalism, he rejected his physical interpretation, the latter becoming a major contention in modern physics. As Born states, Schrödinger believed

> that he had accomplished a return to classical thinking; he regarded the electron not as a particle but as a density distribution given by the square of his wavefunction $|\psi|^2$. He argued that the idea of particles and of quantum jumps be given up altogether; he never faltered in this conviction. . . . I, however, was witnessing the fertility of the particle concept every day . . . and was convinced that particles could not be abolished. A way had to be found for reconciling particles and waves.[4]

Born received the Nobel Prize in physics in 1954 "for his fundamental research, especially for his statistical interpretation of the wave function."

However, just as previously it was found that the formalism of Dirac's quantum mechanics was equivalent to Heisenberg's matrix mechanics, now it was determined that the formalism of Schrödinger's wave mechanics was equivalent to the previous two formalisms. As Segrè states:

> For all three the essential relation that produces the quantization is $pq - qp = h/2\pi i$. For Heisenberg p and q are matrices; for Schrödinger q is a number and p is a differential operator: $p = (h/2\pi i \times (\partial/\partial q)$. For Dirac p and q are special numbers obeying a noncommutative algebra. The results of any calculation on a concrete problem done by any of the other three methods are identical.[5]

Thus the physical interpretations vary while the formalisms are equivalent, a completely unexpected development. What today is called quantum mechanics is a formalism that John von Neumann (1903–1957) derived from the preceding theories and presented in his classic book translated as the *Mathematical Foundations of Quantum Mechanics.*

While Heisenberg was still at Bohr's Institute at Copenhagen Bohr invited Schrödinger, whom he had not yet met, to visit his Institute on October 27, 1926. There they engaged in such intense discussion that, even though Schrödinger became ill with a feverish cold, Bohr persisted in his questioning at his bedside. Yet Schrödinger retained the highest regard for Bohr as he wrote:

> In spite of everything I had already heard, the impression of Bohr's personality from a purely human point of view was quite unexpected. There will hardly again be a man who will achieve such enormous external and internal success, who in his sphere of work is honored almost like a demigod by the whole world, and who yet remains—I would not say modest and free of conceit—but rather shy and diffident like a theology student. (Pais, NBT, p. 299)

Dispelling the somewhat common impression of scientists as cold and aloof, the following description conveys their fervent pursuit of physical reality.

> The intensity of the debate at the birth of quantum mechanics was more than a matter of careers, temperaments, and prejudices; the participants were gripped by the conviction, endemic to the science, that they were arguing about the shape of the Universe itself, and that the picture they were forming had profound philosophical resonances. In some respects, the belief that discerning the laws of the quantum world is equivalent to deciphering the most primary code of nature is naïve and reductionistic; but in other ways it is exactly what they were doing, and the physicists of the time—as well as their successors today—have rightly been caught up in the breathtaking implications of their quest. (Crease and Mann, p. 59)

The inevitable introduction of probabilities into quantum mechanics stems from the fact that for the measurement of the position and momentum of a particle, such as an electron, there must be an interaction between a detector like a light ray and the electron, but for an exact position determination the ray must have the shortest possible wavelength, like a gamma ray, but such a short wavelength has a high frequency or energy, as indicated in Planck's formula $e = h\nu$. To minimize the resultant momentum displacement due to the high energy, a ray with a longer wavelength and less energy is required, but this results in a less accurate position measurement. An analogous situation occurs regarding energy and time.

In another formulation of the problem, the interaction of a macroscopic instrument with the evolving quantum system collapses what is called the superposition of *possible* states of the wave packet into a *definite* state. It is the existence of these possible states, only one of which can be actualized during the measurement interaction, that also introduces the unexpected uncertainties or indeterminacies since they are not antecedently determinate. Thus the uncertainties are not due to the imprecision of the measuring device, but to the inherent opposition of the wave properties of the light rays required to measure the conjugate properties of the particle.

HEISENBERG'S UNCERTAINTY PRINCIPLE

Disturbed and stressed by his intense discussions with Heisenberg and Schrödinger, Bohr left on a skiing trip to Norway leaving Heisenberg at the Institute to work out his famous uncertainty or indeterminacy relations. First described in a letter to Pauli in February 1927 and then to Bohr in the following March, Heisenberg wrote that "I believe that I have succeeded in treating the case where both [the momentum] p and [the coordinate] q are given to a certain accuracy . . ." (Pais, NBT, p. 304; brackets in original). What he finally concluded was that while either the momentum or the position coordinate, along with the energy and the time coordinates, can be individually measured accurately, the more precisely one determines the one the less accurate is the determination of the other, accounting for the indeterminacy in the measurements.

As Crease and Mann clearly state: "If the error in the position (that is, Δq) is very small, then the error in the momentum (that is, Δq) must increase to keep their product, $\Delta q \times \Delta p$, larger than $\hbar$"

(p. 65). This is true also of the time Δt and the energy Δe. Thus Heisenberg's formulas are: $\Delta p \times \Delta q \geq h/2\pi$ and $\Delta t \times \Delta \varepsilon \geq h/2\pi$. Although these noncommutative uncertainties had been identified by Born, Heisenberg was the first to express them in the precise equations for which he is famous.

The intense discussions between Bohr and Heisenberg were based on their contrasting beliefs as to how the paradoxes of incommensurate quantities and the wave-particle duality could be solved, Heisenberg insisting that the solution lay in the mathematical formalism with Bohr maintaining that it required a clarification of the application of the theoretical concepts. As Heisenberg states in a famous passage reversing the position of classical physics, the resolution meant accepting that only those experimental situations could occur that were permitted by the mathematical formalism.

> Instead of asking: How can one in the known mathematical scheme express a given experimental situation? the other question was put: Is it true, perhaps, that only such experimental situations can arise in nature as can be expressed in the mathematical formalism? The assumption that this was actually true led to limitations in the use of those concepts that had been the basis of classical physics since Newton.[6]

BOHR'S COMPLEMENTARITY AND COPENHAGEN INTERPRETATION

While Heisenberg was working out his equations in Copenhagen, Bohr in Norway was formulating his conceptual clarification of the principle of "complementarity," the term appearing "for the first time in a draft from 10 July 1927," while in "correspondence it shows up for the first time in a letter to Pauli in August" (Pais, NBT, p. 311). One of the clearest expressions of his position occurs in an article titled "Quantum Physics and Philosophy—Causality and Complementarity," written much later in 1958, which reads like a general interpretation of Heisenberg's uncertainty relations. As he states,

> within the scope of classical physics, the interaction between object and apparatus can be neglected or, if necessary, compensated for. . . . In quantum physics, however, evidence about atomic objects obtained by different experimental arrangements exhibits a novel kind of complementary relationship. Indeed, it must be recognized that such evidence which appears contradictory when combination into a single picture is attempted, exhausts all conceivable knowledge about the object. Far from restricting our efforts to put questions to nature in the form of experiments, the notion of *complementarity* simply characterizes the answers we can receive by such inquiry, whenever the interaction between the measuring instruments and the objects forms an integral part of the phenomena.[7]

This "Copenhagen interpretation," as it came to be called, if true is almost as radical a reorientation in knowledge as the Copernican revolution. Just as prior to that revolution our ordinary experience of the world had deceived people into believing that they were living on an unmoving sphere in the center of the universe, so ordinary experience has also misled us into thinking that we directly perceive the world as it is, the Aristotelian view of experience. Owing to the nearly instantaneous velocity of light at close distances and the behind the scenes hidden functioning of our neuronal processes, even today we normally assume that the perceptual world is the real world.

However, when philosophers and scientists in the seventeenth century began to realize that our sensory perceptions preclude us from having a *direct* knowledge of the world as it exists independently of us, they distinguished between the subjective "sensory qualities" of perceptual objects, such as colors, sounds, and smells, from their inherently real "primary qualities," like size, shape, solidity, and mass. The independent absolute status of the latter was reinforced by Newton's conception of absolute space and time. In Immanuel Kant's terminology, there was the "phenomenal world" of appearance and the "noumenal world" of things in themselves, which, however, for him were completely unknowable. Most scientists rejected this extremely skeptical distinction believing that at least the primary qualities of objects and some of their causal effects could be inferred from experience.

EINSTEIN, PODOLSKY, AND ROSEN'S GEDANKEN-EXPERIMENT

It was this realist assumption that before they are measured particles possess inherent properties such as position, momentum, energy, and time and that quantum states such as light packets are determinate, that led Einstein throughout his life to maintain that quantum mechanics, in precluding this, was "incomplete." In 1935 Einstein, Boris Podolsky, and Nathan Rosen devised a *Gedanken-Experiment,* known as the "EPR experiment," to show that if measured successively, rather than concurrently, an electron could be found to have an exact position and momentum, in contrast to Heisenberg's equation which depicted the uncertainties as measured simultaneously. As stated in the *Physical Review* (1935, 47, p. 777), the article claimed that the error in the indeterminacy relations consisted in regarding two physical quantities, such as position and momentum, "as simultaneous elements of reality *only when they can be simultaneously measured or predicted,*" concluding that "[*n*]*o reasonable definition of reality could be expected to permit this.*" If measured successively, it was claimed, each predicted magnitude could be precisely determined because it could be made without disturbing the second magnitude, thereby eliminating the uncertainties.

After a sleepless night Bohr formulated his reply in an article, "Can the Quantum Mechanical Description of Reality be Considered Complete?" which appeared in the next issue of *Physical Review* (1935, 48, p. 700). In rebuttal, Bohr stated that if examined closely,

> we now see that the wording of the above-mentioned criterion of physical reality proposed by Einstein, Podolsky, and Rosen contains an ambiguity as regards the meaning of the expression "without in any way disturbing the system." Of course there is in a case like that just considered no question of a mechanical disturbance of the system under investigation during the last critical stage of the measuring procedure. But even at this stage there is essentially the question of *an influence on the very conditions which define the possible types of predictions regarding the future behavior of the system.* As these conditions constitute an inherent element of the description of any measurement the argumentation of the above-mentioned authors does not justify their conclusion that the quantum mechanical description is incomplete.

Furthermore, Bohr claimed that the Copenhagen interpretation of quantum mechanics that the status of certain quantum properties or states is indeterminate until measured, as radical as that seems, is not so different from relativity theory. According to Einstein's special theory, *measurements* of extension, mass, and time are not invariant among systems in motion, but are *relative* to the velocity of the system from which they are measured. Furthermore, according to the general theory of relativity such magnitudes are not inherently unchangeable as previously believed but vary with their velocities, mass increasing, space contracting, and time dilating in systems approaching the velocity of light. And just as relativistic effects occur only when the velocities approach those of light, so quantum effects take place only when the magnitudes are subatomic.

Thus modern physics suggests that we may have to revise our naïve conception of a preexistent physical world which is there to be discovered to a *conditional* world whose disclosed properties depend upon our interacting with it, as is true of ordinary experience!

As objective and independent as they seem to be, the colors and solidity of ordinary objects appear as they do only to someone with a visual system like ours. As I argued in a previous book, *Contextual Realism*,[8] at least from our position in the world it seems that whatever exists with the properties displayed depends upon a more extensive or underlying physical background, yet as long as that background is stable, so is that portion of the world and our experience and understanding of it. But when we explore domains that exist under radically different physical conditions owing to their dimensions, velocities, temperatures, and so forth, and therefore display unusual limited properties and effects, we cannot expect the same laws or concepts to apply as in our macroscopic domain.

Thus the conditions within which we experience the world are much more limited than we realized, yet they do not represent the absolute barrier to further knowledge that Kant claimed. With telescopes and spectroscopes it has been possible to arrive at the Big Bang theory of the beginning, expansion, diversification, and structure of the universe and by probing the atom with subatomic particles we have learned much about the interior structure of the atom which would be impossible if we were confined to the world of appearance. On the other hand, though we have discovered some constants, such as gravity and the velocity of light, the uncertainties or indeterminacies imposed by the reliance on intermediary instruments of investigation have made us aware of how approximate or provisional all knowledge is. What has been opened up to us is a universe so vast and complex that any claim to final knowledge seems naïve, and yet one cannot but be amazed that a creature with such humble beginnings could have acquired the extensive knowledge that we have, however conditional or approximate.

GOUDSMIT'S AND UHLENBECK'S DISCOVERY OF ELECTRONIC SPIN

Returning to the discussion of the development of quantum mechanics, during the years 1921–1924 two physicists in Hamburg, Otto Stern and W. Gerlach, had performed several experiments that indicated that electrons, in addition to their properties of charge and mass, rotate. Then in 1925 two physics students at the University of Leyden, Samuel A. Goudsmit and George E. Uhlenbeck, confirmed that electrons like planets have an axial rotation in addition to their orbital angular momentum. Calling this axial rotation a "spin," they found it had a fixed axial angular momentum $\frac{1}{2}h/2\pi$, that the direction of the spin has two possibilities, and that the possible opposite directions of its orbital vector is determined by its spin vector (according to the right hand rule).

Furthermore, because of their spins particles act as tiny magnets with the magnetic moment of the electron determined with an accuracy of 1.001159652 μ_e (the μ_e called the Bohr magneton). While it was found that spin is a fundamental property not only of electrons but of all particles, the conception of an actual rotary motion is no longer taken literally.[9]

PAULI'S EXCLUSION PRINCIPLE

Another discovery pertaining to electronic motions was made by Wolfgang Pauli (1900–1958), an Austrian physicist of unusual temperament, bearing, and analytical skills. His peculiar behavior was displayed in an uncontrolled rocking motion when he was concentrating on what someone was saying, while his acerbic temperament and insulting manner were legendary, as when he occasionally signed his communications with "The Wrath of God." Yet scientists acquired enough respect for his analytical acuteness that they could overlook his caustic rudeness.

Since Bohr formulated his model of the hydrogen atom it had been agreed that no two electrons within the atom can occupy the same orbit because they do not have the same set of quantum numbers. As described by Ne'eman and Kirsh, the atomic state of an electron can be characterized by four quantum numbers:

> the first (*n*) indicates the "shell" or energy state which the electron occupies; the second (*l*) determines the orbital angular momentum of the electron; the third (*m*) determines the inclination which the direction of the orbital angular momentum will take up relative to a defined direction (e.g., a magnetic field); the fourth (m_s) determines the direction of the spin (for which there are only two possibilities). (ibid. p. 60)

The exclusiveness of electronic orbits based on their atomic numbers was identified by Pauli in 1925 and thus is called the "Pauli exclusion principle" and is one reason for his being awarded the Nobel Prize in physics in 1945.

It was proven later, however, that his principle is valid for any particle whose spin is fractional, but not for those with integral spins. The particles with fractional spin obeying the Fermi-Dirac statistics are named "fermions," while those with integral spin following the Bose-Einstein statistics are called "bosons." No fermions having the same quantum numbers can occupy the same orbital state while any number of bosons with the same number can occur in the same region of space, the difference due to the connection between their spin and the symmetry of the wave function.

DIRAC'S QUANTUM ELECTRODYNAMICS

Following his immensely successful interpretation of quantum mechanics known as "transformation theory" and while he was at Bohr's Institute from September 1926 to February 1927 doing postdoctoral research, Dirac realized that although quantum mechanics dealt with particles whose velocity approached the speed of light, none of the existing theories had taken into account relativistic consequences—Schrödinger had ignored them. With his usual ingenuity influenced by Einstein's adoption of Maxwell's field theory, Dirac created a quantum field theory which would take into account relativistic effects, such as the field associated with the electron. Thus he succeeded in combining electrodynamics with quantum mechanics which he called "quantum electrodynamics." (QED)

The paper presenting the theory, "The Quantum Theory of the Emission and Absorption of Radiation" submitted to the *Proceedings of the Royal Society* and published in March 1927, is described by Crease and Mann as "One of the most influential papers in the history of twentieth-century physics" (p. 82). In it Dirac states:

> *The new quantum theory, based on the assumption that the dynamical variables do not obey the commutative law of multiplication, has by now been developed sufficiently to form a fairly complete theory of dynamics. . . . On the other hand hardly anything has been done up to the present on quantum electrodynamics. The questions of the correct treatment of a system in which the forces are propagated with the velocity of light instead of instantaneously, of the production of an electromagnetic field by a moving electron, and the reaction of this field on the electron have not yet been touched.* (p. 82; italics in the original)

This paper was followed by two others in 1928 also published in *The Proceedings of the Royal Society* which contained his well-known relativistic wave equation of the electron, known as the "Dirac equation," describing the motion of an electron with "four components, two associated with spin and two with the energy [the Hamiltonian] of the particle" (Ibid., p. 84; brackets added). Following these outstanding achievements he was appointed to Newton's old chair of Lucasian Professor of Mathematics at Cambridge University in 1932 and shared the Nobel Prize in physics with Schrödinger in 1933. But though his equation eliminated many puzzles it still led to strange predictions, such as the space in atoms not being empty but filled with oscillators and photons that in a "zero state" could not be detected, but still possessed a vast amount of energy which was equivalent to mass or matter according to Einstein's equation $E = mc^2$.

In addition, solving his equation for the energy or Hamiltonian of a single electron resulted in two values, one positive and one negative. While positive energy was well known, the existence of negative energy and therefore negative matter (again according to Einstein's equation) was unheard of and yet the formalism predicted that space was filled with this undetectable negative energy. To add to the perplexity, in 1930 when Carl D. Anderson built an improved version of Wilson's cloud chamber to investigate cosmic rays, he discovered unusual tracts of light particles with a mass similar to electrons but having a positive charge. He named the positively charged electrons "positrons" and the name stuck. It was also considered a new form of unstable matter, "antimatter," which does not appear to exist in our universe. When an electron and a positron strike they annihilate one another producing two photons. For his experimental discovery of positrons and antimatter he received the Nobel Prize in physics in 1936 (cf. Ibid., p. 90).

Based on Dirac's initial quantum electrodynamics in the succeeding decades, QED was developed into what is now called "the standard model." It was completed by the Japanese physicist Sin-iteo Tomonaga, Richard Feynman, and Julian Schwinger who shared the Nobel Prize in physics in 1965 for their joint achievement. According to Chris Quigg at the Fermi National Acceleration Laboratory (Fermilab):

> QED is the most successful of physical theories. Using calculation methods developed in the 1940s by Richard P. Feynman and others, it has achieved predictions of enormous accuracy, such as the infinitesimal effect of the photons radiated and absorbed by the electron on the magnetic moment generated by the electron's innate spin. Moreover, QED's descriptions of the electromagnetic interaction have been verified over an extraordinary range of distances, varying from less than 10^{-18} to more than 10^{8} meters.[10]

MORE RECENT DEVELOPMENTS

Here our journey through the historical maze of scientific inquiry will end with a brief summary of some recent achievements, as obscure as they are, to indicate the various views about the future prospects of scientific research.[11] With the creation of increasingly powerful particle accelerators the succeeding decades have witnessed the discovery of a plethora of subatomic particles with unforeseen properties. Predicted by theorists such as Eugene Wigner, Hideki Yukawa, Murray Gell-Mann, Sheldon Glashow, Julian Swinger, Steven Weinberg, Samuel Ting, Burton Richter, *et al.,* these particles are created out of mass-energy transformations and their existence experimentally confirmed. So varied are the particles that they have been compared to a zoo with the construction of a new kind of periodic table composed of leptons and hadrons, with hadrons further divided into baryons and mesons.

In addition to this new world of particles and associated antiparticles with similar properties but opposite charge, two new atomic forces were discovered, a strong force binding the nucleons (protons and neutrons) and a weak force controlling the radioactivity within the nucleus. These new forces operate only within extremely short distances and by the exchange of "virtual" particles: photons in electromagnetic interactions, gluons in strong interactions, and vector bosons W^-, W^+, and Z^0 in weak interactions. The hadrons react to both strong and weak forces while leptons react only to the weak force. Initially, it was thought that the leptons and hadrons, along with the photons and the postulated gravitons (to explain gravitational attraction), exhausted the basic number of particles, but in 1964 Murray Gell-Mann and George Zweig proposed additional particles to explain the composition of the hadrons.

Gell-Mann whimsically called these new particles "quarks" derived from a passage in James Joyce's *Finnegan's Wake* ("Three quarks for Muster mark!") and the term caught on. The quarks are unusual in that unlike previous particles that have integral charges, they have fractional charges of plus two-thirds or minus one-third and in addition never exist separately but are joined as pairs or triplets to form the hadrons, a quark and antiquark pair composing a meson and three quarks constituting a baryon. Although at first purely conjectural to explain the interactions of the hadrons, there now is experimental evidence for their existence.

Continuing the whimsical classifications, Gell-Mann postulated a property or charge called "strangeness" while Sheldon Glashow added "charm" so that physicists now explain the hadrons and their interactions as caused by combinations of quarks, classified as "flavors". up (u)-down (d), charm (c)-strange (s), bottom/beauty (b)-top/truth (t). Thus matter consists of the two groups of six leptons and six quarks, plus the basic particle forces. Eventually a new quantum theory supplementing QED was introduced by Gell-Mann called "quantum Chromodynamics," so named because it consisted of a strong force or charge called "color" binding the quarks in the hadrons.

Yet not satisfied with two separate categories of basic particles, the quarks and the leptons, as well as three independent exchange particles accounting for the strong, weak, and electromagnetic forces, there began a search for a grand unified theory (GUTs). But rather than strain the reader with more detail regarding the developments leading to the formation of GUTs, I will conclude by quoting the summary of Crease and Mann, which is complex enough.

> The result is a ladder of theories. Firmly on the bottom is SU(3) × SU(2) × U(1) [the SU stands for special unity group based on Cartan's group theory], whose predictions have been confirmed ("to the point of boredom," Georgi says). . . . The W and Z particles were discovered at CERN [European Center for Nuclear Research] . . . but the theory was so well established . . . the event was—for theorists at least—an anticlimax. The GUTs proposed by Georgi and Glashow and other physicists, which fully unite the strong, weak, and electromagnetic forces, are the next rung on the ladder. Although as yet unconfirmed, these theories are considered compelling by most physicists. Finally, at the top of the ladder . . .

are supersymmetry and its cousins, which are organized according to a principle somewhat different from SU(5), though, like that model, they put apparently different particles together in groups. Supersymmetry groups are large enough to include gravity, but are so speculative that many experimenters doubt that they can ever be tested.[12]

FUTURE PROSPECTS OF PHYSICS AND COSMOLOGY

The last statement is significant. Written in 1984 this conclusion is somewhat dated, yet it indicates how scientific theories have become so abstract and complex that they tend to outstrip experimental evidence. Several decades ago physicists had high expectations of developing "a theory of everything," one that would reduce quantum mechanical particles to waves of vibrating strings, analogous to the notes produced by plucked violin strings, recalling and vindicating the Pythagorean dream of a musical harmony pervading the universe. But not only did this initial "string theory" and the subsequently revised "superstring theory" encounter the bizarre requirement that the universe have ten dimensions, after over ten years it has not produced a single prediction that could be tested empirically, the *sine qua non* of a scientific theory.

Cosmology is another area that had made such remarkable progress that cosmologists were tempted to predict a final theory of the universe. For example, in his inaugural lecture (delivered by a student because he has amyotrophic lateral sclerosis) when he assumed the Lucasian chair in physics in April 1980, Stephen Hawking, one of the world's best known cosmologists, made the following prediction.

In this lecture I want to discuss the possibility that the goal of theoretical physics might be achieved in the not-too-distant future, say, by the end of

> *the century. By this I mean that we might have a complete, consistent, and unified theory of the physical interactions which would describe all possible observations. Of course one has to be very cautious about making such predictions. . . . Nevertheless, we have made a lot of progress in recent years and . . . there are some grounds for cautious optimism that we may see a complete theory within the lifetime of those present here.* (Crease and Mann, *The Second Creation*, p. 410; italics in the original)

Having turned the century and not yet attained a final, complete theory, there would seem to be more grounds for caution than for optimism. Recently, however, Michio Kaku, based on the data derived from the Wilkinson Microwave Anisotropy Probe (WMAP), expressed considerable optimism regarding the "inflationary theory" of the bizarre cosmology of "Parallel Worlds."

> With the flood of new data we are receiving today, with new tools such as space satellites which can scan the heavens, with new gravity wave detectors, and with new city-size atom smashers nearing completion, physicists feel that we are entering what may be the golden age of cosmology. It is, in short, a great time to be a physicist and a voyager on this quest to understand our origins and the fate of the universe.[13]

Although being bizarre is no criterion of a scientific theory's promise as we have seen throughout the history of science, since revolutionary scientific theories initially have invariably seemed strange because of being counterintuitive, the theory of multiple or parallel worlds does read like science fiction.

> In this radically new interpretation . . . at each quantum juncture, the universe splits in half, in a never-ending sequence of splitting universes. All universes are possible in this scenario, each as real as the other. People living in each universe might vigorously protest that their universe is the real one, and that all the others are imaginary or fake. These parallel universes are not ghost worlds with an

> ephemeral existence; within each universe, we have the appearance of solid objects and concrete events as real and as objective as any. (ibid., p. 168)

Still, it seems to me that a universe in which "all possible worlds coexist with [ours]" would be so complex that any final conception of the universe would be impossible.

There are now some books on science that address the extraordinary implications of these recent scientific developments, one by a theoretical physicist, titled *The End of Physics: The Myth of a Unified Theory*, the other by a former senior staff writer at *Scientific American,* titled *The End of Science: Facing the Limits of Knowledge in the Twilight of the Scientific Age.* The author of the first book, the scientist David Lindley, describes the history of physics leading up to this pessimistic dénouement as follows:

> Modern physics was set on its present course by the pragmatic methods of Newton and Galileo and their many successors, and this effort, three centuries old, has led to the elaborate physical understanding we now possess. But this kind of physics seems to have run its course. Experiments to test fundamental physics are at the point of impossibility, and what is deemed progress now is something very different from what Newton imagined.[14]

He then goes on to classify unified theories, such as a theory of everything touted today, as mythical because of their being untestable.

> The theory of everything will be, in precise terms, a myth. A myth is a story that makes sense within its own terms, offers explanations for everything we can see around us, but can be neither tested nor disproved. A myth is an explanation that everyone agrees on because it is convenient to agree on it, not because its truth can be demonstrated. This theory of everything, this myth, will indeed spell the end of physics. It will be the end not because physics has

> at last been able to explain everything in the universe, but because physics has reached the end of all the things it has the power to explain. (ibid., p. 255)

It comes as an ironic surprise to read, after arguing throughout the book that Christianity is an extension of mythology because its origin, based on allegedly revealed scripture, presents a worldview that has no evidential support in terms of what we have since learned about the universe and human existence, that some scientists believe that science itself, for all its past achievements, could have reached a similar denouement. This is not to claim that previous science was mythological, but that a "grand unified theory" explaining everything presupposes that contemporary science already possesses all the information needed to achieve this goal, which seems implausible. It is even questionable whether such an achievement would be desirable since that would put an end to basic research, even though there would remain innumerable practical problems to solve with the existing knowledge and technology.

The questioning of leading scientists regarding the future prospects of science by John Horgan reinforces this conclusion, though I doubt that it is the most prevalent view among scientists. What he found is that some scientists no longer believe that a final theory is attainable, either at the present or in the foreseeable future, and that perhaps it is not even possible. Horgan calls this view of science, "ironic science," and refers to the scientists who hold it "strong scientists" because they are aware that seeking a final answer could be an illusion and yet still pursue science as a speculative endeavor.

> The most common strategy of the strong scientist is to point to all the shortcomings of current scientific knowledge, to all the questions left unanswered. But the questions tend to be ones that may *never* be definitively answered given the limits of human science. How, exactly, was the universe created? Could our universe be just one of an infinite number of universes? Could quarks and electrons be composed of still smaller particles, ad infinitum. What does

> quantum mechanics really mean? . . . Biology has its own slew of life's insoluble riddles. How, exactly did life begin on earth? Just how inevitable was life's origin and its subsequent history?[15]

After describing these current impasses in science he then characterizes what this implies about the current and future attempts to resolve them.

> Ironic science, by raising unanswerable questions, reminds us that all our knowledge is half-knowledge; it reminds us of how little we know. But ironic science does not make any significant contributions to knowledge itself. Ironic science is thus less akin to science in the traditional sense than to literary criticism—or to philosophy. (p. 31)

This conception of the limits of scientific explanations has received formal reinforcement by Gödel's "incompleteness theorem" which Horgan defines as asserting that "any consistent system of axioms beyond a certain basic level of complexity yields statements that can be neither proved nor disproved with those axioms; hence the system is always incomplete" (p. 174). It seems to me this theorem supports my proposed theory of "contextual realism" described previously, that known physical systems, though seemingly autonomous within their own domains, are always embedded in larger physical contexts on which their ultimate existence and further explanation depend, *ad infinitum*. According to Gödel's theorem all explanations are incomplete because they ultimately depend upon explanatory principles that lie beyond the horizon of the system being considered. Yet I would suggest that this does not imply that they are mythical or not relatively or conditionally real and true within their limiting conditions, as exemplified by the ordinary world of experience which we do not regard as unreal nor our descriptive statements referring to it as untrue because that world is not ultimately or unconditionally real.

A close friend, Dr. Lionel J. Skidmore, a gifted mathematician who studied with Richard Feynman among others and who spent his professional life plotting and predicting the orbital trajectories of earth satellites and space shuttles, has described to me the method he used to determine how close to being correct his computed predictions were. The method, called "least squares," was first introduced by the great German mathematician Carl Friedrich Gauss to calculate orbital dimensions. He developed the method in 1795 when he was eighteen years old, but it was not published until six years later in his two volume work on celestial mechanics. Another famous mathematician, the Frenchman Adrien-Marie Legendre, independently developed the same method in 1805, as did the American Robert Adrain in 1808.

The formula used in the method of least squares is known as the Gauss-Markov theorem, but a precise verbal description of the method can be found on the Internet.

> **Least squares** . . . is a mathematical optimization technique which, when given a series of measured data, attempts to find a function which closely approximates the data (a "best fit"). It attempts to minimize the sum of the **squares** of the ordinate differences (called *residuals*) between points generated by the function and corresponding points in the data. Specifically, it is called ***least*** mean ***squares*** (LMS) when the number of measured data is 1 and the gradient descent method is used to minimize the squared residual. LMS is known to minimize the expectation of the squared residual, with the smallest operations (per iteration). But it requires a large number of iterations to converge.

Dr. Skidmore has described to me how essential this method was using data such as the vehicle parameters, launch data (winds, pressures, etc.), engine model specifications, and known post-flight anomalies, all of which could only be given probable values based on observations, in *determining how close his calculations were to an exact measurement and therefore how reliable his predictions were.*

My reason for describing this is that if it were possible to apply this method to the data used in formulating a unified theory of everything, then we would know how close we were to a correct final theory, which in turn would eliminate all the speculation as to whether such a theory were even possible. But since the necessary data are inaccessible or beyond our present reach, we can only speculate about the answer which presumably will be decided in the future. Yet I just read that the situation may not be as dismal or as final as the pessimists have predicted.

This is due to the construction by the European Organization for Nuclear Research (CERN) of a colossal particle accelerator, the Large Hadron Collider also known as "Atlas" located on farmland at the boarder of Switzerland and France. The most ambitious and expensive research endeavor since the Manhattan Project, it will cost more than 8 billion dollars. Consisting of a 27-kilometer or 17-mile circular tunnel buried 300 feet in the ground, it will circulate two streams of protons composed of quarks at nearly the speed of light around the tunnel. Beamed in opposite directions, when ready it will cause 600 million collisions per second with sophisticated detectors measuring the effects at four major intersections. If all goes as predicted, sometime in 2009 the collisions will produce "tiny fireballs of primordial energy, creating conditions that last prevailed when the universe was less than a trillionth of a second old."[16]

On the morning of 10 September 2008 the first beam completed its momentous journey much to the relief and delight of the physicists and engineers who have worked years on the project. Presumably these fireballs will momentarily replicate the forms of matter, kinds of energy, and laws that characterized the state of the universe following the Big Bang 13.7 billion years ago. Resolving many of the questions facing theoretical physicists today, it could lead to explanations of the origin of black holes and of the nature of dark matter and dark energy, the latter believed to be causing the expansion of the universe. It also may help explain supersymmetry, antimatter, and perhaps provide some evidence, pro or con, for string theory.

Furthermore, physicists hope to detect evidence of the illusive Higgs boson, named after the Scottish physicist Peter Higgs who predicted the existence of such a particle in 1964. It is especially significant because it is thought to impart mass to the elementary particles produced by the Big Bang that are believed to have been originally massless. Because of its explanatory importance, the Higgs boson has been referred to as the "God particle." Thus scientists anxiously await the outcome of the CERN experiment, a few even fearful, as during the test of the nuclear reaction at Los Alamos, that it could get out of control and devastate the planet. Yet most scientists are not anxious out of fear, but because they hope its success could provide data that would open up a whole new era and areas of scientific research, while its failure would confirm the pessimistic prognosis of the end of physics and cosmology.

Recently, data derived from NASA's orbiting Chandra Ex-Ray observatory (named for a Brahmin scholar Subrahmanyan Chandrasekhar who in the thirties posited black holes to explain imploding stars leaving a dark hole in space) of a collision between two galaxies that some scientists claim provide "direct evidence" of the creation of dark matter and energy; however, other physicists believe that we must await the experimental results of the Large Hadron Collider before any decision can be made. Thus one can understand the anxiety that has been typical of scientific inquiry from the beginning.

NOTES

1. Peter Gibbins, *Particles and Paradoxes* (Cambridge: Cambridge University Press, 1987), pp. 23–24. Future citations will be preceded by the author's name followed by the page number.

2. Werner Heisenberg, *Physics and Beyond* (New York: Harper Torchbooks, 1972), p. 8. Further citations of his works will be preceded by his name followed by the acronym of the title and page number.

3. Cf. Robert P. Crease and Charles C. Mann, *The Second Creation: Makers of the Revolution in Twentieth-Century Physics* (New York: Macmillan Publishers, 1986), pp. 50–51. Future citations will be indicated by their names followed by the page number.

4. Abraham Pais, *Niels Bohr's Times* (Oxford Clarendon Press, 1991), p. 288. Further quotations will be identified as Pais, NBT.

5. Emilio Segrè, *From X-Rays to Quarks,* (San Francisco: W. H. Freeman and Company, 1980), p. 165.

6. Werner Heisenberg, *Physics and Philosophy: The Revolution in Modern Science*, ed. by Ruth Nanda Anshen (New York: Harper & Brothers, 1958), p. 42.

7. Niels Bohr, *The Philosophical Writings of Niels Bohr*, Vol. III (Woodbridge, Connecticut: Ox Bow Press, 1963), p. 4.

8. Cf. Richard H. Schlagel, *Contextual Realism: A Meta-physical Framework for Modern Science* (New York: Paragon House, 1986).

9. Yuval Ne'eman and Yoram Kirsh, *The Particle Hunters* (Cambridge: Cambridge University Press, 1983), pp. 52–59.

10. Chris Quigg, "Elementary Particles and Forces," *Scientific American*, April 1985, p. 89.

11. This discussion is based on Richard H. Schlagel, *From Myth to Modern Mind*, Vol. II, *Copernicus through Quantum Mechanics*, op. cit., chap. 11, which in turn is largely based on Crease and Mann, op. cit. above, chaps. 15, 16.

12. Robert P. Crease and Charles C. Mann, "How the Universe Works," *Atlantic Monthly*, August 1984, p. 84; brackets added.

13. Michio Kaku, *Parallel Worlds: A Journey through Creation, Higher Dimensions, and the Future of the Cosmos* (New York: Doubleday, 2005), p. xvii.

14. David Lindley, *The End of Physics: The Myth of a Unified Theory* (New York: Basic Books, 1993), p. 255.

15. John Horgan, *The End of Science: Facing the Limits of Knowledge in the Twilight of the Scientific Age* (New York: Broadway Books, 1997), p. 7.

16. Dennis Overbye, "Replaying the universe's birth," *International Herald Tribune,* May15, 2007, p. 2.

Chapter 16

CONCLUSION

I hope that religionists do not conclude that because some, though by no means all, scientists claim that scientific inquiry might have reached its explanatory limits this requires or justifies a religious 'explanation,' such as Henry Morris's claim quoted previously that "there is no *sure* way (*except by divine revelation*)" of settling empirical questions such as "the true age of any geological formation" (p. 435). What evidence is there that any religion has ever provided a plausible answer to an empirical question? Genesis is a mythological explanation while the recourse to God to explain the origin of the universe or of life on earth is futile because the concept of God itself is so amorphous that nothing of empirical significance can be deduced from it to explain anything.

What can be inferred about the universe that is explanatory and testable from the conception of God as an "all loving, omniscient, and omnipotent spirit"? While the attributes are meaningful, they have no more referential significance than words like "angel," "unicorn," or the "devil." The disparity between the primitive anthropomorphic conception of God and the requirements of a cosmic explanation is blatant. Just attempting to imagine what the nature of God would have to be to have created the entire universe with its infinite complexity defies comprehension. If we had examples of previous universes having been created by a God and evidence that our universe resembled those, then we might infer that ours was similarly created, but not otherwise.

As indicated previously, Western civilization was formed by two contrasting traditions, the empirical-rationalistic perspective of the ancient Greek philosophers and the mystical-revelatory approach of the Hebraic-Christian prophets. Beginning with the Edict of Milan issued by the co-Emperors Constantine and Licinius in 313 CE that decreed the freedom of worship of all religions in the Roman Empire and eventually led to Christianity becoming the state religion and gradually spreading to be the dominant ideological, moral, and cultural influence in the West. Then the reintroduction of ancient Greek manuscripts to Europe owing to the Arab conquests rekindled an interest in mathematical and empirical inquiries that had been quelled by the otherworldly outlook of Christianity. Their progression has established the present scientific worldview and technological advances evident in much of the world threatening the Muslim culture and arousing the hostility of terrorist groups, such as the Taliban and al-Qaeda, toward the Western "infidels." Even countries like China and India are now adopting Western methodologies to improve their economies and living conditions.

The purpose of this book has been to reconstruct the advances of science, in contrast to the anachronistic biblical worldview, to show how and why scientific inquiry emerged contesting the religious outlook, describing the empirical discoveries and theoretical interpretations that brought civilization to its current level of knowledge and technological achievements. With this résumé as background it should be apparent that religiously motivated arguments, such as special creation and intelligent design, have no evidential justification. Unless one's thinking has been predetermined by the religious worldview, nothing in our present understanding of the universe and of human existence indicates that it was created by an all loving, all powerful, all knowing God. All of our knowledge is in terms of natural processes that are too causally determined, self-contained, conciliatory, and physicalistic to be attributed to a spiritual being.

Moreover, unlike the Abrahamic religions that claim to find final answers in revelation, scientific inquiry is an ongoing process in which further discoveries requiring radical conceptual revisions, institutional changes, and cultural modifications seem inevitable. Despite Einstein's persistent belief in a *determinate* universe in contrast to quantum mechanics, he nonetheless stated in 1938 that "Science is not and will never be a closed book. Every important advance brings new questions. Every development reveals, in the long run, new and deeper difficulties."[1]

Thus my critique of Christianity (and religions in general) differs from atheists such as Richard Dawkins, Sam Harris, and Christopher Hitchens in exposing the erosion of religion's evidential basis, rather than maligning religions for their obvious shortcomings. Though it is true that religions have been dogmatic, repressive, and the source at times of terrifying cruelties and wars, this is due to the omnipresent defects of human nature, not to perverse religious doctrines or practices as such. Religions have been too influential in supporting compassionate values, advocating humane social legislation, and promoting philanthropic missions throughout the world for that caricature to be justified.

Vilifying religion ignores the moral teachings, exhortations, and inspiring influence of religious leaders like Buddha, Moses, Jesus, Paul, and to some extent Muhammad, that have contributed immeasurably to subduing the weaknesses, selfishness, cruelties, and perversions of human nature by extolling the values of love, forgiveness, charity, kindness, and self- sacrifice. Nor should one overlook or disparage the tremendous consolation religions have provided in this predominantly perilous, conflicted, and despairing world.

That this recourse has diminished in more advanced civilizations owing to a greater understanding and control of natural disasters, more humane social and political institutions, and improved living conditions does not erase religion's crucial supportive role in the past—as well as in the present where the need and cultural context still persists. My criticism of religions is not that they have had a deleterious influence on society—one only has to witness worshipers in solemn prayer asking for forgiveness of their sins and seeking strength to lead better lives or to overcome adversities to see otherwise—but their continued advocacy and defense of a supernatural system of belief established millennia ago when ignorance and superstition were equated with knowledge and truth.

As we have seen, religions arose at a time and in a culture when there was almost no understanding of the world or control over one's life, and since the most familiar form of causality was intentional human acts, it was natural to attribute the reigning forces of natural events and the reasons for the human plight to the malicious, capricious, or benign interventions of supernatural beings, such as demons, evil spirits, guardian angels, or Gods. But scientific inquiry gradually replaced occult, vitalistic, or supernatural agencies with naturalistic causes and forces owing to the latter's greater explanatory efficacy and predictive controls. Nor can the threatening contingencies of life and wrenching moral choices be mitigated by the purported absolutes of religion. As science attests, eventually everyone will have to accept that we live in a vast, enormously complex, and possibly unfathomable universe in the sense of never being finally or intentionally explainable as to why, rather than how, it is as it is. In the poignant words of Erich Fromm, man must

> acknowledge his fundamental aloneness and solitude in a universe indifferent to his fate, to recognize that there is no power transcending him which can solve his problem for him. Man must accept the responsibility for himself and the fact that . . . *there is no meaning to life except the meaning man gives his life by the unfolding of his powers* . . . within the limitations of our existence.[2]

Thus wonder at the mystery of existence and how it all came about should replace futile attempts at supernatural explanations. Only hubris or naiveté could lead one to think, like the proponents of the anthropic principle, that the universe, so vast in extent and time in contrast to the finitude and frailty of human existence, was intentionally preprogrammed to produce human beings, as if our existence was preordained in the original Big Bang, analogous to the astrological belief that a person's destiny lay in the configuration of the zodiac or in the constellations of the stars. Just the capriciousness and cruelty of the evolutionary process and the calamitous nature of the human condition should be enough to disabuse us of this fantasy. Our existence has no more cosmic relevance than that of the Neanderthals, Cro-Magnons, or that poor young girl who lived 3.3 million years ago whose skeleton depicts the partial transition between apes and humans. However emotionally difficult it may be to resign ourselves to such an inauspicious status, the evidence for it is overwhelming and wishing it were otherwise will not make it so. The only things supporting the Abrahamic religions are tradition, indoctrination, ignorance, and inclination, not empirical evidence or rational argument. Monotheism is no more credible than the polytheism it superseded and just as vulnerable to eventual replacement. Accepting the truth and facing reality are far worthier than chasing mystical or mythical illusions.

NOTES

1. Albert Einstein and Leopold Infeld, *The Evolution of Physics: The Growth of Ideas From Early Concepts to Relativity and Quanta* (New York: Simon and Schuster, 14th ed., 1951), p. 308.

2. Erich Fromm, *Man for Himself: An Inquiry into the Psychology of Ethics* (New York: Rinehart, 1947), p. 45. Quoted from John F. Schumaker, *Wings of Illusion* (Buffalo, New York: Prometheus Books, 1990), p. 158. Italics in original.

BIBLIOGRAPHY

Achinstein, Peter. *Particles and Waves*. Oxford: Oxford University Press, 1991.

Appleman, Philip. *Darwin*, selected and ed., 2nd ed. New York: W. W. Norton, 1970.

Aristotle. *Metaphysics*, *The Basic Works of Aristotle,* ed. by Richard McKeon. New York: Random House, 1941.

———. *De Anima.*

———. *De Cealo.*

———. *De Generatione Animalium.*

———. *De Generatione et Corruptione.*

———. *De Partibus Animalium.*

———. *Metaphysica.*

———. *Physica.*

———. *Posteriora Analytica*

Arnaldez, R., and I. Massignon. "Arabic Science," René Taton, ed. *History of Science: Ancient and Medieval Science*, Vol. I, trans. by A. J. Pomerans. New York: Basic Books, 1963.

Artz, Frederick B. *The Mind of the Middle Ages,* 3rd ed. rev. New York: Alfred A. Knopf, 1962.

Augustine. *The Confessions of St. Augustine*, trans. by J. C. Pilkington. New York: The Heritage Press, 1963.

Beare, F. W. *The Earliest Records of Jesus*. Oxford: Blackwell, 1964.

Bolles, Edmund Blair. *The Ice Finders*. Washington, DC: Counterpoint, 1999.

Bohr, Niels. *The Philosophical Writings of Niels Bohr,* Vol. III. Woodbridge, Connecticut: Ox Bow Press, 1963.

———."Can the Quantum Mechanical Description of Reality be Considered Complete?" *Physical Review*, 1935, p. 48.

Bronowski, J. and Bruce Mazlish. *The Western Intellectual Tradition*. New York: Harper & Brothers, 1960.

Butterfield, Herbert. *The Origins of Modern Science*, rev. ed. New York: Collier Books, 1962.

Cantore, Enrico. *Atomic Order*. Cambridge: The MIT Press, 1969.

Caspar, Max. *Kepler*, trans. by C. Doris Hellmann. New York: Dover Publications, 1993.

Clagett, Marshall. *The Science of Mechanic in the Middle Ages*. Madison: University of Wisconsin Press, 1959.

Clavelin, Maurice. *The Natural Philosophy of Galileo: Essay on the Origins and Formation of Classical Mechanics*, trans. by A. J. Pomerans. Cambridge, Mass.: The MIT Press, 1974.

Cohen, I. Bernard. *Franklin and Newton*. Cambridge: Harvard University Press, 1966.

Cohen, M. R. and I. E. Drabkin. *A Source Book in Greek Science*. Cambridge: Harvard University Press, 1948.

Conant, James Bryant and Leonard K. Nash. *Harvard Case Histories in Experimental Science*, Vols. 1 & 2. Cambridge: Harvard University Press, 1948.

Copernicus, Nicolaus. *On the Revolutions of Heavenly Spheres*, trans. by Charles G. Wallis. Amherst, New York: Prometheus Books,1995.

Cornford, F. M. *Principium Sapientiae,* ed. by W. K. C. Guthrie. New York: Harper & Row Publishers, 1952, 1965.

———. *Plato's Cosmology*. New York: The Library of Liberal Arts, 1957.

Crease, Robert P. and Charles C. Mann. *The Second Creation: Makers of the Revolution in Twentieth-Century Physics*. New York: Macmillan Publishers, 1986.

———. "How the universe Works." *Atlantic Monthly,* August, 1984.

Crombie, A. C. *Robert Grosseteste and the Origins of Experimental Science*. Oxford: At The Clarendon Press, 1953, 1962.

Crossan, John Dominic. *Jesus: A Revolutionary Biography*. New York: HarperCollins, 1994.

Dennett, Daniel C. *Darwin's Dangerous Idea: Evolution and the Meanings of Life*. New York: Simon and Schuster, 1995.

Denton, Michael. *Nature's Destiny: How the Laws of Biology Reveal Purpose in the Universe*. New York: Free Press, 1998.

Diogenes Laertius. *Lives of Eminent Philosophers*, 2 Vols., ed. by H. S. Long. Oxford: Oxford University Press, 1964.

Dionysus. "Festival Letters," quoted by Eusebius, in *Ecclesiastical History*, by G. A. Williamson. Middlesex: Penguin Books, 1965.

Durant, Will. *The Story of Civilization*, Part III. *Caesar and Christ*. New York: Simon and Schuster, 1944.

———. *The Story of Civilization*, Part IV. *The Age of Faith*. New York: Simon and Schuster, 1950.

Edds, Kimberly. "At Grand Canyon Park, a Rift over Creationist Book." *Washington Post*, January 20, 2004, A17.

Einstein, Albert. *Relativity: The Special and General Theory,* trans. by Robert W. Lawson. New York: Crown Publishers, 15th ed., 1961.

———. "Autobiographical Notes," Paul A. Schlipp, ed., *Albert Einstein: Philosopher-Scientist*. Evanston: Library of Living Philosophers, Vol. VII, 1949.

Einstein, Albert, and Leopold Infeld. *The Evolution of Physics*. New York: Simon and Schuster, 14th ed., 1951.

Einstein, Albert, Boris Podolsky, and Nathan Rosen. *Gedanken-Experiment (EPR Experiment)*, *Physical Review*, 1935, 47.

Ehrman, Bart D. *Misquoting Jesus: The Story Behind Who Changed the Bible and Why*. New York: HarperSanFrancisco, 1995.

Field, J. V. *Kepler's Geometrical Cosmology*. Chicago: The University of Chicago Press, 1988.

Freeman, Charles. *The Closing of the Western Mind: The Rise of Faith and the Fall of Reason*. New York: Alfred A. Knopf, 2003.

Freeman, Kathleen. *Ancilla to the Pre-Socratic Philosophers*. Oxford: Basil Blackwell, 1962.

Funk, Robert W., Roy Hoover, and the Jesus Seminar. *The Five Gospels*. New York: Macmillan, 1993.

Galileo, Galilei. *Dialogue Concerning the Two Chief World Systems*, trans. by Stillman Drake, 2nd ed. Berkeley: University of California Press, 1967.

———. *Dialogues Concerning Two New Sciences,* trans. by Henry Crew and Alfonso de Salvio. New York: McGraw Hill, 1914, 1963.

Gibbins, Peter. *Particles and Paradoxes*. Cambridge: Cambridge University Press, 1987.

Gilbert, William. *De Magnete*, trans. by P. Fleury Mottelay. New York: Dover Publications, 1948.

Greenaway, Frank. *John Dalton and the Atom*. Ithaca: Cornell University Press, 1966.

Greene, John C. *The Death of Adam: Evolution and Its Impact on Western Thought*. Ames, Iowa: The Iowa State University Press, 1959.

Gugliotta, Gay. "For Noah's Flood, a New Wave of Evidence," *Washington Post*, November 18, 1999, pp. A1, A28.

Guth, Alan H. *The Inflationary Universe*. Reading, Massachusetts: Helix Books, 1997.

Hanson, R. P. C. and A. T. Hanson. *The Bible Without Illusions*. London: SCM Press, 1989.

Hanson, Victor Davis. *A War Like No Other*. New York: Random House, 2005.

Heath, Sir Thomas L. *Aristarchus of Samos: The Ancient Copernicus*. Oxford: At The Clarendon Press, 1966.

———. *A History of Greek Mathematica,* Vol. I. Oxford: Clarendon Press, 1921.

———. *The Works of Archimedes with the Method of Archimedes*, ed. New York: Dover Publications, 1953.

Heisenberg, Werner. *Physics and Beyond: Encounters and Conversations*, trans. by Arnold J. Pomerans. New York: Harper Torchbooks, 1972.

———. *Physics and Philosophy: The Revolution in Modern Science,* ed. by Ruth Nanda Anshen. New York: Harper & Brothers, 1958.

Hibbert, Christopher. *Rome: The Biography of a City*. London: Penguin Books, 1987.

The Holy Bible: Revised Standard Version. New York: Thomas Nelson & Sons, 1953.

Horgan, John. *The End of Science: Facing the Limits of Knowledge in the Twilight of the Scientific Age*. New York: Broadway Books, 1997.

Hyde, W. W. *Paganism to Christianity in the Roman Empire*. Philadelphia: The University of Pennsylvania Press, 1946.

Jaeger, Werner. *Aristotle*, 2nd ed., trans. by Richard Robinson. Oxford: Clarendon Press, 1948.

Kaku, Michio. *Parallel Worlds: A Journey Through Creation, Higher Dimension, and the Future of the Cosmos*. New York: Doubleday, 2005).

Kaufman, Marc. "2 in U.S. Win Nobel Prize for Research of Universe's Origin." *Washington Post*, October 4, 2006, A3.

Kee, Howard Clark. *Understanding the New Testament*, 5th ed. New Jersey: Prentice Hall, 1993.Kepler, Johannes. *Astronomia nova*, 1604.

Kepler, Johannes. *Astronomia nova*, 1604.

———. *Epitome Astronomiae Copernicanae*, 1618–1621.

———. *Harmonice mundi*, 1619.

———. *Mysterium Cosmographicum*, trans. by A. M. Duncan. New York: Abaris Books, 1981.

———. *Tabulae Rudophinae,* 1627.

Kirk, G. S. & J. E. Raven. *The Presocratic Philosophers*. Cambridge: At The University Press, 1962.

Koestler, Arthur. *The Sleepwalkers*. New York: Grosset's Universal Library, 1959.

Koyré, Alexandre. *Galileo Studies*. New Jersey: Humanities Press, 1978.

———. *Newtonian Studies.* Cambridge: Harvard University Press, 1965.

Kuhn, Thomas. *The Copernican Revolution.* Cambridge: Harvard University Press, 1957.

Lear, F. "Medieval Attitude Toward History," *Rice Institute Pamphlets*, 1933.

Lindley, David. *The End of Physics: The Myth of a Unified Theory.* New York: Basic Books, 1993.

Long, A. A. *Hellenistic Philosophy.* Berkeley: University of California Press, 1986.

Lorentz, H. A., A. Einstein, H. Minkowski, and H. Weyl. *The Principles of Relativity,* trans. by W. Perrett and G. B, Jeffery. New York: Dover Publications, 1923.

Lucretius. *On Nature*, trans. by Russell M. Greer. New York: Bobbs Merrill, 1965.

Mack, Burton L. *Who Wrote The New Testament? The Making of the Christian Myth*. New York: HarperCollins, 1996.

McKenzie, Norman, *Dreams and Dreaming*. New York: The Vanguard Press, 1965.

McNeill, W. H. *Plague and Peoples*. New York: Doubleday, 1976.

Marcus Aurelius. *Meditations*, trans. by A. S. L. Grube, The library of Liberal Arts. New York: Bobbs-Merrill, 1963.

Morris, Henry, ed. *Scientific Creationism*. San Diego: Creation-Life Publishers, 1974.

Nash, Leonard K. "The Atomic-Molecular Theory," in James Conant and Leonard K. Nash, eds., *Harvard Case Histories in Experimental Science,* Vol. 2. Cambridge: Harvard University Press, 1948.

Ne'eman, Yuval and Yoram Kirsh. *The Particle Hunters*. Cambridge: Cambridge University Press, 1986.

Newton, Sir Isaac. *Opticks,* based on fourth ed. London, 1730, prepared by Duane H. D. Roller. New York: Dover Publications, 1952.

———. *Principia*, Vol. I. *The Motion of Bodies* and Vol. II, *The System of the World*, Motte's trans. of 1729 rev. by Cajori. Berkeley and Los Angeles: University of California Press, 1962.

Norwich, John J. *Byzantium: The Early Centuries*. New York: Alfred A. Knopf, 1999.

Pais, Abraham, *'Subtle is the Lord . . .'* Oxford: Oxford University Press, 1982.

———. *Inward Bound: Of Matter and Forces in the Physical World*. New York: Oxford University Press, 1986.

———. *Niels Bohr's Times: In Physics, Philosophy and Polity.* Oxford: Clarendon Press, 1991.

Partington, J. R. *A Short History of Chemistry*, 3rd ed. rev. and enlarged. New York: Harper Torchbooks, 1960.

Patterson, Elizabeth C. *John Dalton and the Atom.* New York: Anchor Books, 1970.

Persinger, Michael. *Neurophysiological Basis of God Beliefs*. New York: Praeger, 1987.

Plato. *The Collected Dialogues of Plato*, ed. by Edith Hamilton and Huntington Cairns. New York: Pantheon Books, 1964.

———. *Cratylus*, in *The Collected Dialogues of Plato*, ed. by Edith Hamilton and Huntington Cairns. New York: Pantheon Books, 1964.

———. *Epinomis*, in *The Collected Dialogues of Plato*, ed. by Edith Hamilton and Huntington Cairns. New York: Pantheon Books, 1964.

———. *Phaedo*, in *The Collected Dialogues of Plato*, ed. by Edith Hamilton and Huntington Cairns. New York: Pantheon Books, 1964.

———. *Republic,* trans. by F. M. Cornford. Oxford: Oxford University Press, 1945.

———. *Sophist*, in *The Collected Dialogues of Plato*, ed. by Edith Hamilton and Huntington Cairns. New York: Pantheon Books, 1964.

Plutarch. *Life of Marcellus,* trans. by Bernadotte Perris, London, 1917.

Quigg, Chris. "Elementary Particles and Forces," *Scientific American*, April 1985, p. 89.

Randall, John Herman. *Aristotle.* New York: Columbia University Press, 1960.

Ridley, Matt. *Genome.* New York: Harper Collins, 1999.

Robertson, Andrew. *Einstein: A Hundred Years of Relativity*. New York: Harry N. Abrams, Publishers, 2005.

Ross, W. D. *Aristotle*, 5th ed. rev. London: Methuen and Co., 1949.

Sarton, George. *A History of Science*, Vol. II. Cambridge: Harvard University Press, 1959.

Schlagel, Richard. H., *Contextual Realism: A Meta-physical Framework for Modern Science.* New York: Paragon House, 1986.

———. *From Myth to Modern Mind: A Study of the Origins and Growth of Scientific Thought,* Vol. I, *Theogony through Ptolemy.* New York: Peter Lang, 1995.

———. *From Myth to Modern Mind: A Study of the Origins and Growth of Scientific Thought*, Vol. II, *Copernicus through Quantum Mechanics*. New York: Peter Lang, 1996.

———. *The Vanquished Gods*. Amherst, New York: Prometheus Books, 2001.

Schweitzer, Albert. *The Quest for the Historical Jesus,* trans. by W. Montgomery. Great Britain: A & C. Black, Ltd., 1910.

Scott, Eugenie C. *Evolution vs. Creationism*. Berkeley and Los Angeles: University of California Press, 2005.

Segrè, Emilio. *From X-Rays to Quarks*. San Francisco: W. H. Freeman, 1980.

Singer, Charles. *A Short History of Scientific Ideas to 1900.* London: Oxford University Press, 1959.

Slater, E. and A. W. Beard. "The Schizophrenic-like Psychoses of Epilepsy. I Psychiatric Aspects." *British Journal of Psychiatry,* 1963.

Small, Robert. *An Account of the Astronomical Discoveries of Kepler,* a reprinting of the 1804 text. Madison: The University of Wisconsin Press, 1963.

Sprong, John Shelby. *Liberating the Gospels*. New York: HarperCollins, 1996.

Stark, Rodney. *The Rise of Christianity*. San Francisco: HarperCollins, 1997.

Strauss, David Friedrich. *Life of Jesus Critically Examined,* 2 Vols., 1835.

Stein, Robert. "3.3 Million Years Later, Skeleton of Girl Found," *Washington Post*, November 21, 2006, A9.

Struik, Dirk J. A. *A Concise History of Mathematics*, 3rd ed. New York: Dover Publishing, 1948, 1967.

Taton, René, ed. *History of Science: Ancient and Medieval Science*, trans. by A. J. Pomerans, 3 Vols. New York: Basic Books, 1963.

Taylor, H. O. *The Medieval Mind,* Vol. I, 4th ed.Cambridge: Harvard University Press, 1962.

Vedantam, Shanker. "Near-Death Experiences Linked to Sleep Cycles," *Washington Post*, April 11, 1006, A 2.

Voltaire. *Eléments de philosophie of Newton*, 1783.

Weigall, A. *The Paganism in Our Christianity*. New York, Kessinger Publishing, 1928.

Weins, Robert C. *Radiometric Dating: A Christian Perspective*. ASA Resources, 2002.

Wells, G. A. *A Critique of the New Testament Record*. La Salle, Ill.: Open Court, 1989.

———. *Who Was Jesus?* La Salle, Ill.: Open Court, 1989.

Westfall, Richard S. *Never at Rest: A Biography of Isaac Newton*. Cambridge: Cambridge University Press, 1980.

White, Andrew D. *A History of the Warfare of Science with Theology in Christendom*. New York: Appleton Press, 1896.

Whittaker, Sir Edmund. *A History of the Theories of Aether & Electricity*, Vol. I, *The Classical Theories*. New York: Harper Torchbooks, 1960.

INDEX